Human Ancestry

LONDON : GEOFFREY CUMBERLEGE

OXFORD UNIVERSITY PRESS

Man-Origin

Human Ancestry

From a Genetical Point of View

By

R. RUGGLES GATES, F.R.S.

EMERITUS PROFESSOR OF BOTANY IN THE
UNIVERSITY OF LONDON
RESEARCH FELLOW IN BIOLOGY
HARVARD UNIVERSITY

HARVARD UNIVERSITY PRESS
CAMBRIDGE, MASSACHUSETTS
1948

COPYRIGHT, 1948
BY THE PRESIDENT AND FELLOWS OF HARVARD COLLEGE

PRINTED AND BOUND BY NORWOOD PRESS
J. S. CUSHING CO.; BERWICK AND SMITH CO.
NORWOOD, MASSACHUSETTS, U.S.A.

Preface

This book is a study of the origin and history of the races of mankind, based on an evolutionary background. Some of the principles of evolution which have been dealt with mainly by the paleontologists are here considered in relation to the evolution of mammals, the Primates, and man. Many of the important discoveries of human remains, especially in South and East Africa, have not yet been incorporated into the general fund of knowledge regarding fossil man.

While large gaps still remain in our knowledge of human evolution, the important discoveries in China and Africa and elsewhere in the last quarter-century necessitate many reorientations and make it possible to write a tentative anthropological history of man from a genetical point of view.

It is only by learning how man has come to his present state that we may hope intelligently to control in any measure his future development. This book is offered as a small contribution to that end. In the present disturbed state of the world it is important that questions of race and population be recognized as the fundamental problems they are.

In the earlier chapters such problems as parallel evolution and the origin of the mammals have been emphasized. The paleontological background leads clearly to the conclusion that the modern so-called races of man are not simultaneous divergents from a single stock, but that there has been much independent evolution (partly parallel and partly divergent) on the different continents, from diverse ancestral stocks.

In the last two chapters the question of species is considered from a genetical point of view. The futility of erecting intersterility as the sole criterion of species is made clear, since it would lead to chaos, not only in regard to human types but in the taxonomy of all animals and plants. The Hominidae therefore take their place with other organisms as a succession of species and genera in time and space, with several species surviving into modern times.

Much of this book has been written at the Marine Biological Laboratory, Woods Hole, with the facilities of its excellent library. I am much indebted to Dr. A. S. Romer for reading Chapters 2 and 3 and making some emendations. I have also had the benefit of a critical reading of the whole book with valuable suggestions by Dr. Carleton S. Coon, from his wide and exact knowledge of human types. For the original drawings in Figures 5 and 6 and also for the originals of Plate XVI (a, b, c, d) and Plate XVII (a, b), I have to express my thanks to Dr. R. Broom, F.R.S., of Pretoria, whose fundamental contributions to South African paleontology and human ancestry are well known. The Harvard Libraries, especially in Anthropology and in Comparative Zoology, have been the source of many references. I should like to add that without the stimulus of Professor E. A. Hooton, a leader among anthropologists, this evolutionary study of man might not have been written.

The secretarial assistance in connection with this book has been done mostly by Mrs. Genevieve C. Doyle, and I am indebted to Mr. Claude Ronnie, of the Oceanographic Laboratory, Woods Hole, for the photographic work. To the American Philosophical Society I am indebted for a grant while this book was being written.

I have to acknowledge permission to publish the following illustrations:

FIGURE 1.	LeGros Clark (after Gregory and Hellman) in *Biol. Revs.* 15:205. Fig. 1.
FIGURE 3.	Gear in *S. Afr. J. Sci.* 26:688. Fig. 1.
FIGURE 5.	Galloway in *Am. J. Phys. Anthrop.* 23:46. Pl. 1.
FIGURE 6.	FitzSimons in *S. Afr. J. Sci.* 23:817. Pl. 17, fig. 4.
PLATE I (a, b).	Dart in *S. Afr. J. Sci.* 26:658. Pl. VIII.
PLATE I (c).	Weidenreich in *Anthrop. Papers, Am. Mus. Nat. Hist.* Vol. 40, pt. 1. Pl. 4b.
PLATE I (d).	Gregory and Hellman in *Am. J. Phys. Anthrop.* N.S. 3:267. Pl. 2a.
PLATE II.	Koenigswald in *Proc. Roy. Acad. Amsterdam.* 41:192. Pl. 1.
PLATE III (a).	Weidenreich in *Anthrop. Papers, Am. Mus. Nat. Hist.* Vol. 40, pt. 1. Pl. 10 (3, 5, c).
PLATE III (b).	Black in *Palaeont. Sinica.* Ser. D, vol. 7, fasc. 1. Pl. 2.
PLATE IV (a).	Black in *Palaeont. Sinica.* Ser. D, vol. 7, fasc. 2. Pl. XI.

PLATE IV (b).	Black in *Palaeont. Sinica*. Ser. D, vol. 7, fasc. 2. Pl. I.
PLATE V (a).	Black in *Bull. Geol. Soc. China*. Vol. 9, no. 1. Pl. 2.
PLATE V (b).	Weidenreich in *J. Roy. Anthrop. Inst.* 67:51. Pl. XI.
PLATE VI (a).	Weidenreich in *Palaeont. Sinica*. Vol. 7, fasc. 3. Pl. 12.
PLATE VI (b).	Weidenreich in *Palaeont. Sinica*. Vol. 7, fasc. 3. Pl. 10, fig. 8.
PLATE VII (a).	Sollas in *Phil. Trans. Roy. Soc.* 199B:281. Pl. 29.
PLATE VII (b).	Morant in *Ann. Eugen.* 3:337. Pl. 2.
PLATE VIII.	Gear in *S. Afr. J. Sci.* 26:697. Pl. IX.
PLATE IX (a).	Weidenreich in *Palaeont. Sinica*. No. 127. Figs. 264, 265.
PLATE IX (b).	Smith in *Phil. Trans. Roy. Soc.* 208B:351. Pl. 12.
PLATE X.	Ogilvie in *Malayan Nature J.* 1:23. Fig.
PLATE XIII (a).	Shellshear in *Phil. Trans. Roy. Soc.* 223B:1. Pl. 2, fig. 12.
PLATE XIII (b).	Shellshear in *Phil. Trans. Roy. Soc.* 223B:1. Pl. 2, fig. 9.
PLATE XIV (a).	Broom in *Ann. Transv. Mus.* 20:243. Fig. 15.
PLATE XIV (b).	Broom in *Ann. Transv. Mus.* 20:217. Pl. IX.
PLATE XV.	Broom in *Ann. Transv. Mus.* 20:217. Pl. VI.
PLATE XVI (a).	Keith in *Trans. Roy. Soc. S. Afr.* 21:153. Fig. 3.
PLATE XIX (b).	Hrdlička (after Schoetensack) in *Smithson. Misc. Coll.* 83:99. Pl. 16.
PLATE XX (b).	Hrdlička in *Smithson. Misc. Coll.* 78:63. Fig. 64.
PLATE XXI (b).	Stewart in *Smithson. Misc. Coll.* 100:26. Fig. 2.

Certain others are acknowledged in the text.

Biological Laboratories
Harvard University

R. Ruggles Gates

CONTENTS

Foreword by Earnest Hooton   xv

1 *Introduction*   3

2 *Principle of Parallel Evolution*   13

     Gigantism in birds · Horses and Litopterna · Rodent-like incisors · Parallel lineages in fossil Echinoderms and in Cretaceous Polyzoa · Amphibian and reptilian evolution · Transition from reptiles to mammals · South African Therapsida · Notoungulata · Lemurs · Proboscidea

3 *Evolution of the Mammals*   44

     Parallel origins and adaptive radiation · Marsupials and placentals · Molar tooth patterns · Parallelisms in the Primates · Notharctus and Tarsius · Dryopithecinae · Australopithecinae · Anthropoid apes and Man

4 *Evolution of the Hominidae*   78

     Pithecanthropus · Meganthropus · Gigantopithecus · Sinanthropus · Skulls · Mandible · Dentition · Long bones · Javanthropus · Africanthropus · Rhodesian skull · Neanderthal man (Paleoanthropus) · Pleistocene successions · Mount Carmel

5 *Head Shapes and Their Inheritance*   119

     Dolichocephaly · Origin and spread of brachycephaly · Sergi's classification applied to Bushmen and Bantu · Egyptian craniology · Chamaecephaly and hypsicephaly · Loss of superciliary and occipital tori · Gorilloid and orangoid lines of evolution

6 *Local Evolution of Modern Racial Types: From Pithecanthropus to the Australian Aborigines*   144

     Homo australicus · Wadjak man · The Keilor skull · Talgai skull · Modern Australoid types · Indian jungle tribes · The Tasmanians · Veddoid type in Arnhem Land · Australian compared with Bushman brain · Scheme of human phylogeny

Contents

7 Evolution of Man in South and East Africa   166

Florisbad skull · Boskop man · Zitzikama · Pleistocene climates in Africa · Bushmen and Hottentots · Strandloopers · Korana · Springbok Flats skeleton · Matjes River type · Archeological types · East African climates and cultures · Racial mixtures · Bushman race · Chinese influence in East Africa · Pygmies in Africa and elsewhere · Asselar man · Alfalou skeletons · Bantus · Nilotes · Bushman morphology · Oldoway man · Gamble's cave · *Homo kanamensis* · Kanjera skulls · Elmenteita

8 Human Evolution in Europe   236

Eoanthropus · Galley Hill · Swanscombe · Heidelberg jaw · Neanderthal man · Gibraltar, Spy, Krāpina, Steinheim · Aurignacian man · Cro-Magnon, Chancelade, Grimaldi · Brachycephaly · Solutré · Magdalenian man · Lautsch · Canary Islands skulls · Ofnet · The Mediterraneans

9 From Sinanthropus to the American Indians   274

Chinese origins · Blood groups · Routes into America · Dixon's skull types · Co-existence of man and Pleistocene animals in America · Florida remains · Calaveras skull · Folsom man · Paleo-Indian cultures · Sandia cave · The Minnesota girl · Lagoa Santa · Punin skull · Skeletons in Argentina · Chapadmalensian man · Blood groups of Fuegians · Gruta de Cadonga · Eskimo · Siberian cultures · Thule culture

10 Polynesians, Melanesians, and Negroes   335

Population increase · Blood groups in the Pacific · Polynesian origin and spread · Maoris · Indonesians · Madagascar · Hawaii · Sumatra · New Guinea · Melanesians · Negritos · Blood groups in India

11 Some Principles of Speciation in Primates   360

Darwin's views · Secondary races · Origin of Dinarics · Crossing between species · Human species · Species in gorilla and chimpanzee · Intermediates between them · Geographical variation in monkeys · No single criterion of species

12 Paleontology, Speciation, and Sterility   382

Species in paleontology · Choroclines and chronoclines · Intersterility in species and subspecies, examples from Drosophila, Crepis, Corvus · F_2 interspecific sterility in salamanders and Gallus · Geographic strains of frogs and toads · Examples from Cobaya and Peromyscus · Breakdown of intersterility as a sole criterion of species

ILLUSTRATIONS

Plates

FOLLOWING PAGE 80

I   a. Restoration of the head of Australopithecus by A. Forestier, under the direction of Professor G. Elliot Smith (by permission of the *Illustrated London News;* from Dart in *Natural History,* 26:327)
b. The Taungs skull (after Dart)
c. Reconstruction of *Pithecanthropus robustus* (skull IV), with the mandible B of *P. erectus* (after Weidenreich, 1945)
d. Reconstruction of *Plesianthropus transvaalensis* Broom (after Gregory and Hellman, 1945); from a copy kindly supplied by Dr. W. K. Gregory

II   Side view and vertical view of *Pithecanthropus erectus,* skull II (after Koenigswald, 1938)

III   a. Molars of (1) Sinanthropus, (2) modern Amerind, (3) Gigantopithecus (after Weidenreich, 1945)
b. Molars of (1) chimpanzee, (2) Sinanthropus, (3) Chinese child (after Davidson Black, 1927)

IV   a. Sinanthropus skull in preparation
b. Vertical view of the same skull (both after Black, 1931)

V   a. Sinanthropus skull in side view (after Black, 1930)
b. Sinanthropus skull III, showing frontal and occipital tori (after Weidenreich, 1936)

VI   a. Mandibles of (a) female gorilla, (b) Sinanthropus male, (c) Chinese male
b. Sinanthropus mandibles, (1) adult male, (2) adult female, (3) female child (all after Weidenreich, 1936)

VII   a. The Gibraltar (Neanderthal) skull (after Sollas, 1908)
b. The Rhodesian skull (after Morant, 1928)

VIII   Three shapes of skull in vertical and lateral aspect (after Gear, 1929)

FOLLOWING PAGE 176

IX   a. The undeveloped Talgai skull (after Smith, 1918)
b. Head of an Australian man (after Weidenreich)
c. Skull of an Australian aboriginal woman (after Weidenreich)

Plates

X A group of Che Wong, a primitive tribe in Perak, Malaya (after Ogilvie, 1940)

XI a. Kanikar jungle tribe in Travancore
b. A group of Kanikar women and a child (photos by the author)

XII a. A group of Pulayas in Travancore
b. Urali jungle tribe in the Nilgiri Hills (photos by the author)

XIII a. Line drawing showing the sulcal pattern in b
b. Marshall's photograph of the brain of a Bushwoman (both after Shellshear, 1934)

XIV a. Two old Korana women at Kimberley (after Broom, 1941)
b. Boskop type of skull in a modern Hottentot (after Galloway, 1937)

XV Skeletons (a) from Zitzikama, (b) inland Cape Bushman (after FitzSimons, 1926)

XVI a. Typical old Bushman at Postmasburg
b. Young Bushman from Langeberg
c. Typical Korana at Kimberley
d. Old Korana woman (all photographs by A. M. Duggan-Cronin)

FOLLOWING PAGE 304

XVII a. Korana girl (photo by A. M. Duggan-Cronin)
b. Pseudo-Australoid Korana (after Broom, 1941)
c. Barotse Negroes at Victoria Falls, Northern Rhodesia (photo by the author)

XVIII a. Young Basuto women in Pietersburg, northern Transvaal
b. Native mine workers near Pretoria
c. Young Zulu woman at Durban, East Africa
d. Basuto family in the Kruger Game Reserve, Transvaal (photos by the author)

XIX a. Typical kraal in Southern Rhodesia
b. Bavenda women at Pietersburg, northern Transvaal (photos by the author)

XX a. Young Sesutos in the northern Transvaal
b. Kaffir women and children in the northern Transvaal (photos by the author)

XXI a. Two Matabele women at Bulawayo, Southern Rhodesia
b. Two women near Mafeking, Bechuanaland, showing evidence of Hottentot ancestry
c. Negro children in the northern Transvaal (photos by the author)

XXII a. and b. The Heidelberg or Mauer jaw (from Hrdlička, after Schoetensack)
c. The Steinheim skull (after Weinert, 1936)

Illustrations xiii

Plates

XXIII a. A Tibetan woman at Darjeeling (after Hrdlička)
       b. Slave Indians on the Mackenzie River (photo by the author)
XXIV   Tierra del Fuegians (after Lipschutz, Mostyn, and Robin)

FOLLOWING PAGE 368

XXV   a. and b. Eskimo girls
       c. Loucheux Indian girl
       d. Eskimo boy (photos by the author)
XXVI   a. and b. Eskimos at Aklavik
       c. Loucheux Indians at Fort MacPherson (photos by the author)

FACING PAGE 400

XXVII Jurua (Negrito) woman and five children from the Andaman Islands

Figures in Text

1. Restoration of the mandible of Dryopithecus, showing the simian character of the teeth and the simian shelf (from LeGros Clark after Gregory and Hellman) — 54
2. Scheme of higher primate evolution — 56
2a. Side view of skull of *Paranthropus robustus* Broom. The shaded portions are those preserved (after Broom) — 68
3. Six forms of skull in vertical view (from Sergi after Gear, 1929) — 129
4. Scheme of human phylogeny — 161
5. Side view of (a) Florisbad, (b) Cape Flats skull (from original drawings kindly sent by Dr. R. Broom) — 168
6. Vertical views of (a) Florisbad skull, (b) Korana male, (c) Bush female skull (from drawings made by Dr. R. Broom) — 168
7. Outline of a Matjes River skull, showing the trigonocephaly (after Keith, 1933) — 187
8. Comparison of hypsicephalic Eskimo and chamaecephalic Aleut skull (after Stewart, 1940) — 283

Map of South Africa, showing archaeological sites — 170

Tables

1. Frequency of torus mandibularis and torus palatinus — 92
2. Measurements of Pithecanthropus and related skulls — 97
3. Divisions of the Ice Age in Europe — 104
4. Neanderthal remains in relation to glacial periods — 105

5. Shapes of head 128
6. Skull shapes of Bushmen 130
7. Development of skull shapes of Bushmen 130
8. Skull shapes of Bantu 131
9. Analysis of Egyptian skulls of all periods 135
10. Classification of skull shapes from Russian kurgans 140
11. Measurements of Australoid tribes in India 152
12. Blood groups and types of Australian Aborigines 158
13. Brain weights of different races 162
14. Cranial capacity of Boskop skulls 171
15. Comparison of European and South African archaeological ages 189
16. Correlation of European climates and climates and cultures in South Africa 191
17. Blood groups of Bantu and Arabs 201
18. Measurements of three tribes of Nilotes 214
19. Analysis of facial elements in the Bantu 218
20. Association of cranial and facial type in Bushmen 219
21. Blood groups of Tierra del Fuegians 319
22. Blood groups of some Polynesians, Melanesians, and Negroes 338
23. Blood groups in India 356

Foreword

New discoveries of fossil man and of fossil apes have come thick and fast since 1935, but no comprehensive work that describes all of these finds in detail has appeared. The few specialists in the study of human ancestry mostly have been too busy writing technical monographs and papers on the numerous new specimens and types that have come to light. The revised edition of Keith's great *Antiquity of Man* (1929) and a supplementary volume, *New Discoveries Relating to the Antiquity of Man,* published a few years later, have remained the only detailed works in English that cover the entire subject.

In the present volume, Professor R. Ruggles Gates has assembled the new evidence and combined it with the old to make an up-to-date account of human paleontology, together with elaborate summaries of new studies of the skeletal remains of geologically recent man in various parts of the world. He has attempted to relate the history and analysis of contemporary "races" of man to our knowledge of ancient human types, so as to fill the gap that is ordinarily left by specialists dealing with the two fields of studies as separate entities. *Human Ancestry* is thus an unusual, if not unique, book. The mere compilation of all of this information in one volume results in a most useful reference work.

Nearly all authoritative works on fossil man have been written by specialists in human anatomy or in physical anthropology, with some valuable contributions by vertebrate paleontologists. Interpretations of the evidence have been based principally upon anatomical considerations. Most students of the skeletal remains of contemporary races have this same anatomical viewpoint, but, since they usually work with series rather than isolated specimens, they employ also biometric methods in their analyses. Professor Gates is primarily a geneticist rather than an anatomist or a physical anthropologist. Hence he brings to the present work a fresh and different outlook. He is particularly concerned, for example, with the taxonomy, or classification in the zoological hierarchy, of the various fossil and existing types of man. Most physical anthropologists are more interested in trying to establish human groups on the basis of anatomical resemblances and differences, and in tracing lines of descent by these means, than in the precise zoological status in systems of classification that those groups should have. The question of the unity or diversity of species in modern man has unfortunate political implications in which the physical anthropologist usually does not wish to involve himself. Professor Ruggles Gates does not subscribe to the dogma that all races of modern man belong to the same species. He argues powerfully for specific diversity. I myself have always thought that detailed discussions of this subject are unprofitable, because of the arbitrary character of taxonomic nomenclature and the apparent lack of consistency of zoologists in assigning taxonomic rank to animals. There seems to be no general agreement upon the definitions of orders, genera, species, and so on, so that semantic confusion arises to complicate the political issues that grow out of attempts at applying systematic classification to man. However, I am glad to see Professor Gates tackle the problem so courageously, even if I myself am indifferent, hesitant, or pusillanimous.

Of course I do not agree with all of the interpretations of the studies of fossil and contemporary man that are offered by Professor Gates in *Human Ancestry*. Over a considerable period of years, I have written on these subjects myself and I cannot consistently agree with my own conclusions and interpretations. Sometimes I change my mind because I have decided that I was simply wrong; oftener, new evidence is uncovered that demonstrates the error of my earlier views. So I would not urge upon the reader of this useful book, *Human Ancestry*, that he accept its contents as eternal verities. He may well "read, mark, learn, and inwardly digest," but he would better not regard the matter as "sacred truth," because all that science can offer is approximations to truth, and that goes for all of us lesser fry and for Professor Gates, too.

Earnest Hooton

Harvard University,
September 28, 1947

Human Ancestry

1

Introduction

The ancestry of man has its roots in the distant past. The vicissitudes of the Ice Age undoubtedly played a part in human evolution, but the beginnings of man take us back to the Pliocene or perhaps even the Miocene period of geology. In this length of time much evolution of other animals has taken place, so that the changes in man need to be considered in connection with the contemporary fauna and flora.

In order to have some understanding of the evolution of man it is necessary to consider first some of the general principles of evolution. In this connection paleontology is of prime importance as showing the course which evolution has actually taken in the past. In the many books on evolution the paleontological aspects are frequently so overlaid by discussions of the importance of natural selection that the general laws of paleontological evolution are frequently lost sight of. In the present work some of these laws will be emphasized, especially in their relation to the evolution of man. Many of the differentiations in man appear to be non-adaptational. It is difficult to believe that a round head has any advantage over a long head, yet brachycephali have more or less supplanted dolichocephali in various parts of the world. Exactly how this has taken place is a matter of great interest, and it is probably of evolutionary significance.

The principle of parallel evolution has been greatly neglected, except by paleontologists. This principle is therefore considered in the early chapters, not only in mammals but in some groups of invertebrates. It will be seen that parallelisms occur plentifully in whatever direction we look. Their abundance also involves recognition of the fact that groups, such as the mammals, which we now regard as uniform, have had a polyphyletic origin.

In the Primates we approach nearer to man's ancestry, and here again parallelisms stud the phylogenetic lines, however we read them. Two phylogenetic schemes have been drawn up, one for higher Primates and one for man, but these need not be taken too seriously. They are only meant as a tentative attempt to bring into a connected picture the known facts of primate and human paleontology, although these are still fragmentary in many respects. The widest gap in man's ancestry appears to be between the Dryopithecus level of Miocene and Pliocene times, and the known types of humanoid and anthropoid as they are found in the Lower Pleistocene. When more complete remains from the Late Pliocene are discovered we shall probably know much more about human ancestry in relation to the Primates than we know now.

Notwithstanding these lacunae in our knowledge, the discoveries of the last twenty-five years have been such as to throw a flood of light on the origin of the modern so-called races of man. They show that the old idea that man as we know him on the different continents diverged simultaneously from a single common ancestry is no longer tenable. The majority of anthropologists recognize that this simple picture does not correspond with fact, but that there have been multiple centers of man's evolution, involving again a certain amount of parallelism. These parallelisms all go back ultimately to parallel mutations, occurring repeatedly, as all mutations do, whether or not they have any selective value.

This point of view, which was first expressed in relation to the human blood groups in 1936, is now seen to be the way in which genes arise and spread in the human races, although hybridization also plays a part in some cases. The new facts which show that, even in animals, new species can occasionally arise by hybridization between old ones, are significant in this connection. The spread of brachycephaly in different races of man since the Lower Paleolithic is a remarkable phenomenon which, in the last analysis genetically, probably depends upon repeated mutations from dolichocephaly or mesocephaly. Here is evolution at work, and the fact that man as an organism is better known than any animal should make it possible for anthropological studies of mankind to contribute directly to the principles of evolution. In fact such contributions have already been made (see Gates, *Human Genetics*, Macmillan, 1946).

Introduction

The intention of this work is to examine the facts of race in their anthropological and paleontological aspects, because the status of the modern races of mankind can only be understood on a paleontological background. History does not take us back far enough, and most races have little or no history in the conventional sense of written records. Instead of beginning with civilization, at say 5000 B.C., we shall have to go back at least 100,000 years, probably 500,000, or possibly even 1,000,000 years, to the roots of human ancestry. This knowledge of human and humanoid fossils only began in the 1860's, but has developed at a constantly accelerating pace. Discoveries in the last twenty-five years have, as already mentioned, thrown a flood of light on early human types. Much remains to be done in the elucidation of their relationships, but some features of human evolution have already emerged from obscurity, and these have a direct bearing on questions regarding the relations of modern living races.

Comparison of skulls and teeth has played a large part in anthropology. They are the parts most frequently preserved in human, and indeed in all primate, fossils, and as they contain many of the features which differentiate human types it is natural that they have been more extensively and intensively studied than any other part of the human skeleton. They also contain the most complicated skeletal structures. Reference will therefore be made to many different skulls which have been excavated in Java, China, South Africa, and other parts of the world in recent years. The developments in South Africa, where many whole skeletons have been found in good preservation, are relatively little known, but they furnish much important evidence regarding human evolution in that continent. The pedomorphic degeneration of the Bushmen in South Africa has been one of the striking episodes of human development, and their curious relationship to the Hottentots —a co-descendant from the Boskop race—bears some similarity to the "symbiotic" relations between the Congo Pygmies and the Negroes, but is of a more intimate character.

In this book, and in my previous work, *Human Genetics,* considerable attention has been given to the genetics of head shape, not only as regards the relative dimensions of length, breadth, and height, but concerning the shape as seen in vertical view—an aspect of skull shape first emphasized by Sergi. A genetic study of the inheritance of the ellipsoid, pentagonoid, and other shapes in families is much needed, but some data are already available regarding the occurrence and genetic segregation of these types in certain races. The distribution of hypsicephalic crania, with a high vault, and chamaecephalic with a low vault, is a racial distinction of much significance. The skulls of Eskimos and Aleuts are very similar except that the former are high-headed and the latter low-headed. More complete investigations of the inheritance of head height should yield results of great interest. The Egyptian population, with large numbers of skulls and skull measurements available over a period of nearly seven thousand years in the history and prehistory of that country, shows how skull shapes have altered, new types coming in but failing to hold their own in competition and crossing with autochthonous types.

Another feature of skull variation which is of great significance in human evolution is the presence of heavy superciliary and occipital tori in such types as Pithecanthropus, Sinanthropus, and Rhodesian man, and to a lesser extent in Neanderthal man. This heavy skull scaffolding persists to some extent in the Australoids, as well as in some of the Korana in South Africa and certain early skulls in South America. It is thus an excellent example of parallel development, in this case parallel loss, in different races. It was formerly assumed that all mankind had evolved through this gorilloid stage, the tori being gradually lost in most modern races of man. The discovery, in southern England and eastern Africa, of types lacking these ledges and therefore belonging to the genus Homo, although as early as or earlier than Neanderthal man or Pithecanthropus, has led many anthropologists to the conclusion that man in some lines of descent evolved without ever developing the gorilloid type of brow ridges.

It appears that the evolution of the orangutans in eastern Asia, Borneo, and Sumatra took place also in the absence of these cranial tori which are so highly developed in the African gorillas and chimpanzees, so the writer suggests that in man also an orangoid strain or line of evolution developed, lacking brow ridges but occasionally crossing with the gorilloid type of contemporaneous man, for instance, at the Neanderthal level. This is only a suggestion at the present time. The future will determine whether it has a permanent basis in fact.

Fossil remains of man have now become numerous enough in the different continents to trace independent lines of descent. One of the clearest of these, accepted by all anthropologists, is from Pithecanthropus to the Australian aborigines, with remnants of the Australoid type surviving in some of the jungle tribes of southern India. Traces of the type are also found in parts of Malaya and New Guinea. The relation of the Tasmanians to the Australian aboriginals has been much discussed. Recent investigations show the survival of a Tasmanoid type in northeastern Australia, which is significant in elucidating the history of the Tasmanians. Intercrossing between such nearly related types has of course occurred throughout the history of man, when they have come in contact or invaded each other's territory.

An immense amount of work has been done in recent years on human paleontology and archeology in South and eastern Africa. Not only the astonishing Rhodesian skull, but many other skulls and skeletons from different periods serve to give a vivid picture of earlier man in South Africa, but when linked with the East African discoveries they connect the ancestral Boskop type with North African types, and these in turn are closely related to types which were found in Europe in the Upper Paleolithic. When we remember that in the Pleistocene the Sahara was a vast fertile and well-watered country which was occupied by man, probably in large numbers, it is natural to suppose that some human evolution as well as migration took place here. The land bridge which formerly existed across the central Mediterranean made the Sahara region much more accessible to Europe than now.

A feature of eastern Africa in relation to human populations which has been scarcely recognized by anthropologists is that the Chinese visited this region in large numbers in the Middle Ages, after skirting the Indian Ocean in their ships. Some of the Mongoloid traits appearing in Africa are probably from this late source. The yellowish-brown skin of the Bushmen and Hottentots appears, however, to be a phenomenon of pedomorphy; quite independent of any Mongoloid ancestry, but representing instead a remarkable tendency of the Boskop race and its descendants, the Bush and Hottentots, to undergo pedomorphic or infantile degeneration. This is but one expression of the tendency to pedomorphy in human evolution.

It is well known that Pygmy races of man, and of several other mammals, exist in various parts of the world. Some anthropologists are inclined to derive them all from one ancestry, the Negrito race, having short stature and kinky hair. However, some human dwarf types are clearly outside these bounds. The writer is inclined to regard the

Congo Pygmies as achondroplastic dwarfs, derived as a mutation from tall ancestors. On the other hand, the Akkas of Uganda and the Jurua of the Andaman Islands are well-proportioned dwarfs or miniatures, apparently of quite different type and independent origin from the Congo Pygmies of the Ituri forest. Many other dwarfs appear to be independent derivatives from different ancestral races. Dwarfism in mankind, like gigantism in birds, thus appears to have arisen many times independently. The relation of the Negritos to other races of man remains a problem.

Many views have been expressed regarding the relation between Neanderthal man and his successors, the Upper Paleolithic races of Europe. Hrdlicka believed that Neanderthal man evolved directly into modern man in Europe, but this view is no longer tenable. On the other hand, we know that when one primitive tribe invades the territory of another they usually kill the men and keep the women as wives. That probably happened in parts of Europe with sufficient frequency to account for the persistence of occasional Neanderthalian characters in later descendants. As we shall see, this does not necessarily mean that they both belong to the same species, or even to the same genus.

Sinanthropus or Chinese man, whose correct name is now recognized as *Pithecanthropus pekinensis,* represents one of the most extensive and important of all discoveries of early man. We shall find that some of his genes for particular skull characters have been transmitted to the modern Mongoloids, and some to the American Indians. The origin of the latter and their route into America will be discussed at some length. The recent discoveries regarding the culture of Folsom man have led to the recognition of Paleo-Indians contemporary with extinct animals and either of late Pleistocene or early post-glacial times. Similar records in South America refer to a type of Indian skull with somewhat Australoid features. This is regarded, not as a sign of Australian ancestry in America, but rather as a parallel stage in the reduction of brow ridges from the heavy Pithecanthropus ancestral type.

The Eskimos, as the last to cross the Bering Straits and establish themselves in Alaska, acted as a deterrent of any later immigrants to America by that route. While the earlier Indian immigrations may have been down the Mackenzie River valley as a corridor, probably beginning during the last interglacial, when ice fields remained east and west of them, the Eskimos, being already adapted to an ice culture, continued eastwards along the Arctic coast until they reached Greenland and Labrador. The absence of the B blood group from most Indians (except through crossing with whites) and its presence with quite high frequency in modern Mongoloids is a clear indication that the Indian ancestry left northeastern Asia before the B had spread to that region.

While most Indian tribes approach 100 per cent O blood group, their A and B probably coming in through crossing with Europeans, the Blackfeet and the related Blood Indians are remarkably high (up to 80 per cent) in A. This has led to the suggestion that they represent a separate (later) migration from Asia. A fuller study of the blood groups of Indians in Alaska and elsewhere should help to settle this point. It is well known that some of the Tibetans closely resemble the American Indians. The blood groups of typical Tibetans from the interior of Tibet would help to show whether this resemblance has a deeper significance. These cases, and others mentioned throughout this book, show that the blood groups and other serological differences, owing to their simple unitary inheritance, can be of great value as an additional anthropological character in the interpretation of racial relationships.

The remarkable spread of the Polynesians in the Pacific came so late that it can be treated almost as a series of historical events. Their origin previous to their "home" in Java is still quite uncertain. They are evidently nearly related to the Caucasians in physical features as well as blood groups (having little B), yet they are not typically Mediterranean except in their dark skin color. According to some writers they were first located in India. If their ancestry were partly derived from the Aryan invaders of India the problem of their origin might be solved. Some of their relatives, as Indonesians, proceeded westwards, skirting the Indian Ocean in their canoes and finally arriving on the African coast, whence they have largely peopled Madagascar. Apparently these westward migrations were motivated by an attempt to escape malaria, to which they were more susceptible than some other peoples. The possible relation of malarial susceptibility to blood groups is worthy of more complete investigation.

The African Negro appears to be a purely African product, most closely related to the Bushmen and Hottentots. His relation to the Melanesians and Papuans, sometimes called the Negroes of the Pacific, remains problematical. They are similar in skin color and in hair characters, but there appears to be little evidence of similar peoples in the intervening areas and it seems possible that their similarities may be partly at least the result of parallelisms. Some of the Papuans have the "Jewish" or Armenoid nose in an extreme form, and it is clear that this at any rate is a case of parallel development.

Views at present differ widely regarding species in anthropoids and in man. Geneticists are almost equally at variance regarding the criteria by which species should be defined. The subject is discussed at some length in the last two chapters of this book, and the endeavor is made to reach a balanced usage which will harmonize the methods of the paleontologists and geneticists without adopting the more extreme views of either. The point of view of the paleontologist is of great value in that they have to examine species in both time and space, whereas the neontologist is apt to think of them only in space. It is well known that related contemporary species frequently develop intersterility as they diverge. Some geneticists have even attempted to make intersterility an absolute criterion of species. This hard and fast rule, however, leads to such absurd results that its application would end in taxonomic chaos. It is evidently sounder to recognize that intersterility is one measure of specific differentiation, but that it frequently bears little or no relation to morphological difference, which is the usual measure of taxonomic distinction.

Darwin discussed at length the status to be accorded to the primary races of man; considering the possibility of regarding them as species or subspecies, he finally decided in favor of the latter, although recognizing that good reasons could be adduced for ranking the main types as species. Since his time the whole field of human paleontology has developed, and anthropologists now recognize that there have been different centers of human evolution, so that the main surviving races have arisen from different ancestors in widely separated areas. This gives a very different picture from the old conception that white, black, and yellow diverged simultaneously by merely spreading out geographically from a single common ancestral stock. In this book reasons are adduced for ranking the main types of living man (about five) as species.

Introduction

The objection has often been raised that neither race nor species always means the same thing. This is of course true. Nature, in her manifold variety, produces larger and smaller species, larger and smaller races, meaning by larger or smaller the relative size of the gap with related species, or the amount of difference from the most nearly related forms. Attempts have also been made to do away with the term race, because it can be applied to larger or smaller groups. This is, however, no drawback, provided the particular use is understood in every case. Some writers have endeavored to revive Deniker's old term "ethnic groups," but this is hardly an advantage, since ethnic groups are also of various sizes or degrees of difference, large and small. We cannot put nature into a strait jacket, but on the contrary, we have to recognize that species are much too multifarious to fit into any one universal definition.

One of the chief advantages in using the term species for the primary races of mankind is that of convenience, and the same argument applies to all taxonomic work. There is also the more cogent argument, arising from facts which were unknown in Darwin's time; namely, that if Pleistocene species A gives rise to modern species B, and if Pleistocene species C gives rise on another continent to a very different species D, then it is absurd to include B and D in one species. In fact, Pithecanthropus in Java is recognized to have given rise to *Homo australicus* in Australia, and quite independently in Africa Africanthropus has produced Boskop man (*Homo capensis*), which has since developed into the modern Bushmen and Hottentots. To contend that *Homo capensis* and *H. australicus* must belong to one species because Linnaeus said so in the eighteenth century is to nullify the great increase in knowledge of the subject which has taken place since that time. If the evolution from A to B and from C to D had been parallel developments, then B and D would be rightly placed in the same species. But every anthropologist knows that the Australian aborigines and the Cape Bushmen are unlike in practically every feature, including not only stature, skin color, facial features, and hair characters, but even the sex organs.

The question of sterility in relation to species in various groups of animals will be discussed, and it will be pointed out that sterility cannot arise in the time-passage from one species to its descendant species, or the latter would never appear. Between contemporaneous diverging species it arises as an incident of the divergence, but is by no means a measure of that divergence. Indeed, it is clear that the development of intersterility may be the first step in speciation or the last or it may make its appearance at any intermediate stage.

The relations between speciation and sterility are therefore by no means simple or uniform, but they vary from species to species and from group to group of organisms. Examples from frogs and toads, salamanders, birds, guinea pigs, and the deer mice, Peromyscus, as well as Drosophila, show how varied these conditions are.

Some recent work with interspecific hybrids in Salamanders shows that the F_1 can be perfectly fertile yet the F_2 produces monsters which fail to develop. Are such species to be considered interfertile or not? In certain other cases a single gene present in some members of one species may act as a lethal in crosses with certain other species but has no effect whatever in the species in which it must originally have arisen as a mutation. Such instances show how incidental is the occurrence of intersterility between any two species.

In plants the origin of a new species as a result of crossing between two other species is well known and many cases are on record. In animals the possibility of such an occurrence has often been denied, but Patterson has described such a case in Drosophila. This is important as showing that in animals too hybridization has played a part in the origin of species. In man, cases will be cited in which secondary races with new characters arise as a result of crosses between other races. All these facts go to show that neither sterility nor any other single feature can be used as a universal criterion of species.

2

Principle of Parallel Evolution

In his *Descent of Man* Darwin traced human ancestry back to the fishes. This line of descent was based on the relatively meager paleontology of his time. Enormous accumulations of paleontological knowledge of all the animal groups now makes possible a much more complete picture of the course of evolution. The transitions from group to group especially can now be seen in much greater detail, and their study illuminates certain aspects of evolutionary change which can scarcely be touched upon in the short-time experiments of genetics. They thus serve in part as a complement and foil to the results achieved by geneticists.

A fundamental evolutionary principle, the recognition of which was impossible in Darwin's time, has thus emerged. This is the principle of parallel development, which has been expressed in various ways by paleontologists and has been recognized in the form of parallel mutations and parallel variations by geneticists, but has received far less emphasis than its importance deserves. In the older literature it is frequently referred to as "convergence," although not all cases of convergence are true instances of parallelism. Haas and Simpson (1946) have an extended discussion of usage regarding convergence, parallelism, homoplasy, and similar terms. They define parallel adaptations and parallel evolution, but make no mention of parallel mutations, a term first used by Gates in 1912.

These parallelisms are of every degree and many kinds. Thus the fact that many reptile groups showed increasing gigantism ending in extinction in the Mesozoic is well known, and many attempts have been made to explain it. These are parallelisms, whatever their explanation. The tendency to great increase in size is equally marked in the mammals, where the whales, the elephants, and the horses are perhaps the most conspicuous examples. In birds, the fossil and recently extinct types, as well as the ostriches, emus, and cassowaries, show that gigantism has probably arisen independently perhaps a dozen times, including cases recently extinct among the geese, penguins,[1] pigeons, and rails. Lowe (1929), in a detailed study, recognized seven families of struthious birds—the moas, rocs (Aepyornithidae), kiwis, emus, cassowaries, ostriches, and rheas—but maintained that they all came from one ancestry, which left the avian stem before flight had been attained. This, however, is highly improbable, as many of them have developed on isolated islands. The dodo, for instance, was a giant pigeon of the Mascarene Islands, extinguished by man in the seventeenth century, and the solitaire of Rodriguez was a related member of the Columbae. These flightless giants are characteristic of islands, where the absence of mammalian predators rendered their evolution safe, for example, Aepyornis in Madagascar, extinct within historical times. A tendency to increase in size, until gigantism ends in extinction, must indeed be recognized as one of the most widespread laws of animal evolution. The same tendency can be seen in some at least of the plant groups. For instance, the giant lycopods and selaginellas of the Carboniferous coal measures became extinct, while their smaller relatives have survived to the present day.

Owen (1869) described the osteology of the dodo. The Newtons (1870), in a full account of the flightless solitaire, show that it was abundant on Rodriguez in 1693, the males weighing up to forty-five pounds. It still existed as late as 1761. There was formerly another species of solitaire on the island of Bourbon. These and the related dodo were all derived from a common ancestry in the pigeons at a time when the Mascarene Islands formed part of a single land mass in the Indian Ocean. Thousands of bones of the solitaire were dug from caves in Rodriguez, and the Newtons show their extreme variability.

On the other hand, the giant Diatryma, from the Eocene of New Mexico and Wyoming, is not nearly related to any bird living or extinct (Matthew and Granger, 1917), but may be distantly related to

---

[1] The giant penguins of the Miocene (Simpson, 1946), known from Patagonia, New Zealand, and Antarctica (Seymour Island), are probably an evolutionary reaction to cold climate.

the cranes. It was larger than any bird except the extinct Moas of New Zealand, but was a true carinate bird with a huge head and a thick neck, the skull being seventeen inches in length, the wings vestigial, the height about seven feet. Phororhachos, of equal size, developed independently much later in South America, and Gastornis was an Eocene giant of France, Belgium, and England. That these continental flightless birds could evolve implies an absence of effective enemies, as with the modern rheas and ostriches.

There is reason for believing, however, that the rhea of South America, the ostriches of North and South Africa, and the cassowary of Australia had a common ancestor, although now so widely separated. Studies of Kellogg and of Harrison on the Mallophaga or birdlice of these struthious birds, as well as the similarity in their nematode and cestode parasites (see Gates, 1946, p. 1419), strongly support this conclusion. The ostrich formerly extended as far as Mongolia, and fragments of ostrich eggs have been found beneath the hearths of Paleolithic man in northern China. There are several fossil emus (Dromaeus) in Australia, and a generalized ostrich in the Lower Eocene (Fayum) of Egypt. The giant Moas (Dinornithidae) of New Zealand have been divided by some authors into seven genera, some extending back to the Pliocene. They were hunted by early man and apparently became extinct only within the period covered by the last three or four centuries.

Lowe (1928) maintains that all these giant birds were derived from one ancestry, which never attained flight. In any case, the gigantism on islands must have been a parallel development. Lowe's view is in harmony with the conception of Broom (1910) that some primitive relatives of the Dinosaurs (Thecodonts) which began walking on their hind legs became warm-blooded and developed into birds, the mammals being derived from those which remained tetrapods. Mayr (1942, p. 278) supports the view, now widely accepted, which divides the Ratitae (ostrich-like, flightless birds) into five groups, their similarities being "secondary isomorphisms" resulting from the loss of flight. In other birds (Carinatae) the loss of flight accompanied by varying degrees of gigantism has occurred independently at least in the rails (Aptornis of Pleistocene New Zealand), pigeons (*Didus ineptus* in the Mascarenes, Dronte in Mauritius, and Pezophaps in Rodriguez), Carianae (Diatryma in the Eocene of Wyoming and Phororhacos in the

Miocene of South America), and geese (Cnemiornis in the Pleistocene of New Zealand).

As another example of parallelism, Watson (1921) finds in the Coelacanth fishes complete loss of the hyomandibular as a supporting element of the jaw. "This loss is an exact parallel to that which has occurred in Tetrapods and Dipnoi." Bather (1927) recognized the importance of evolutionary parallelisms when, after citing numerous examples, he wrote, "The whole of our system . . . is riddled through and through with polyphyly and convergence." Thus Abel maintains that the genus Equus arose independently in the Old and New Worlds, the European horses from Hipparion and those in North America from Protohippus by way of Pliohippus. He therefore calls the American horses Neohippus. However, his view is generally discredited by later work, and Hipparion is believed to have migrated from America to the Siwalik Hills of India. McGrew has recently given reasons for the belief that the zebras developed from a different subgenus of Pliohippus than other members of the genus Equus.

The parallel evolution of the proterothere Litopterna in South America to the horses in North America is well known. The former became extinct in the Pleistocene, about the time the true horses reached South America (Simpson, 1940). Matthew (1926) has an excellent account of the evolution of all these groups, showing many parallelisms. In the Cervidae, Abel similarly traced distinct lines of descent, through four successive genera or grades of evolution in Miocene and Pliocene times.

In these cases it is generally recognized that the parallelism has arisen and continued in a long series because in each case a group of herbivorous animals was undergoing evolutionary adaptation to rapid flight from predatory enemies on open, grassy plains. In some other cases there is at least a doubt regarding the adaptational advantage to be derived from the variations in question, yet parallel developments have occurred in separate families of animals. Thus an extinct goat, *Myotragus balearicus,* in one of the Balearic Islands (Minorca) developed long lower incisors like those of a rodent (Andrews, 1915). The lower leg-bones (metacarpus and metatarsus) were also short, probably as an adaptation for rock climbing. The same condition of the incisors has arisen (Miller, 1924) in the vicuña, a member of the camel family in South America, although the nearly related guanaco shows no such condition. Andrews suggests that the long, permanently growing lower median incisors of this Balearic goat (the upper incisors and canines are missing in all the Bovidae) may be an adaptation for scraping lichens off the arid rocks. However, this would hardly furnish an adequate diet even for a goat, and it seems more likely that the condition has arisen as a parallel mutation in both the goat and the vicuña, the animals then having to make use of the instruments with which nature provided them.[2]

The Aye-Aye (*Daubentonia* (*Cheiromys*) *madagascariensis*) is now (Romer, 1936) thought to be descended from Plesiadapids but may be a parallel development within the Madagascar group of lemurs. It is the size of a small cat. By Owen it was regarded as the most primitive living primate, but is now recognized as a specialized form—the only primate to have rodent-like incisors. It was originally regarded as an arboreal rodent. It is nocturnal and lives on insects, including woodborers, and this element in its diet seems to have so conditioned its survival as to lead to the development of rodent incisors by natural selection. However, its long middle finger is more slender than the others, and this appears clearly to be an extraordinary specialization, with a claw for dragging the grubs from their hole. In the description of Owen (1866), the great ears were regarded as a specialization to enable the animal to hear the larvae boring.

To take a similar example from domesticated animals, the bulldog in Europe has been independently paralleled by an extinct type of dog developed by the Incas in northern Peru (Hilzheimer and Wegner, 1937). Skulls and pottery from the Chimu period before the Spanish conquest show that the features were the same but the shortening of the skull less extreme than in the French bulldog. The selection by man of independent parallel mutations obviously accounts for both.

These cases have been selected almost at random, but they are typical

---

[2] In a recent comparative study of the evolution of hypsodonty in the molars of Equidae (horses), Stirton 1947 (*Evolution* 1: 32–41. Figs. 9) shows that the molar heights develop at different rates in different lines of descent. Similarly in the camels, antilocaprids, beavers, and rhinoceroses the rates may vary in different orders, families, genera, and even species. He finds no evidence of macromutations but recognizes hypsodonty as an adaptation to feeding on siliceous grasses (generally mingled with some sand) which developed in the Tertiary (Stipidium, Stipa). The different rates of wear of the teeth are related to differences in the food (ecological niches) of the various animals. The increase in tooth height occurs too slowly to be of selective value, but in three or four million years the gene or genes for hypsodonty could become linked to other factors which are of selective value.

of the innumerable instances of longer and shorter parallelisms of variation to be found throughout the animal and plant kingdoms. Many such cases from both plants and animals were cited in an earlier work (Gates, 1920) in which this evolutionary principle was emphasized. Vavilov, later (1922), under the term homologous variations, pointed out the widespread occurrence of such parallel variations in plants. The principle is, of course, a natural extension of the conception of mutations. If a variety of discrete variations can take place spontaneously in the germ plasm of one species, the same mutations may be expected to occur in more or less distantly related forms having a similar germinal constitution. The deep significance of this principle of parallelism on the development of phylogenies has still received relatively little attention from others than paleontologists, so it is necessary to restate and emphasize its importance in the evolution of all organisms, including man. This is even more necessary because some writers have supposed that time and space can be disregarded in the consideration of human evolution.

It is clear that there is abundant basis for applying the principle of parallelism universally, along with that of divergence or radiation, not only to living species and genera but also to the longer series found by paleontologists. A real difficulty in the construction of phylogenetic "trees" is that the diverging branches and twigs of a tree inadequately represent what takes place in the evolution of any group or phylum of organisms. They represent the divergent variations, but take no account of the equally numerous parallel mutations. A comprehensive review of the phylogeny of any group must keep both in mind as well as the convergencies. Phylogenies in the past have too frequently emphasized one at the expense of the other. Bather, in his presidential address to the Geological Society, cites examples of parallel lineages in the fossil Echinoderms. He shows how the Pentacrinine stem, on which the genus Balanocrinus is based, has arisen several times from the genus Isocrinus during Mesozoic and Caenozoic times and apparently in various seas. In the same way the Lamellibranch genus Ostrea has changed repeatedly into the giant Gryphaea in parallel lineages. As these lineages may evolve at various rates in different lines of descent and may also anastomose, in other words, intercross, the result may give a highly complicated picture to unravel.

A detailed study of the evolution of the starfishes in the Cretaceous, by Spencer (1914), is illuminating in this connection. Because of slow, placid, and uninterrupted sedimentation over vast periods of time, the course of evolution of these starfishes is perhaps as clear as any in paleontology. Spencer shows that isolated ossicles in the white chalk of England can be identified. Many thousands of specimens, mainly from one zone, *Micraster cor-anguinum,* were assigned to their species. The starfish lineages ("species series") show unbroken continuity. Variation is usually of the continuous type, the variations in different lineages being frequently independent of each other. The types of variation in each lineage, Spencer concludes, are "predetermined and limited by innate causes." Environment affects the course of variation, since the evolution of a lineage in the depths of the Cretaceous seas differs from that nearer shore. Each zone or group of zones has its own distinctive fauna, the Upper Cretaceous being divided into three periods by using the ossicles for zonal determination.

There is clearly parallelism, because a number of lineages appear simultaneously at the beginning of the Cenomanian and "pass through corresponding approximately synchronous stages." Evolution was traced in the characteristic genus Metopaster. The ossicles of the Cretaceous Asteroidea were found to undergo parallel changes in shape and ornament in the various "species series." Nine parallel series are recognized, with few breaks. "The evolutionary changes in the forms are so very gradual that it is difficult to find forms which are distinctive of the various horizons." Variation in the ossicles was continuous as regards height, length, and other characters. Saltations occur, but are lost after a comparatively short life. Periodicity or rhythm was found in the character of the ossicles. There may be a short period of katagenesis or regression, followed by a long period of anagenesis. In the Ammonites of the Inferior Oolite the ornamentation may similarly develop, regress, and later develop again. Forms lacking the necessary plasticity (variations) become extinct, but measurements do not support the view that selection moulds the low into high forms of ossicles, and the regression period is unexplained.

Spencer concludes that the lineage history is predetermined. "Parallel stages are passed through by all the races" of *M. cor-anguinum,* as with the Ammonites. The course of some variations at least is not fortuitous but predetermined. Closely allied races differ in their capacity for elaboration of ossicles. Step mutations account for the origin of

Metopaster uncatus and *Mitraster Hunteri* as branches from the *Metopaster Parkinsoni* stem, which continues. The hereditary changes are in definite directions which are, as a whole, parallel in character for the majority of the lineages. But Spencer concludes that the "rate of elaboration" of the ossicles is determined in some way by the environment. Whatever the predetermination may mean, there is no doubt of the abundant parallelism.

It is obvious that in any such scheme both time and space are involved. Successive forms arise in a lineage along a time scale. The parallel lineages in some phyla of animals may be widely separated in space. The relation of these lineages in space and time, as well as their breakup and classification into species and genera thus becomes a matter of practical moment. There are various possible solutions of this problem, but a uniform treatment is generally impossible because of the incompleteness of the geological record. This often, in fact generally, makes it impossible to know whether two related fossil forms belong to the same lineage or to different ones. Only in such groups as Echinoderms, where for limited periods the fossil record of past life in certain areas is reasonably complete, can the relations between lineages be clearly determined. Nevertheless, the fact has to be recognized in all paleontological records that the time series and the space series have both existed in the past, with generally only occasional fossilization taking place. The more incomplete the record, the more difficult it is to determine whether any two fossils belong in the same or in different lineages. This is now a basic difficulty in determining the phylogeny of man.

The simplest and natural method where the fossil record is abundant is to regard the time scale as proceeding along the lineage, different levels or grades of evolutionary advance being recognized as separate genera. The parallel lineages at any one level or horizon will then represent different species of the same genus if all are changing at the same rate. This is the simplest condition that can be expected. It will generally be complicated not only by branching and anastomosis of lineages but also by gaps of varying extent, rendering difficult the determination of how a given lineage continues beyond a gap, or whether it disappears. Another complication is the fact that in related parallel lineages the evolution, at least of certain characters, may proceed at different rates. The migrations of all organisms, even man, were limited in the past, and so any given lineage may be expected to show limitations in space. This is now being recognized as true of human fossils. In the matter of classification it is therefore natural to regard different levels of evolution in any lineage as genera, and the diversities arising at any level within parallel lineages as species. The genera will thus be determined on a time basis and the species will be classified on the basis of their spatial distribution. Thus in the paleontological record a species belonging to one genus often passes directly into a species of another genus (but belonging to the same lineage), whereas the branching of one twig or lineage may result in the production of several contemporaneous species of the same genus.

These remarks about lineages of course involve considerable simplification. The history of any lineage will be further complicated by such factors as shifting in populations, variations in numbers, and invasions from other lineages; but these will often disturb without disrupting the lineage. It must also be remembered that in fossils as a rule the skeleton only is preserved, whereas many specific characters of every living species in higher organisms are based on external features of color and epidermal structure in the widest sense, which perish in the process of fossilization. Most fossil "species" based on skeletal characters in the higher animals thus probably represent differences which are really of generic value.

Lang made a similar elaborate study of the Cretaceous Polyzoa. In a further elaboration and explanation of evolution in the subfamily Pelmatoporinae, Lang (1920) states that other subfamilies of Cribrimorph Cretaceous Polyzoa show the same principles, which are remarkably similar to those found by Spencer in Cretaceous starfishes. He finds that "various Cribrimorph stocks had arisen independently over and over again from Membranimorph ancestors, had run through a more or less similar evolution, and, finally, become extinct; so that the many forms described under *Cribrilina, Membraniporella,* and other Recent genera were really in no way closely related to these, and that the Cretaceous Cribrimorph forms, in consequence, needed at least a generic nomenclature of their own."

The Cretaceous forms fell into ten main stocks, between which no direct relationship could be discovered, and whose common ancestor must be sought far back among the primitive Membranimorphs, so each is given the status of a family. Without going into details, we may quote his conclusion (p. 217): "The whole tendency of this paper is to show that parallelism of evolution occurs in the characters of the *Pelmatoporinae* and that the method of formation of the secondary aperture differentiates the various genera; that a similar parallelism occurs within each genus, making the diagnostic features of every lineage very difficult to determine." The Pelmatoporinae are thus a natural group in which the earlier and simpler forms pass gradually into species which are later in time and more specialized in structure. This specialization consists fundamentally in the laying down of more calcium carbonate along definite tracts. "This results in a parallelism in the evolution of each lineage tending to produce homoeomorphic forms, and suggesting a predisposition in the parental stock to evolve along determined lines." There are thus certain characteristic differences in each lineage, and excessive elaboration in each has led to extinction. The Tertiary and Recent Cribrimorph genera must then have developed independently from less specialized forms.

It is necessary to conclude that the principle of parallel variations in related groups and the principle of evolving gigantism in each phylum are two of the most general laws of paleontology.

In order to show how general is this principle of parallel lineages in evolution, we may consider a few further cases from the fossil record. First let me remark that it is in the transitions from one phylum to another that paleontology can furnish the greatest analytical aid. These transitions are like the turning of a corner; they are marked by rapid changes of direction. Some think they take place only in small populations, in which Sewall Wright has shown mathematically that evolution should proceed more rapidly than in large populations. However, some of the modern studies indicate that large numbers of individuals in many parallel lineages were involved to a certain extent in some transitions.

In an early study of the origin of land Tetrapods (Amphibians of the Coal Measures) from the Crossopterygian fishes, Gregory (1915) assumes "convergence" in many characters of the fishes. The otic notch of the skull was probably inherited from the Rhipidistia but may have developed independently (as a parallel) in the Stegocephali. There is in many cases no doubt of the existence of parallelisms in this transition. The problem is rather to determine from which group a particular character has been inherited and in which it has appeared as a parallel. The whole phylum of Angiosperms is shot through with similar problems in the cross-relationships of particular family characters. This is brought out, for instance, in the study of Hayata (1921) on the flora of Formosa.

The transformation of fishes to amphibians took place by the modification of many sets of organs, these modifications occurring independently over a long period of time. Thus a lung was generally present in primitive bony fishes, and, contrary to older views, if lungs and swim-bladder are homologues, the latter arose through modification of the former. In all these cases selection of parallel variations seems to have been the guiding factor in adaptation to a new environment. Since the transformation of fishes into amphibians took place so early (in the Devonian), the steps in the transition are naturally less well documented than in later paleontology. This long drawn out metamorphosis resembles in some respects that of Mammalia (to be considered shortly) but differs greatly from that of Reptilia, whose earliest members differ only slightly from the Embolomerous Amphibia. It adds another to the many instances in paleontology of "evolution proceeding steadily in definite directions with time" (Watson).

The situation is even clearer in Amphibian evolution. The large Labyrinodonts are in three morphological stages, (1) the primitive transitional Carboniferous Embolomeri, (2) the Rachitomi, occurring throughout the Permian, and (3) the Stereospondyli, from the lower Triassic to the Rhaetic. Watson (1920) finds that the Rachitomi were derived from the Embolomeri and must have given rise to the Stereospondyli, the two latest groups being related by many characters. In this long period of evolution there were many changes from 2 to 3. These changes included (a) a remarkable increase in the size of the openings in the palatal bones; (b) a steady reduction in the length of the quadrate ramus of the pterygoid, so that the lower jaw articulation, at first far behind the connection of the skull with the neck, comes finally to be in front of that plane. These changes are found to be common to all the larger Amphibia, but they proceeded at very diverse rates in different stocks; (c) gradual flattening of the dorsal surface of the skull and a decrease in depth has occurred, but some aquatic members were already considerably flattened in the lower Permian; (d) reduction in the basi-occipital and basi-sphenoid, so that the tripartite condyle of Eryops is converted into the double condyle of Capitosaurus. Watson refers to seven other transformations in the skull.

Lydekkerina is a specialized type, "far off the line of descent of Capitosaurus," but intermediate between Eryops and Capitosaurus in a series of characters. In the whole group he finds "a definite trend of evolutionary change in the basicranial and otic regions of the skull, which continues throughout the history of the group." The Brachopidae, a very distinct natural family, shows "an exactly similar series of changes," each step in this family being taken rather early.

The pair of occipital condyles are an adaptation, allowing greater dorsiventral flexibility in opening the mouth. They have developed independently in an exactly similar manner in the Cynognathidae, in which the skull is not flattened. Watson concludes that the six characteristic features of the skull which developed in these Amphibia must have evolved independently in the Urodela (tailed Amphibians) and the Ecaudata (frogs and toads), which have a different ancestry. "This case of the parallel evolution of diverse branches of the same class is perhaps the most striking that paleontology has yet revealed to us." Some further parallels to these changes can also be found in the Reptiles.

In a later account, Watson (1926) recognizes that the Labyrinthodonts show at least six and probably many more developments in a time sequence, occurring "independently in a strictly parallel manner in many distinct phyla" of this group. In the cases that have been followed, these changes progress regularly. Other evidence renders it certain that the Lower Permian Labyrinthodonts were terrestrial, the Upper Triassic aquatic. Hence these Triassic Amphibia were descended from terrestrial forms which had taken to the water, just as living Perennibranchiate Urodela (salamanders and their relatives) are secondarily aquatic. In the twenty species of Embolomeri in the British Carboniferous there is a wide range of adaptive radiation. Some were primitively aquatic with no terrestrial ancestors, others were land-living, and (if Watson's interpretation is correct) a large group had returned to the water.

In a recent monograph which includes all the Stegocephalia, Case (1946) recognizes twelve different non-adaptive trends in the skeleton, and seven adaptive changes which appear repeatedly in the different lines of descent. The latter affect the shape and other features of the skull, the number and form of the teeth, form of the body and presence or absence of dermal armour. The very recent monograph of Romer (1947) on the Labyrinthodonts recognizes parallelisms in their development and that of other tetrapods from Crossopterygian ancestry, with an early separation of Labyrinthodonts into the Embolomere and Rhachitome types.

Watson (1921) concludes that the Lower Permian Pelycosaurs gave place to the Triassic Cynodonts through an evolutionary series in which the Cynodont skull was derived from a primitive reptile like Seymouria (Cynodectes).[3] The Cynodonts were carnivorous therapsids with a well-developed secondary palate. Broom regards their development as a result of the need to capture the more active reptiles for food. This necessitated a rapid gait with the body off the ground. The same evolutionary trends were followed in many allied branches in the Pelycosaurs, Gorgonopsids, Therocephalia, and Deinocephalia, the changes being often adaptive. The Pelycosauria, which are found mainly in the Carboniferous and Lower Permian of Europe and America, resemble the living monotremes and are closely related to the ancestral therapsids. The Anomodontia (Broom, 1914) are another group of mammal-like reptiles, off the main line (i.e., a parallel development), from the Middle Permian to the Triassic in the South African Karroo beds. The genus Dicynodon, with about forty species ranging in size from a rat to a tapir, is typical. They have no incisors, but some have canines, some molars, some both, and they show many mammal resemblances.

In a recent study, Romer (1946) concludes that Limnoscelis is "close to an ideal ancestor for most if not all of the reptilian groups." This primitive reptile was described by Williston in 1911 from the early Permian or late Carboniferous rocks of New Mexico. It cannot be at the base of the reptile phylogenetic tree because of the time factor, but it is a relict type, little modified from the real ancestor. It is "in almost every regard an ideal ancestor for the Pelycosaurs and, through them for therapsids and mammals." Romer suggests that the early reptiles were still essentially aquatic, like their amphibian an-

---

[3] Broom (1930), however, believes that Seymouria is an amphibian which resembles reptiles in certain characters, through parallel development. It is a remarkable "mosaic" of amphibian and reptilian characters with very few intermediate features, indicating rapid independent development of separate body regions.

cestors. These early amphibious reptiles may have laid their eggs on land like the modern turtles.

In animal evolution we see group after group increasing in size until they become gigantic and end in extinction. Cowles (1945) attempts to explain the extinction of giant reptiles by assuming that climates became warmer until the high temperatures produced sterility of their male germ cells. He points out that in the endothermic mammals Bergmann's law holds, that in higher latitudes (colder climates) the animals of many mammal species are larger, whereas in the reptiles (exothermic) the smaller species or varieties occur in higher latitudes. However, the evolutionary tendency to increase in size appears to be much too widespread both in time and space, and in the animal kingdom, to be accounted for by any such principle, either of climatic cooling as applied to the mammals or of warming as applied to the reptiles. The increase in size of the horses and other groups of mammals began long before the climatic cooling of the Pleistocene, and there is no evidence of increasing temperature while the reptile groups were reaching their maximum size. Similarly, gigantism in various invertebrate groups develops without any apparent reference to temperature changes.

The origin of the mammals, having occurred much more recently, can be examined in greater detail. It offers perhaps the clearest instances of parallel lineages in the transition period from reptiles to mammals. Sir Richard Owen, beginning as early as 1844, studied the Dicynodont reptiles from South Africa in which the teeth were already differentiated into incisors, canines, and molars. By 1880 he had concluded that the mammals were descended from this group of reptiles. The Pelycosauria, as already mentioned, are closely related to the therapsid ancestry. Great numbers of fossil skeletons of Therapsida have since been obtained from the Karroo beds of South Africa, of Permian and Triassic age, and Broom showed in 1910 that the North American Permian Pelycosaurs agree essentially with the South African Therapsids, but are rather more primitive, in correlation with their greater age.

Broom (1914), in discussing the origin of mammals, recognized several sub-groups of Therapsida in South Africa, mammal-like reptiles, the lowest of which were nearly related to the Texan forms. He shows how in each of these groups there was progression from small to large size. All the carnivorous therapsids from the later Triassic of South Africa differ from the Permian types in having a well-developed secondary palate and were formerly grouped together in the Cynodontia, having mammal-like dentition. He regards it as uncertain whether some of them are ancestral aberrant forms or whether they acquired cynodont characters by convergence (that is to say, parallel variations). More recently, Broom (1938) shows that the skull of a newly described genus, Millerina, from South Africa, is remarkably like that of Eothyris, a pelycosaur from America described by Romer. He believes that these two genera form a link between the higher cotylosaurs of Africa and the most primitive pelycosaurs of America. Both these groups crawled like reptiles and did not walk like the mammals.

The earliest mammals, in the Upper Triassic and the Jurassic, fall into several distinct groups even at their first appearance, and this can best be accounted for through parallel variations in the transition period. Broom lists thirteen mammalian skeletal characters found in the Cynodontia and not in the Therocephalia. He shows peculiarities of the Therapsida in shoulder girdle, pelvis, and limbs, all related to walking with the body off the ground, which distinguishes mammals from reptiles. The Therapsida were regarded as a group showing as much variation and as many orders as the mammals themselves.

Matthew (1927), in a summary of his great knowledge of fossil mammals, states that Dromatherium and Microconodon, represented by two lower jaws of Upper Triassic age found in 1857 in a coal mine in North Carolina, are really pro-mammalian reptiles, while Microlestes, from the Upper Triassic of Germany, consists only of a few isolated teeth whose affinities cannot be determined. Records of true mammals begin with minute forms in the Jurassic, the jaws already representing ancestors of several living orders as well as others that became extinct. Of the latter, the Multituberculata, extending from the Jurassic to the Eocene, were regarded as a separate stock from reptile ancestry, coördinate with the Prototheria and Eutheria and having a cynodont pelvis. In 1924, in the Cretaceous of Mongolia among dinosaur remains were found skulls of two small mammals, one of them multituberculate (referring to the peculiar character of the teeth), the other (Zalambdolestes) a placental having tooth characters of the insectivore-creodont pattern. The other Mesozoic mammals were apparently arboreal.

In the Paleocene, following the extinction of the dinosaurs, the ancient Cretaceous orders of marsupial mammals and Insectivora developed. The Paleocene thus saw the culmination of the archaic mammals, the Multituberculatae becoming extinct in the basal Eocene. The only modern orders are marsupials (related to the opossum), insectivores of early Mesozoic origin, and Carnivora represented by primitive types. At the end of the Paleocene small Primates and primate-like Insectivora appear, showing relations to the tarsioids and tree-shrews to be discussed later. Chiromys appears to be a true lemur with rodent-like incisors, a parallel specialization to the earlier Plesiadapis.

With the Eocene a new mammalian fauna suddenly appears, the same genera being represented in North America (Wyoming and New Mexico) and Europe (the London Clay, Paris Basin, Belgium). They must have migrated to these areas, presumably from some Asiatic region of origin, perhaps Mongolia. The principal orders of modern mammals thus began in the Eocene and diverged into families during this period. The European and American forms also diverge, but evolve in lines which are more or less parallel. The Eocene Carnivora all belong to the primitive Creodonta, giving rise in the Upper Eocene to forms which are ancestral to the modern Canidae. The Tillodonts, represented by the Lower Eocene Esthonyx, developed gnawing incisors parallel to those of the rodents, but unrelated to them. At the end of the Eocene the older orders of mammals became extinct. Many of the Eocene phyla were also replaced by more modern forms and the Oligocene begins a great faunal invasion of new forms from elsewhere, probably connected with climatic changes.

In South America the carnivorous marsupials developed as beasts of prey parallel to the Creodonts of North America. Some suggest that these and other marsupials passed over Antarctica as a land bridge to Australia where they became the modern Thylacines, but Matthew (1912) thinks they developed in Australia as a parallel from Paleoarctic Cretaceous marsupials similar to the American. The South American Edentata and Platyrrhine monkeys also developed in South America in relative isolation. Simpson (1940) and others have shown that the Tertiary fauna of South America had a North American origin but developed an extraordinary assemblage of unique mammals in isolation until the two continents were reunited in late Pliocene times. This connection began early in the Tertiary and was probably at first a series of stepping stones (islands) before it became a continuous land bridge. Simpson finds no evidence that any considerable part of the South American fauna came from Australia over the Antarctic land bridge and he believes that the composition of the South American land fauna is inconsistent with the theory of continental drift. He believes that the Australian Dasyuridae are a parallel development to the South American Borhyaenidae, and similarly the Peramelidae, are parallel to the Diprotodontidae. With this brief summary of some features in mammal evolution we may return to the problem of mammal origin.

In connection with the comprehensive studies of Broom and Watson there has been much further increase in knowledge of the South African Therapsida, and Olson (1944) has recently made a still more detailed approach to the problem of therapsid phylogeny and the origin of mammals. He regards the Gorgonopsia as probable ancestors of the Cynodontia and finds that the repeated independent development of suites of characters in the suborders of Therapsida is demonstrated. He concludes that each suborder of this group approached the mammalian threshold independently. In a detailed study he concludes that the mammals had a polyphyletic ancestry, at least three lines of mammals arising independently from different therapsid stocks. By the end of the Permian a vast assemblage of animals existed in South Africa, mammal-like in varying degree and with an almost infinite variety of association of mammalian and reptilian characters.

Nothing appears to be known of the epidermal or pigmental characters of these animals, or of their methods of gestation or feeding of the young, all of which are very different in the two groups today. But there is no reason to suppose that the present distinctions between reptiles and mammals were in any way adhered to in these respects, since they were completely disregarded in the skeletal variations. One can consider this a transition period only in the sense that the separate phyla, reptiles and mammals, have since become sharply demarcated through extinction of the intermediates. Before this extinction the population of species seems to have included every combination of reptilian and mammalian characters which had resulted from free variation over a long period. In the early Triassic these therapsid faunas were impoverished by increasingly harsh climatic conditions, only the more mammal-like being able to survive. Four of these lines were potential mammal ancestors and from them the mammals arose in at least three distinct lines and perhaps more, but the exact therapsid source of each mammalian group remains to be determined. The survival of the Monotremes shows what an extraordinary assemblage of characters some of these extinct animals must have possessed.

This impoverishment of the Permian-Triassic transition was survived by the therapsid groups, Cynodontia, the advanced Anomodontia, and the Therocephalia, all of which approached the mammalian threshold more closely than their less successful contemporaries. Elimination in the early Triassic thus acted as a filter, permitting only the most advanced (mammal-like) therapsids to pass. This surviving assemblage lay very close to the base of the mammal stock. Those with mainly reptilian or mammalian characters survived in their different ecological niches, while those with mixed characters and habits were unable to do so.

Among the mammal-like reptiles were the Dicynodonts. Broom recognized in Dicynodon certain mammalian features but, from a study of the endocranial cast, Schepers (1937) found by comparison with the brain of living reptiles that the Dicynodon brain was of a distinctive type. It shows evidence of specialization but has primitive features linking it with the Chelonia and Amphibia. This he regards as a case of parallel evolution in the same geological period and under similar conditions, since the Chelonia and Amphibia are not related to Dicynodon. Schepers remarks that the nervous system "on account of its innate peculiarity of construction, and the mechanical nature of the neurobiotactic force operative within it" can permit of only a limited range of variation. Parallelisms in the nervous system might thus be more frequent than in other organ systems. Dicynodon thus shows similarities to the mammal skeleton and to the chelonian and amphibian brain. There are probably many such cases of cross-relationships in animals. They are certainly abundant in the flowering plants.

The mammals, as indicated above, appear to have originated not from one line of therapsids but from several, probably three to five, and the monotremes may have come independently from a very primitive stock of Cynodonts. The characteristic double occipital condyle and other features of the mammal skull developed independently in the Cynodontia, the Anomodontia, and certain bauriamorphs. The variations in this great plexus of "transitional" forms were not adaptational except in so far as they were pre-adapted for the harsher conditions of a future climate. Many of the characters seem to have appeared not in one locality but over large parts of the earth's surface. This was not selective adaptation, because the mammal-like variations appeared in some lines while in other contemporary lines, such as the pro-lizards, these characters never appeared. Hence we have progressive development in limited lines, controlled by similar gene complexes and for this reason parallel. The Tritylodontoidea are another reptile group which independently approached the mammalian level but never reached it.

Viewing this transition from reptiles to mammals as a whole, we have to recognize a riot of variations giving rise to large groups with many combinations of what we now consider reptilian or mammalian characters. These characters were non-adaptive in their origin and involved many parallel variations. Later, when the pinch of a hard climate put them to the test, those with incongruous combinations of characters were extinguished and only those with more mammalian or more reptilian features were able to survive in their various ecological niches. None of these were as yet mammals, but from several different lines of survivors the true mammals arose through parallel variations.

Simpson (1945) has recently expressed a monophyletic view. Discussing the development of the mammals from reptiles, he says (p. 17), "They certainly arose from a unified group of reptiles of much smaller scope than a class, perhaps a family or perhaps a superfamily." But he says (p. 165) regarding the Upper Jurassic Multituberculata, "Everything points to their having been distinct from all other mammals since the very beginnings of the Mammalia." He also recognizes the monotremes as a subclass, equally distinct from the rest of the mammals. The Tritylodontoidea, one of the two subclasses of Microtuberculata, have recently been found to be reptiles and not mammals, showing how tenuous and indeed arbitrary is the line between these two groups in their fossil ancestry. Simpson similarly concludes (p. 169) that the Jurassic Triconodonta are a parallel evolution from triconodont reptiles.

The principle of parallelism, so broadly exemplified here, has continued in the later evolution of mammals and man, accompanied of course by the principle of adaptive radiation. The latter type of variation has led groups of mammals into every habitat, from the marine seals and whales to the parachuting phalangers and squirrels and the flying bats. The former type of variation comprises the similar but independent mutations in different groups of mammals. Such marsupials as the thylacine "wolf" of Australia are remarkably similar to the placental wolf, and most of these similarities must be parallel developments. Bensley (1903) regarded the Australian marsupial fauna as of South American origin, having come over the Antarctic land bridge, since there is a special resemblance to the Patagonian Miocene fauna. He thinks Thylacinus may have migrated from South America, and that the opossums (Didelphyidae), now found mainly in South America, took the same route in the early Eocene or later Cretaceous. As already mentioned, Simpson does not accept this view. A *late* isolation in the Australian continent is indicated by the fact that the Australian marsupials, in their radiative evolution, have only differentiated into eight families, not orders. On the other hand, it is possible that evolution was slower in the absence (until developed within the group) of predatory enemies. As an example, Myrmecobius, the marsupial anteater, has its teeth specialized, but not obliterated as in the placental anteaters. Incidentally, this conception of the part played by the Antarctic continent appears to be in harmony with the theory of continental drift, which has developed much more recently.

Gregory (1920–21), in his great monograph on mammal (and earlier) dentition, points out that Myrmecobius was the culmination of a line of marsupials specialized for insect diet in which the incisors, canines, and premolars were less specialized than the molars. He says (p. 160), "In the straightness of the upper and lower tooth rows, which form nearly parallel lines, in the wide spacing between the teeth and in the flatness and length of the palate, the dentition of Myrmecobius parallels that of the armadilloes," although they are far from being related.

From the many other cases of parallelism in the evolution of dentition cited in Gregory's monograph, the following may be mentioned. The crescent-shaped pattern of the molars has often evolved among placental mammals in adaptation to leaf-eating habits. The same pattern is found in the Australian (marsupial) Koala (Phascolarctos).

In the Kangaroos, bilophodont molars are found, in which opposite cusps are linked by cross-crests. This development is parallel to that in tapirs and other ungulates, also in the Old World monkeys. The Australian marsupial, Phascolomys, has one pair of gnawing incisors growing from persistent pulps, followed by a wide diastema in the jaw, a condition parallel to that in the beaver. Thus the marsupials show parallel development to the placentals, not only in dentition but also in their body form and habits (adaptive radiation).

In modern Carnivora (Eucreodi) the first lower molar (m_2) and the last upper premolar (m^1) form carnassial teeth for tearing flesh, while in the fossil Pseudocreodi m^1 and m_2 or m^2 and m_3 are carnassial. Butler (1946) finds that extreme carnassial dentition developed at least four times in the extinct Pseudocreodi, twice in the Marsupials, and more than once in the modern Carnivora. He indicates other parallelisms in the dentition of the mammals.

Several of the evolutionary principles we have been considering are well exemplified in a study by Simpson (1937) of the Notoungulata, a group of herbivores, mainly hoofed, in the Tertiary of South America, which became extinct during the Pleistocene. The first fossils of this group were discovered by Darwin in South America and described by Owen. By 1890 some ten genera were recognized. Toxodon, the first described, was of Pleistocene age; Nesodon was Miocene. As a parallelism, some of the group developed incisors like the rodents—an adaptational specialization which has evidently happened in mammals many times. Three groups of Notoungulata were soon recognized. In one of these—the entelonychians—"successive genera that are almost identical in dentition are profoundly different in the limbs, just as some successive genera of typotheres are hardly distinguishable in the limbs, but very distinct in the dentition." Similarly, in the typotheres one set of structures (the limbs) may be relatively invariable while another set (the teeth) in the same group are highly variable and evolving rapidly. This principle in relation to human evolution will be referred to in Chapter 11. In two related groups of Notoungulata, certain parts (a) will become specialized and certain others (b) remain primitive in one group, whereas in the other group (a) will remain primitive and (b) become specialized. In the typothere dentition, a limited period of progressive change was preceded and followed by long periods of relative stability. Different rates of evolution in the group were very common. The Notoungulata show many adaptive parallelisms to the North American and Holarctic Ungulata. The dentition of the family Notohippidae developed closely parallel to that of the Equidae, similar mutations appearing independently in both families; but the parallel development of the two families differed widely in time as well as in space. The notohippids began these developments in the Eocene, but the corresponding development of the horses took place only during the Oligocene and Miocene.

The lemurs are included by most zoologists in the Primates, although Wood Jones (1918) excludes them. He ascribes their primate characters, not to ancestry but to parallelism; but this seems to be extending the idea of parallelism too far. The more advanced and later lemurs closely parallel the higher Primates in some features; some lemurs are so monkey-like that it has been difficult to prove their lemuroid origin. The Nycticebidae closely parallel the tarsoids in skull, dentition, and limbs. There is a wealth of fossil lemurs in Madagascar, of Pleistocene and earlier age. Some of them have become gigantic, such as Megaladapis with a skull 31.5 cm. long. Some of the North American Eocene lemuroids appear to have been evolving towards (parallel to) the South American monkeys. Simpson (1940) considers it probable that these South American Primates are descended from early Tertiary lemuroids of North America.

Owing to the dentition of the American Eocene Hypsodontidae they were formerly placed in the Primates, but Matthew showed that their limb bones placed them in a totally different order, as primitive ungulates. Several modern families of lemurs have a specialization of the incisors, canines, and first premolars which is associated with an enlarged tongue and connected with the habit of cleaning the fur with the lower front teeth. These conditions appear to have been acquired independently in different families.

The Archaeolemuridae are an extinct Madagascar family of lemurs, in which the upper molars were bilophodont, in other words, the summits of opposite cusps are linked by cross-crests. The same condition has developed independently in the tapirs and other ungulates, as well as in the Cercopithecidae (a family of Old World monkeys). They also have, like these monkeys, a short face and expanded brain. In the family Indrisidae the premolars are reduced from four to two, the first and second disappearing. This has occurred independently in many primate groups.

The evolution of molar patterns in marsupials closely parallels that in placentals, from the primitive triangular (trituberculate) type in the oppossums to the diverse shearing, cutting, crushing, and grinding molars of modern marsupials. Probably nowhere are adaptive mechanisms better exhibited than in the teeth, yet the means by which these adaptations arise and complement each other in the two jaws is by no means easy to understand in the ordinary evolutionary terms.

The lorises of Asia and Africa and the galagos of Africa are small, large-brained, short-jawed forms derived from a primitive lemuroid stock. They are parallel to the tarsioids in these features, as well as in the large eyes (nocturnal) and large internal ears. The Tarsioidea are separated from the lemurs, as a suborder coördinate with the Lemuroidea and Anthropoidea. The only surviving species, *Tarsius spectrum,* is a shy nocturnal tree-dweller in Borneo and the Philippines. These three groups apparently have a remote common ancestry. There are numerous Eocene tarsioid genera in North America, all small or minute, with large orbits and wide brain case, and seven Eocene genera (different from the American) in Europe. They are derived from a common Lower Eocene stock, but both groups are polyphyletic. The upper molars have evolved from trituberculate to quadrangular, as in other mammals. Parapithecus, from the Lower Eocene of Egypt, is one of the most primitive known anthropoids and is near the tarsioid stem, some authors considering the tarsioids ancestral to the anthropoids. The Catarrhine (Old World) and Platyrrhine (New World) monkeys form a remarkable parallel series derived from primitive stocks which probably began with quite different genera of tarsioids. The Notharctidae are probably ancestors of the Platyrrhinae. The marmosets (Hapalidae) of tropical America are apparently derived from ancestors of the Cebidae, some of whose characters they parallel. It will thus be seen that the evolution of the mammals is shot through with parallelisms, some between nearly related and others between more distantly related groups.

That there is much parallel evolution in the development of the Proboscidea is emphasized by Watson (1946). His views are based mainly on the extensive monograph of Osborn (1936–1942), who recognized six parallel lines. The complicated pattern of the teeth is an evolution in response to the food habits (browsing or grazing). The increasing size of the animals necessitates longer eating periods, up to eighteen hours per day. Beginning with Moeritherium, from the Upper Eocene of Fayum, Egypt—an animal only two feet high and related to the Sirenia—they progress through many types to the Pleistocene mammoths and the modern elephants. Cuvieronius, in the Pleistocene of South America, survived so late that the type specimen was killed, cooked, and eaten by men with obsidian implements, who carved bones and had decorated pottery which can be dated as A.D. 200–400. The increasing weight of the proboscidians has been a driving force in modifying the structure of their legs and other parts. The Pleistocene mastodon of India, Pentalophodon, shows remarkable resemblances to the genus Elephas which are the results of parallel evolution. The Miocene Serridentines, in which the great molar teeth develop serrate spurs which pass from the ridges into the valleys, agree with the Trilophodonts, the two lines pursuing parallel evolutionary courses, but the Serridentines are generally smaller at any time. They probably occupied a different ecological niche; their evolutionary advance was slower and less extreme. The Zygolophodonts were another line, probably forest dwellers, their evolution probably delayed by their retention of browsing habits. Beginning with Miomastodon in the Miocene, they later gave rise to the true elephants, beginning with the Upper Pliocene and Pleistocene Stegodon of India and spreading in the Pleistocene to China and Japan, Java, the Philippines, and central Africa. From the most primitive *Archidiskodon platifrons* of Pleistocene India three subfamilies of true elephants developed, showing parallel evolution, adding to the number of plates and increasing the height of the molar teeth. In all their branches the Proboscidea show the same general course of evolution. At first the lower jaw lengthens by extension of the diastema. Then this elongation of the jaw suddenly stops and it begins to shorten, the lower tusks, which have developed by overgrowth of the incisors, often being lost at this time.

The lower tusks incidentally take a great variety of forms (even becoming shovels) according to the uses to which they are put. The cheek teeth are reduced until only the enormous third molars remain. There are many correlated changes in the skull, to support the huge tusks and increase the efficiency of the grinding apparatus. In the meantime the trunk has been developing as an elongation and specialization of the anterior part of the face and the upper lip. When it has developed sufficiently to be of prime service as a prehensile organ, the lower tusks have lost their uses and sooner or later disappear, presumably through one or more negative mutations.

The parallel changes of the Proboscidea are all related to the increase in weight and consequent necessity to strengthen the legs. The great lengthening of the legs is an example of heterogonic growth. The brain remained relatively small, but the domesticated elephant nevertheless shows many remarkable instances of intelligence, an intelligence which impresses those who have seen elephants at work. Evolution was much more rapid in the elephants than in the horses, as shown by comparing the Lower Pliocene Stegolophodon and the Pleistocene Elephas on the one hand with the Lower Pliocene Pliohippus and the Pleistocene Equus on the other. This may have been because the elephants moved in smaller herds as breeding units so that any advantageous mutations arising could thus be more rapidly incorporated into the strain. This important evolutionary principle, that even in related stocks the rates of evolution may be quite different, regardless of the length of the reproductive cycle, is at present but little understood. Simpson (1944) has discussed it at length, with many examples from paleontology and an explanatory analysis.

The evolution of the elephants in Europe during the Pleistocene, as worked out by Soergel (1912), is also illuminating. From the genus Stegodon arose *E. platifrons,* which gave rise to the two species, *E. meridionalis* and *E. hysudricus. E. meridionalis* is found in the Pliocene in nearly all places where Pleistocene elephants occur. It was frequent in the Mediterranean region, France, southern and middle England, central Germany, the northern Balkans, and southern Russia (Odessa). During the Pleistocene *E. meridionalis* produced *E. antiquus* and *E. trogontherii,* which later spread eastward, and are so nearly related that one is sometimes regarded as a variety of the other. *E. trogontherii* was more northern, extending from southern England to central Germany and southern Russia in the first interglacial, but was not found in Spain, Italy, or Greece. In France it was less frequent than *E. antiquus* and more or less intermediate in character. In the second interglacial it gradually develops into *E. primigenius,* which is found widely distributed in Europe in the third interglacial, from Italy and France to England, Germany, Switzerland, Hungary, Finland, Siberia, in countless herds, and from Alaska to Mexico. By the beginning of the Würm it was infrequent in Europe but survived on the Siberian tundra. In the east, *E. hysudricus* gave rise to the Indian elephant, *E. indicus,* the African elephant coming from a separate ancestry.

The Titanotheres are in some respects an even more instructive group than the elephants. Although they only existed from the Lower Eocene into the Upper Oligocene, yet they were a larger group than the elephants and they evolved much more rapidly than the horses. The difference between the Lower Eocene Eotitanops and the Lower Oligocene Brontotherium is vastly greater than between the Lower Eocene Eohippus and the Lower Oligocene Mesohippus. There were twenty branches of the Titanothere family before they suddenly became extinct, while the other odd-toed mammals, the tapirs, horses, and rhinoceroses, continued into the present time. Osborn (1929), in his great monograph of the group, shows that, so far as known, they were confined to a small area near the 40th parallel of north latitude in western North America, Asia, and Europe. Even in the region of four western United States they evolved along many lines of descent. This polyphyletic evolution showed many parallelisms, but also local, adaptive radiation. From small forms they increased in size until they were second only to the elephants, undergoing at the same time adaptation to aquatic, forest, savanna, and plains life. In the ungulates, the development of multiple lines of descent in the same region was more common than monophyletic evolution.

These different lines of descent, or phyla, in the Titanotheres are distinguished mainly by contrasting proportions of the head (dolichocephaly and brachycephaly), of the limbs (dolichomely and brachymely), the feet (dolichopody and brachypody), and the teeth (hypsodonty and brachydonty). But long or short head, face, limbs, and feet "occur in geographical species and subspecies in their corresponding stages exactly as they occur in the geological phyletic time series" (p. 18). Osborn thus distinguishes between "ascending mutations" of Waagen and "contemporaneous mutations" of de Vries, one being in effect phyletic, the other differentiating. In the genus Brontotherium he recognizes ten successive species, from the Lower Oligocene *B. leidyi* to the Upper Oligocene *B. platyceras,* a form with a bifurcated flat horn on its nose.

The various series of Titanotheres show progressive changes in proportion, in ontogeny, and in the rate (velocity) of progression or retrogression in the phylogeny of each character, for example, in the indices and ratios of teeth to skull, of skull to body, body to limbs, and so on, while from time to time new characters (rectigradations) make their appearance, so that some become heavy and slow, others more cursorial. The proportional (allometric) changes are in part at least adaptational. The molars of the earliest Titanotheres, Lambdotherium and Eotitanops) were similar, cusp for cusp, to those of Eohippus, but whereas the horse dentition became highly specialized for grazing, the Titanotheres remained browsers, and their molars in no case became fully adapted to grazing. Their sudden extinction would appear to have been because the mutations which could have saved them by giving them a grazing dentition never appeared.

Differences in phylogenetic velocity are another feature which is exhibited in this remarkable group. A given character evolves faster in one line of descent than in another, although the environment of both may be the same. In twelve subfamilies of the Titanotheres the cusps on the teeth and the horns on the skull evolve independently (parallel) but at different rates. In one phylum horns appear early, in another late. While new types appeared to develop irrespective of external influences, yet the action of natural selection was continuous and in Osborn's view the germinal potentialities of evolution were "evoked in response to certain environmental and habitudinal conditions." Extinction resulted from failure to cope with changed conditions or (more likely) to compete with other herbivores, as the grassy plains increased and the forest areas diminished. Only part of the abundant parallelism in the evolution of the group can be attributed to adaptational processes.

In concluding this chapter we may quote from Clark's (1934) study of the ancestry of man. He arrives at the conclusion (p. 287) that "evolutionary parallelism has been a much more common phenomenon than is usually recognized, and it must obviously be taken into account in any philosophical study of evolutionary principles."

Also, "The descendants of a common ancestor always tend to evolve along similar lines"; and again, "The evolutionary development of new somatic characters may occur in face of, or at least independently of, direct environmental influence." Finally, evolution is not "merely a matter of action and reaction between the physico-chemical factors of the environment and those of a passive or at least a neutral and completely plastic organism." These views are in harmony with the mutation theory if we assume that the mutations of organisms tend to occur in certain directions and not in others—a view which appears to be abundantly supported by the known facts of genetics.

Mydlarski (1947) has recently expressed views similar in many respects to those developed in this book. He points out that the lemurs differentiated into arboreal and terrestial forms in the Upper Oligocene. He speaks of parallel development of the different races of Homo from different ancestors, regarding both Meganthropus and Gigantopithecus as side branches.

One corollary of the principle of parallel mutations needs to be emphasized. It is, briefly, that morphological similarity is not by itself a perfect measure of degree of relationship between species or higher categories. If two distinct lineages both produce the same mutation, the two new forms will appear to be more nearly related than their ancestors, although in reality this is not so. If they have two or more mutations in common, it may be that only the fossil record can determine whether their similarities are derived through a common ancestry or have arisen later as a "convergence" resulting from parallel changes. This, of course, introduces a complication into all questions of relationship and phylogeny, but one that has to be faced. The paleontological facts show that long parallel series as well as short can occur. Selection in some form must be involved in the longer series, but may be arrested at least temporarily in the short, as we have seen in the case of the transition between reptiles and mammals. If the same variations occur in lineages of any group of organisms in different parts of the world, we are no longer compelled to assume that they have spread from a common center. They may have developed independently as parallel mutations or even parallel adaptations. We shall see that this view is supported by the known facts of mutation.

REFERENCES

Andrews, C. W. 1915. A description of the skull and skeleton of a peculiarly modified rupicaprine antelope (*Myotragus balearicus*, Bate), with a notice if a new variety, *M. balearicus* var. *major*. *Phil. Trans. Roy. Soc.* 206B:281–305. Pls. 4.

Bather, F. A. 1927. Biological classification: past and future. *Quart. J. Geol. Soc. London.* 83:lxii-civ.

Bensley, B. A. 1903. On the evolution of the Australian Marsupialia. *Trans. Linn. Soc. London.* 9:83–217. Pls. 3.

Broom, R. 1910. A comparison of the Permian reptiles of North America with those of South Africa. *Bull. Am. Mus.* 28:197–234. Figs. 20.

———. 1914. On the origin of mammals. *Phil. Trans. Roy. Soc.* 206B:1–48. Pls. 7.

———. 1930. *The Origin of the Human Skeleton*. London: Witherby. pp. 164. Figs. 46.

———. 1938. On recent discoveries throwing light on the origin of the mammal-like reptiles. *Ann. Transv. Mus.* 19:253–255. Fig. 1.

Butler, P. M. 1946. The evolution of carnassial dentitions in the Mammalia. *Proc. Zool. Soc. London.* 116:198–220. Figs. 13.

Case, E. C. 1946. A census of the determinable genera of the Stegocephalia. *Trans. Am. Phil. Soc.* 35:325–420. Figs. 186.

Clark, W. E. L. 1934. *Early Forerunners of Man*. London. pp. 296.

Cowles, R. B. 1945. Surface-mass ratio, paleo-climate and heat sterility. *Am. Nat.* 79:561–567.

Gates, R. R. 1912. Parallel mutations in *Oenothera biennis*. *Nature*. 89:659–660.

———. 1920. *Mutations and Evolution*. Cambridge: Cambridge University Press. pp. 118. Also in *New Phytol*.

———. 1944. Phylogeny and classification of hominids and anthropoids. *Am. J. Phys. Anthrop.* N.S. 2:279–292.

Gregory, W. K. 1915. Present status of the problem of the origin of the Tetrapoda, with special reference to the skull and paired limbs. *Ann. N. Y. Acad. Sci.* 26:317–383. Pl. 1. Figs. 15.

Haas, O., and G. G. Simpson. 1946. Analysis of some phylogenetic terms, with attempts at redefinition. *Proc. Am. Phil. Soc.* 90:319–349.

Hayata, B. 1921. The natural classification of plants according to their dynamic system. *Ic. Plant. Formos.* 10:97–233.

Hilzheimer, M., and R. N. Wegner. 1937. Die Chincha-Bulldogge eine ausgestorbene Hundrasse aus dem alten Peru. *Zeits. f. Hundforsch.* N.F. 7:1–43. Figs. 16.

Lang, W. D. 1920. The Pelmatoporinae, an essay on the evolution of a group of Cretaceous Polyzoa. *Phil. Trans. Roy. Soc.* 209B:191–228. Figs. 72.

Lowe, P. R. 1928. Studies and observations bearing on the phylogeny of the ostrich and its allies. *Proc. Zool. Soc. London.* 185–247. Pl. 1. Figs. 22.

Matthew, W. D. 1912. The ancestry of the Edentates. *Nat. Hist.* 12:300–303.
Matthew, W. D., and W. Granger. 1917. The skeleton of Diatryma, a gigantic bird from the Lower Eocene of Wyoming. *Bull. Am. Mus. Nat. Hist.* 37:307–326. Pls. 13.
———. 1926. The evolution of the horse. *Quart. Rev. Biol.* 1:139–185. Figs. 28.
Mayr, Ernst. 1942. *Systematics and the Origin of Species.* New York: Columbia University Press. pp. 334.
Miller, G. S. 1924. A second instance of the development of rodent-like incisors in an Artiodactyl. *Proc. U. S. Nat. Mus.* 66:Art. 8. pp. 3. Pl. 1.
Mydlarski, Jan. 1947. The mechanism of evolution concerning human phylogeny. *Ann. Univ. Mariae Curie-Sklodowska, Lublin.* 1:71–131. Figs. 8.
Newton, A., and E. Newton. 1870. On the osteology of the solitaire or didine bird of the island of Rodriguez, *Pezophaps solitaria* (Gmel.). *Phil. Trans. Roy. Soc.* 159:327–362. Pls. 10.
Olson, E. C. 1944. Origin of mammals based upon cranial morphology of the Therapsid suborder. *Geol. Soc. Am., Special Papers.* No. 55. pp. 136. Figs. 27.
Osborn, H. F. 1929. The Titanotheres of ancient Wyoming, Dakota and Nebraska. 2 Vols. pp. 1654. Pls. 236. Figs. 1436. *Monograph* 65, *U. S. Geol. Survey.*
———. 1936. Monograph of the Proboscidea. *Am. Mus. Nat. Hist.* Vol. 1. 1942. Vol. 2.
Owen, R. 1866. On the Aye-Aye (*Chiromys madagascariensis* Desm.). *Trans. Zool. Soc. London.* 5:33–101. Pls. 13.
———. 1869. On the osteology of the dodo (*Didus ineptus*, Linn.). *Trans. Zool. Soc. London.* 6:49–85. Pls. 10.
Romer, A. S. 1945. *Vertebrate Paleontology.* Chicago: University of Chicago Press. 2nd Ed. pp. 687. Figs. 377.
———. 1946. The primitive reptile Limnoscelis restudied. *Am. J. Sci.* 244:149–188. Figs. 10.
———. 1947. Review of the Labyrinthodonts. *Bull. Mus. Comp. Zool. Harv.* 99. No. 1. pp. 368. Figs. 48.
Schepers, G. W. H. 1937. The endocranial cast of *Dicynodon dutoiti* sp. nov. *S. Afr. J. Sci.* 33:731–749. Figs. 8.
Simpson, G. G. 1937. Supra-specific variation in nature and in classification from the viewpoint of paleontology. *Am. Nat.* 71:236–267. Figs. 10.
———. 1940. Review of the mammal-bearing Tertiary of South America. *Proc. Am. Phil. Soc.* 83:649–709. Figs. 4.
———. 1944. *Tempo and Mode in Evolution.* New York: Columbia University Press. pp. 237.
———. 1945. The principles of classification and a classification of Mammals. *Bull. Am. Mus. Nat. Hist.* 85:1–350.
———. 1946. Fossil penguins. *Bull. Am. Mus. Nat. Hist.* 87:Art. 1. pp. 99. Figs. 33.

Soergel, W. 1912. *Elephas trogontherii* Pohl, und *Elephas antiquus* Falc., ihre Stammesgeschichte und ihre Bedeutung für die Gliederung des deutschen Diluviums. *Palaeontogr. Stuttgart.* 60:1–144. Pls. 3. Figs. 14.

Spencer, W. K. 1914. The evolution of the Cretaceous Asteroidea. *Phil. Trans. Roy. Soc.* 204B:99–177. Pls. 7.

Vavilov, N. I. 1922. The law of homologous series in variation. *J. Genet.* 12:47–89. Pls. 2.

Watson, D. M. S. 1920. The structure, evolution and origin of the Amphibia. *Phil. Trans. Roy. Soc.* 209B:1–73. Pls. 2. Figs. 31.

———. 1921. The bases of classification of the Theriodontia. *Proc. Zool. Soc. London.* 35–98. Figs. 29.

———. 1921. On the Coelacanth fish. *Ann. and Mag. Nat. Hist.* 8:320–337. Figs. 5.

———. 1926. The evolution and origin of the Amphibia. *Phil. Trans. Roy. Soc.* 214B:189–257. Figs. 39.

———. 1946. The evolution of the Proboscidea. *Biol. Revs.* 21:15–29. Figs. 3.

3

Evolution of the Mammals

Having considered the frequency and importance of parallel variations in the evolution of various groups of vertebrates and invertebrates, we may now examine their incidence and significance in the more immediate ancestry of mammals and man. Klaatsch went to the absurd extreme of supposing that different races of man had descended respectively from the gorilla, the chimpanzee, and the orangutan. This would not only imply an excessive amount of parallel development in human evolution, but it throws everything out of focus by assuming that living anthropoid types could be ancestral to living human races. This hypothesis never received any acceptance, but it has served at least to show that the principle of parallel development has its limitations. However, some writers would assume even longer parallel series in the immediate ancestry of man, but we shall see that the facts are against them.

Simpson (1935) agrees that the mammals probably arose polyphyletically from mammal-like reptiles at the end of the Triassic, but he regards the recent mammals, except the monotremes, as monophyletic in origin. The small Multituberculata and Triconodonta in the Jurassic were probably of independent origin from basal mammalian stocks. Simpson pictures four successive waves of mammal radiation. By the end of the Jurassic the small early mammals were world-wide in distribution. In addition to the above two groups were the Symmetrodonta and the Pantotheria, but none of them survived the Jurassic except the Multituberculata. This group lived on into the Lower Eocene, only becoming extinct after a period of perhaps ninety million years. The Pantotheria came nearer to the common ancestry of the marsupials and placentals. In the second mammalian radiation, in the Cretaceous, the Multituberculata developed in more advanced form, but the giant reptiles were still the lords of the earth. The marsupials and placentals began at this time. These marsupials were like the modern opossums, which are essentially living Cretaceous fossils, while the placental-like mammals were insectivores. A third radiation, which may have been in the Cretaceous, consisted in the development of the condylarths and creodonts (archaic placentals). The fourth radiation, in the Eocene, introduces the modern placental mammals. These successive waves of mammals appear abruptly in the fossil record. Where they came from is a mystery. It is assumed that they migrated from their place of origin to the regions where their bones are found. This is something of a difficulty to paleontologists. If these placental mammals were in small numbers in their places of origin, the failure to find records of them before they began to spread is not so surprising.

Before considering the more immediate ancestry of man in the Primates we need to examine briefly further the evidence regarding evolution in the lower mammals. We have seen in the last chapter that the evidence favors the view that several groups of therapsids independently crossed the line from reptiles to mammals. These two groups finally became distinct during a period of climatic severity in which much extinction took place, the survivors being adapted by a series of characters—epidermal, physiological, and reproductive—to quite different ecological niches.

Next there is the question of the relations between the higher and lower mammals. It is now agreed that the placental mammals were not derived directly from a marsupial ancestry, but were they a polyphyletic group? Matthew (1927) considered this question in a memoir on the Paleocene faunas. He found that probably all the orders of placentals had been derived from related families of pre-placental mammals, which were in turn remotely allied with the ancestors of Upper Cretaceous marsupials. Gregory (1920–21), in his extensive monograph on the origin and evolution of dentition in man, agrees with this conclusion. Thus the line between marsupial and placental mammals, like that between reptiles and the earliest mammals, appears to have been crossed on a broad front. Before the end of the Cretaceous the marsupials had become more or less cosmopolitan, having undergone adaptive radiation, especially in the southern centers, Australia and South America. In the meantime the placentals were evolving and dispersing in Holarctica. By the end of the Cretaceous the marsupials had been displaced from the northern land areas, finding refuge almost entirely in the southern hemisphere.

In the latest comprehensive treatment of the mammals, Simpson (1945) concludes that the view according to which the placentals derive from the marsupials is untenable in the light of recent research. The less progressive of the placentals retain primitive characters lost in marsupials, while marsupial characters are in large part specializations not found in the common ancestry. The Jurassic Pantotheria are nearly related to the common ancestry from which both placentals and marsupials diverged. The chromosome numbers (Gates, 1946, p. 52), in so far as they have been counted, are much lower in marsupials than in mammals, the former ranging from 12 to 28, the latter from 32 to 86(?). The basic number appears to be 60 in ungulates and edentates, 48 in insectivores, bats, and Primates, including man. Whether these high numbers have been produced in part by doubling of the chromosomes or by other methods is not yet known.

As the teeth and jaws have been most frequently preserved in all the higher vertebrates, it follows that they have been most intensively studied. Such studies make it clear that the teeth are highly adapted structures, their form being closely related to the food of the animal and its method of using its jaws. How nature determines that the grinding surfaces of the teeth in the upper and lower jaws shall correspond and fit together to produce the most advantageous grinding mechanism is not clear, but of the innumerable adaptations involved in the evolution of different types of teeth, there is no doubt.

Gregory strongly reaffirms the Cope-Osborn theory that the diverse molar patterns of higher mammals have all been derived from a primitive dentition with tritubercular molars in the Paleocene. From this beginning has been evolved an elaborate grinding mechanism in herbivorous animals and an efficient shearing and crushing apparatus in carnivores. Similar adaptations for shearing flesh developed independently at different times and places in the Eocene Creodonts of North America, which are carnivorous placentals, and the carnivorous marsupials. These were not derived from a common ancestor. In paleontology these alternatives are constantly presented, namely that certain similarities or characters in common are the result of (a) inheritance from a common ancestor, (b) parallel variations. In some cases the conclusion is clear, in others there may be insufficient evidence to reach a conclusion. But in the former case they will have had the same ancestor and in the latter different ancestors. This dilemma constantly recurs in the tracing of any phylogeny.

LeGros Clark has written most emphatically on the parallelisms in phylogeny of the Primates. He points out (1935) that (a) some Lemurs are astonishingly monkey-like in skeletal and dental characters; (b) in tarsioids some features of brain and skull are parallel to those of the monkeys; (c) the Catarrhine (Old World) and Platyrrhine (New World) monkeys,[1] for example, Macaca and Cebus, are astonishingly similar in skeletal, visceral, and cerebral anatomy as the result of parallel evolution. They must have diverged early in Tertiary times and may even have arisen from independent tarsioid stocks. In another paper, Clark (1936) points out that within the lemurs (now surviving mainly in Madagascar where they developed some remarkably pithecoid forms) there are parallel specializations in the Lemuriformes and the Lorisiformes; also between the Eocene lemur Adapis in Europe and Notharctus in North America. The Adapidae are the best known Eocene Primates, and Simpson regards them as ancestors of all the Anthropoidea. The nearly related Plesiadapidae, of the Upper Paleocene and Lower Eocene in the United States and Europe, are regarded by Simpson (1935) as an offshoot from lemuroid ancestors, closer to the typical lemurs than to the Tupaioidea and roughly parallel to the later Daubentonoidea in the specialization of their incisors, but not closely related to them.

Elsewhere (1939) Clark says, "Comparative anatomists and paleontologists have come to realize that parallelism has played by no means an insignificant part in organic evolution." While this factor complicates phylogeny it demands the closest consideration in the study of primate evolution. The extinct monkey-like lemurs of Madagascar are not ancestors of the true monkeys, but their simian characters were a parallel development. He concludes that man and gorilla must have many parallel characters although they had a common ancestor. This point will be considered later.

[1] For the classical division into Catarrhini and Platyrrhini, Simpson (1945) substitutes a triple grouping, Cercopithecoidea, Ceboidea, and Hominoidea.

From this and other evidence we may conclude that parallel variations have played an important part in the evolution of the Primates. In addition, like so many other groups, they have developed gigantism —at least in the case of the orangutan, chimpanzee, gorilla, and man. It is quite possible that the progressive increase of size recorded in the modern human populations of many countries is a part of this evolutionary movement, although the rate of increase in human stature and weight is much more rapid in modern civilized man than in any known evolutionary animal series. The Jurassic mammals were all of small or minute size, adapted to a forest habitat and probably mainly arboreal. Only in the Eocene did they become diversified, specialized, and dominant in all habitats, this being accompanied by relatively great increase in size. Between early and late Eocene there were great changes in the mammalian fauna.

From the studies of Gregory (1916), Clark (1940), and others, it appears that the main groups of anthropoids were defined in the Miocene; Clark says as early as the beginning of the Miocene. Osborn's idea that man as we know him already existed in the Pliocene appears to be definitely given up. Rather, many appear to be in agreement that the group of anthropoids represented by Dryopithecus and related genera is a probable source from which the living and fossil Hominidae and later anthropoids descended. However, the South African Australopithecidae (see below) now have equal claims to be ancestral, at least to the Hominidae. Unfortunately, the Miocene and Pliocene Dryopithecus and their relatives are represented only by teeth and fragments of jaws, so that many uncertainties remain regarding these animals, although many species in several genera have been described. It can hardly be doubted that this is a key group from which many later Primates and perhaps man are descended.

One of the starting points of this group is the small *Parapithecus Fraasi* Schlosser from the Lower Oligocene of Fayum, Egypt. It consists of a complete mandible with teeth (Werth, 1919). The jaw is very pointed and has no "simian shelf." The two halves of the mandible converge at an angle of 33° to give the most primitive type of jaw in the Primates. In modern gibbons this angle is 16°–21°, and in man the two rows of molars are of course roughly parallel. The dentition is evidently $\frac{2.1.2.3}{2.1.2.3}$ (2 incisors, 1 canine, 2 premolars, and 3 molars in each half of each jaw), but the outer incisors are much larger than the inner. This led Schlosser to interpret them as canines and the dentition as $\frac{1.1.3.3}{1.1.3.3}$, a view which the comparative study of dentitions shows to be erroneous. The canines are therefore small—a primitive condition in the group. They have become larger and projecting in all the modern genera of anthropoids, but not in the Hominidae so far as known. Parapithecus thus stands between the Eocene insectivorous Lemuroidea, especially the North American Anaptomorphidae and Propliopithecus (lower jaw), a larger animal, also from the Oligocene of Egypt, on the other. This is more advanced and in turn leads on to the Miocene Pliopithecus and recent Hylobatinae (gibbons). Propliopithecus is a very primitive anthropoid, ancestral to the later anthropoids and nearly related to the gibbon line of descent, the oldest known form of which is Pliopithecus. Parapithecus resembles the Hominidae in (1) the small canine, (2) the equal size of the two premolars, whereas the large outer incisor is like the condition in Platyrrhinae (not in Catarrhinae). The large outer incisor is thus a parallel development in Parapithecus and some Platyrrhinae. LeGros Clark (1940) suggests that Parapithecus was derived from a tarsioid stock. The Upper Paleocene Paramomys of Montana (Simpson, 1937) may have been ancestral. The Eocene Necrolemur may be related. Prohylobates is another Middle Miocene form, both this genus and Pliopithecus being very nearly related to the gibbons. Limnopithecus, from the Miocene of Kenya, is a more specialized gibbon, all the gibbons having a shallow and light mandible. About a dozen "species" of Hylobates, the modern southeast Asiatic gibbon, have been described, but Pocock (1927) reduces them to three. He believes that their extraordinary athletic proclivities in the trees have reduced the gibbons, through selection, to a relatively dwarf size compared with their ancestors. They are the only anthropoids with an essentially bipedal gait on the ground. Symphalangus is another genus of gibbons with one to three species in Sumatra.

Notharctus, another key form, is a North American mid-Eocene genus with at least eight recognized species. According to Gregory (1920a), in his extensive monograph, these are survivors of the primitive lemuroid stock which gave rise to all the higher lines of Primates. They had long been arboreal and were descended from earlier arboreal ancestors more like the opossums than the terrestrial insectivores. All other orders of the placentals became sooner or later terrestrial, losing their grasping power. In an earlier study, Matthew (1904) concluded that the small or minute Mesozoic mammals were arboreal and insectivorous or graminivorous, living in forest-clad river deltas and coastal swamps. The American Notharctinae and the contemporary European Adapinae (lemuroids) have a common ancestor.

Beattie (1927) finds evidence that the marmosets (Hapalidae) are primitive, near the base of the Platyrrhine monkeys, but with some late specializations. The two genera Callithrix (Hapale) and Leontocebus (Midas) show no essential differences. He concludes that Hapale is more advanced than Tarsius in its brain and sense organs (a macula lutea is present) but that they are strikingly similar in the nasal region, the orbital cavity, the structure of the vertebral column, and the proximal segments of the limbs. Hapale is thus the most primitive living monkey, an early stage in the series which should connect the Eocene tarsioids with the platyrrhines. These living survivors are thus closely related to the ancestors of Tarsius and the fossil Eocene tarsioids. While Tarsius has specialized in the direction of the higher Primates and in its mode of progression, Hapale has retained its primitive form of progression and developed a more highly specialized form of primate brain.

Tarsius spectrum, another key animal in relation to human ancestry, is the sole survivor of the Eocene tarsioids, now found in Borneo, Java, Sumatra, and the Philippines. It is a small, nocturnal, arboreal, insectivorous animal, much studied since Burmeister gave an account of its anatomy in 1846. In modern accounts (Woollard, 1925) its characters are listed as (1) primitive, (2) lemurine, (3) anthropoid. A discussion of its zoological position and affinities (*Proceedings of the Zoological Society of London.* 1919. pp. 465-498), including its brain and embryology, shows that the majority of zoologists agree in recognizing three groups of Primates: (1) Lemuroidea, (2) Tarsioidea, (3) Anthropoidea. Tarsius is another "living fossil," agreeing with the Eocene tarsioids except in its jumping feet. The name, *Anaptomorphus homunculus,* given by Cope in 1882 to a lemuroid from the Lower Eocene of Wyoming, already recognized the human relationships of the tarsioids. Its very large orbits facing forwards, with reduction of the snout, indicate that binocular (not stereoscopic) vision and nocturnal adaptation had already been attained. But something, perhaps the specialization for nocturnal activity, prevented further evolutionary advance, and now the last of its line is on the verge of extinction. Some of the Eocene bones resembling Tarsius may belong to primitive insectivores, carnivores, or even ungulates, as these orders were still small animals and not yet clearly differentiated.

Skulls of Anaptomorphus and Notharctus from the American Eocene and of Necrolemur from the Eocene of France resemble in essentials the skull of Tarsius, and the hind feet of Notharctus differ only in having a shorter heel bone. The position of Tarsius thus seems clear, with its near ancestral relations to the lemur stock from which it differentiated and its nearer relations to the higher Primates. In certain scattered features it even shows resemblances to man. It has advanced beyond the lemurs in such features as the reduction of the face and the expanded visual cortex of the brain, thus providing a basis for stereoscopic vision and movements of greater skill and precision, and in its habitually upright position. The expansion of the visual cortex, by which the olfactory as the primary sense in the lower Mammalia gave place to stereoscopic vision, made possible the great development of the higher function of the brain, as Elliot Smith has emphasized. But the size of the brain in Tarsius is on the lemur level, much less developed than in the anthropoids, although large compared with that of the other mammals. While Tarsius can be regarded as "the most primitive living primate" (Clark, 1930)—with the exception of the Aye-Aye in Madagascar—it can not be supposed that man is directly descended from a Tarsius-like form (Wood Jones) without doing violence to man's much closer relationships with the monkeys and apes.

However, Clark (1940) suggests that the higher Primates may have originated from the Microchoeridae, an extinct family of Old World tarsioids, the Eocene genus Caenopithecus being regarded as intermediate. The earliest Primates became arboreal and in this environment the eye became the primary sense organ, smell taking second place, whereas among small animals on the ground it had been of first importance. This change in emphasis on the sense organs necessitated corresponding changes in the brain.

Many writers, for example Gregory (1927), regard Upper Cretaceous and Paleocene forms of Insectivora resembling the modern tree-shrews (Tupaiidae) as ancestral to the lemurs, tarsioids, and early anthropoids or an offshoot from near the base of the Primates. These tree-shrews of the order Menotyphla would then represent the group from which the Primates took their rise. In a study of the skull of Tupaia, LeGros Clark (1925) concludes that the tree-shrews combine primitive features with characters only paralleled in the Primates. He formerly regarded them as generalized Insectivora, arising from the base of the primate phylum after it was differentiated from other mammal phyla. Clark (1932), in a later study of the brains of Insectivora, finds the Tupaioidea so similar to the Malagasy lemurs in brain and skull as well as other characters as to class the tree-shrews with primitive Primates rather than with the Insectivora. He controverts the view that these similarities are parallelisms and considers them based on relationship. Tracing the ancestry backwards, Broom (1930) concludes that in early Cretaceous times a primitive marsupial, perhaps related to the bandicoot (Parameles), gave rise to the Menotyphla, one of which produced a primitive lemur in Paleocene times.

Having seen something of the origin and relationships of the Primates, we may now come to the evolution of the Anthropoidea, to which man belongs. The key genus here is Dryopithecus. Ten genera have been described, with over twenty species. These have since been reduced to four genera with ten species (Lewis, 1937). The group of Dryopithecinae almost certainly contains forms which are nearly related to the ancestors of the Hominidae. Within this group, which ranges in time from Middle Miocene to Middle Pliocene and in space from Spain and Germany to India, from Egypt to South Africa, different species already show resemblances respectively to the modern orangutan, chimpanzee, and gorilla, as well as others that come nearer to man. Unfortunately only teeth and jaws are available, except for one femur and one humerus. Many species were described by Pilgrim from the Siwalik Hills of northern India, but the earliest finds were in Europe. Clark (1940) has given a clear summary of our knowledge of these forms, with indications of their probable relationships. The group is an exceedingly variable one. Colbert (1935) concludes that the Siwalik fauna ranges only from Upper Miocene to Lower Pleistocene.

Dryopithecus fontani was discovered in France in 1837. It is of mid-Miocene age and consists of a mandible and part of a humerus. This with a femur from the Lower Pliocene of Germany provides evidence that the Dryopithecinae were ground-apes, with a more erect bipedal carriage than the gorilla and chimpanzee. The German Lower Pliocene species *D. rhenanus* shows similarities to the chimpanzee, although the teeth have human resemblances. These European Pliocene apes died out before the end of that period, and none are found in the Pleistocene, which was probably too cold for them.[2] *D. chinjiensis*, from the mid-Miocene of the Siwalik Hills and *Palaeopithecus sivalensis* from the Lower Pliocene of the same area show resemblances to the gorilla in some dental characters. Gregory (1916) regards the latter as close to the gorilla, differing only in having more primitive characters. Palaeosimia, probably of mid-Miocene age from this area rich in anthropoids, appears to be close to the orangutan. The evidence leads naturally to the conclusion that the Siwaliks were an area from which differentiation and dispersal of early anthropoids took place over a long period, with three genera of anthropoids and one of Hominidae surviving to modern times, the orangutan-like descendants spreading northward into China and eastward to Malaya where it survives, while the other two surviving genera found their modern refuge in Africa. The genus Sivapithecus, with about four species, is also nearer to the orangutan than to other apes or to man. *S. giganteus* was probably as big as the largest modern gorillas. Gregory, Hellman, and Lewis (1938) transfer *Dryopithecus darwini* to Sivapithecus. The teeth are not easily distinguishable from human.

According to the usually accepted view, the primitive trituberculate condition of the molars in mammals gradually changed to the quadrituberculate, in which the four tubercles of the grinding surface are arranged in the form of a square (Gregory and Hellman, 1926). In man, the lower second molar (M_2), for instance, generally has four main cusps in Caucasians, with grooves between them forming a cross (+). On the other hand, M_1 generally has a fifth cusp, which has been more or less retained from the more primitive Dryopithecus pattern. Negroes differ from Caucasians in that M_3 (wisdom teeth)

[2] Incidentally, it is worth mentioning that the chimpanzee and gibbon are very much alike in skull characters and dentition, except that the former is more than twice as large. *Paedopithex rhenanus*, from the Lower Pliocene of Germany, is a femur stated to belong to a large gibbon.

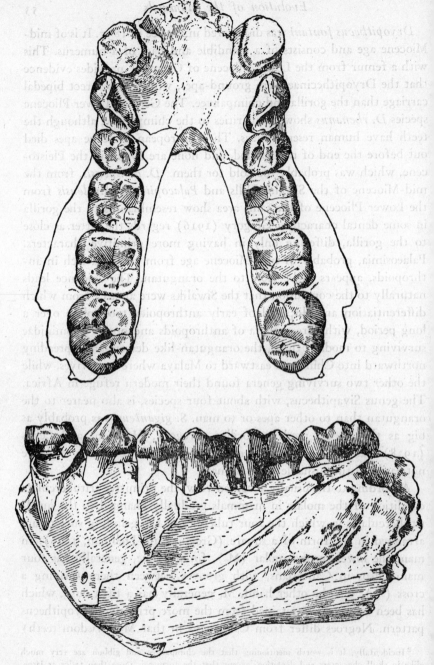

FIGURE 1. Restoration of the mandible of Dryopithecus, showing the simian character of the teeth and the simian shelf.

erupt earlier and frequently have the Dryopithecus rather than the quadrituberculate pattern. In Negroes, Australians, and Indians, M_3 frequently has a sixth cusp. These molar similarities argue for a near relationship between the Dryopithecinae, anthropoids, and modern man.

The canines in Dryopithecus vary a lot, from large in *D. fontani* to weak in *D. pilgrimi* and other species; there is evidence that in hominids the canine has undergone some reduction, while in modern anthropoids it has become large and specialized. *D. giganteus*, represented by a right lower third molar, is regarded by Gregory (1916) as like a chimpanzee but twice as large. Fig. 1 is a restoration of a mandible of Dryopithecus, by Gregory and Hellman (1926), showing the parallel tooth rows, the large canines, and the simian shelf.

The genera Bramapithecus and especially Ramapithecus of Pliocene age (Lewis, 1934) are still nearer to man, the molars being very human, the canines and incisors small, with no diastema in the jaw. The dental arch also is parabolic, as in man, not U-shaped, as in the apes. Ramapithecus might indeed belong in the Hominidae, but as the skull characters are unknown it is impossible to know what its real relationship will prove to be when more evidence is available. These Siwalik apes are, however, very distinctly infra-human, from the evidence available, and there is nothing to connect them with the gibbons, which are recognizable, at least from the Miocene, as a separate group. In the orangutan the mandible is very deep, while in the gibbon it is shallow. In *Dryopithecus fontani* it is very deep and in *Bramapithecus thorpei* quite shallow. In Sivapithecus also the mandible is shallower than in the European species of Dryopithecus, but these differences may be partly sexual, as they are in Sinanthropus. Sugrivapithecus is another genus, of medium size and with small canines, which may be related to man. The size of their brains is not known, nor whether they were arboreal. The excessive variability of the third molar in anthropoids and man reduces its value as an indicator of specific or generic difference. The teeth in most species of Dryopithecus are essentially simian, and in jaws showing the symphysis the simian shelf (lingual torus) is always present. The general relationships of these forms are indicated in the diagram (Fig. 2).

We know nothing of the hands and feet of the Dryopithecinae. Probably some of them were arboreal, brachiating forms and others

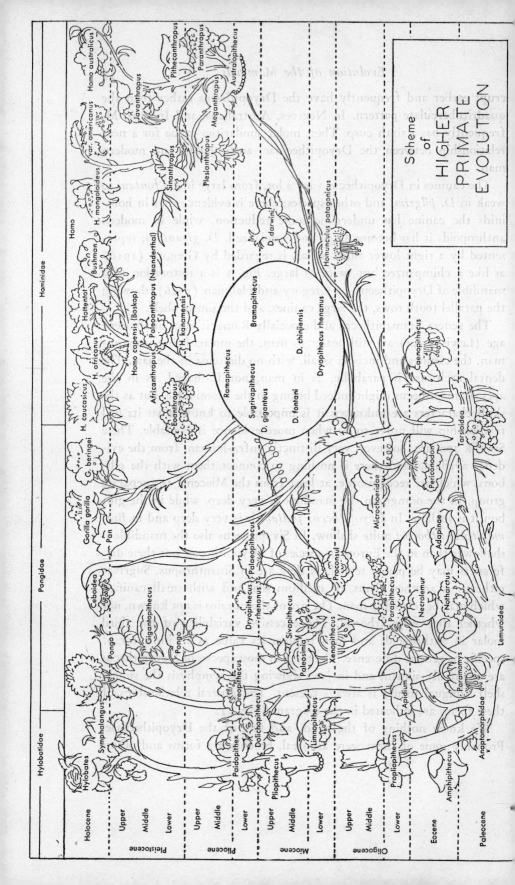

ground-apes. Straus (1940) shows that the chimpanzee, gorilla, and orangutan walk with their flexed fingers touching the ground because the shortness of the long digital flexors automatically bends the fingers as the hand is extended. This condition is present in the newborn chimpanzee but is further exaggerated during development. It is apparently an adaptation to the habit of brachiation. In gibbons, monkeys, and man, on the other hand, the palm is placed on the ground with the fingers extended. In gibbons the brachiation is different from that of the great apes. They are even more expert trapeze performers, but instead of flexing the fingers, they approximate the thumb and forefinger. Straus believes that since man has neither of these specializations, his ancestors never passed through a brachiating stage, although there was an earlier arboreal stage; and that the human hand and arm is nearer to that of a catarrhine monkey than to any other primate. He therefore believes that man is only distantly related to the great apes. The Dryopithecinae are such an extended group in morphology and in time that man and the various anthropoids could all have arisen from different Miocene or Pliocene genera within it.

Recently discovered (Hopwood, 1933) are three new genera belonging with, or near, the Dryopithecinae, from Koru, Kenya. Referred to the Lower Miocene, they are more primitive in some features than the Dryopithecus forms. They were found among Miocene remains of lemuroids, creodonts, carnivores, rodents, and insectivores. Limnopithecus, which is evidently a gibbon, can not be ancestral to Hylobates because of certain specializations. Xenopithecus is an aberrant anthropoid type probably adapted for a different mode of life. *Proconsul africanus* is of greatest interest. It consists of (1) fragments of a left maxilla with a large canine, two premolars, and three molars, (2) a right mandibular ramus with molars and premolars, (3) a right upper first molar, (4) a right lower third molar. The palate was straight-sided and there was a diastema 4 mm. in length. This Miocene anthropoid was more primitive than Dryopithecus in having a strong cingulum on the molars and a prominent hypocone. The dentition shows a mixture of primitive and specialized characters, with many resemblances to the Dryopithecus stock. Hopwood regards it as definitely ancestral to the chimpanzee. It is evidently the most primitive African ancestor of the gorilla-chimpanzee complex. Hopwood concludes that the primary center of anthropoids was in or near northern

Africa. A branch from this group ranged to Europe (*D. rhenanus*) and India (*D. punjabicus*), perhaps giving rise to man somewhere in central Asia. Leakey has found on Rusinga Island in Kavirondo Gulf on Lake Victoria, 160 miles west of Koru, a mandible which occludes perfectly with the maxilla of Proconsul, and a maxilla which occludes with the Koru mandible. Thus Proconsul fills an important place as a Lower Miocene African ancestor of the Middle and Upper Miocene Dryopithecine stock.

These and other finds from the Victoria Nyanza district have since been described (MacInnes, 1943). Another species of Limnopithecus was found, as well as material which probably belongs to Xenopithecus. Most important is the mandible, especially the astragalus and os calcis of *Proconsul africanus*. These bring out hominid characters from which it is concluded that Proconsul could not have been directly ancestral to the chimpanzee. Probably the ancestral chimpanzee diverged from the common stock at a pre-Miocene date. MacInnes regards Proconsul as nearer the main line from which man arose. These three east African genera show no close affinities with the Australopithecinae. It is also significant that the "simian shelf" in the symphysis of the mandible is absent in Proconsul, indistinct in Xenopithecus, and less developed in Limnopithecus than in modern Simiidae. If the African anthropoids are descended from these Miocene genera, the simian shelf must be a later development.

Very recently (*Nature*, October 5, 1946, p. 479) Leakey has found two more mandibles only fifteen feet apart in the Lower Miocene of Rusinga Island. One belongs to Proconsul, the other to Xenopithecus. The Proconsul jaw is nearly perfect (*Illustrated London News*. pp. 197-201. 1946). Not only does it lack the simian shelf, but the chin region is straight and more nearly vertical than in the living anthropoids. Although it lacks a chin, it is not very different from the Heidelberg jaw in this respect. The jaw is less prognathous than in the chimpanzee, and slightly larger. The ascending ramus is narrower (as in man), and the condyles, which articulate with the skull, are small and flat, essentially the same as in man. The rows of cheek-teeth are intermediate in direction between the condition in the anthropoids and man, but they converge forwards as in man. The wear on the molars is also flat, as in man. On the other hand, the huge canine and the monocuspid first premolars are definite anthropoid characters, as well as the third molars, which show the Dryopithecus pattern of cusps. Thus we have again a mixture of anthropoid and human characters, such as we have learned to expect, showing the near relationship of Proconsul to the human ancestral stock. The further evidence regarding Xenopithecus supports the view of Simpson that it is related to the orangutan line of descent. On the basis of all the evidence, Leakey suggests that the Lower Miocene of Kenya may have been the center of primate evolution from which both man and the great apes spread.

Another significant primate genus, Amphipithecus, is based on a fragment of a mandible collected by Barnum Brown from the Upper Eocene of Burma in 1923 and described by Colbert (1937, 1938). This mandible is very deep and heavy, the rami probably not divergent, with a very short vertical symphysis and a heavy simian shelf having a genioglossal pit, as in the advanced Primates. It is probably larger than the Fayum Propliopithecus, the canines were vertical and there was no diastema. The dentition, so far as preserved, was much nearer the anthropoids than the lemuroids, but there were certain lemuroid characters. A peculiar feature is the presence of three premolars, as in the lemurs and South American monkeys, but the second premolar was small and probably in the process of disappearing. This resemblance to the Platyrrhine monkeys is an inheritance from the lemurs. The presence of only two premolars is one of the main distinguishing features of the Catarrhine (Old World) monkeys. Amphipithecus resembles Dryopithecus in having a deep, thick mandibular ramus and a short symphysis, but differs from these and other Fayum forms, such as Parapithecus and Propliopithecus in having three premolars. It is thus near the ancestry of the Old World monkeys and apes, and may be ancestral to Parapithecus. Colbert (1938) points out that if it had only two premolars it could be put in the same class with the gorilla and chimpanzee.

Thus the trail of primate development leads from the Eocene Amphipithecus in Burma to the Oligocene Parapithecus and Propliopithecus in Egypt and thence to Miocene Proconsul in Africa and the Miocene Dryopithecinae of northern India. In Miocene times, tailed monkeys and anthropoid apes were widely distributed in the Eastern Hemisphere. It still remains uncertain whether man's Miocene ancestry was mainly in east Africa or Asia. It seems likely that Proconsul represents an African ancestor of the gorilla and chimpanzee nearly related to man, while the Indian Dryopithecinae and the genus Xenopithecus contain the ancestors of the orangutan. From what is now known, it seems highly improbable that the orangutan ever reached Europe, still less that he survived there into the Pleistocene.

The Australopithecinae are a group of South African man-apes of uncertain but probably Pliocene and Pleistocene age. *Australopithecus africanus* was discovered by Dart (1925) at Taungs, in Bechuanaland in 1924, embedded in a limestone crevice fifty feet below the surface. This area was probably then, as now, a treeless part of the Kalahari Desert. A child's skull with milk-teeth was found, the first permanent molars already present. This skull approaches man more closely than any known anthropoid, living or extinct. The face and mandible are practically complete (Plate I, a, b), and the major portion of the brain pattern can be determined from the natural endocranial cast. Plate I (a) represents a restoration. There is no trace of supra-orbital ridges, the orbits are almost circular (not subquadrate), the teeth humanoid but with a diastema between the canines and incisors. The mandible is massive but there is no simian shelf. In the brain cast the parallel sulcus is pithecoid but the sulcus lunatus is in a posterior position, indicating expansion of the parietal, temporal, and occipital areas. The facial and dental recession and the enlarged brain separate Australopithecus from any anthropoid.

Broom has since discovered two adults evidently belonging to the same group. *Australopithecus transvaalensis* (Broom, 1936) was blasted from a cave deposit at Sterkfontein which was being quarried for its lime, at a point some two hundred and fifty miles east of Taungs and twenty-five miles from Johannesburg. He has since made it the type of another genus, Plesianthropus, and has described the third specimen, from Kromdraai, only two miles from Sterkfontein, as *Paranthropus robustus* (Broom, 1939a). It is possible that the Sterkfontein ape is really an adult of the Taungs type, in which case its first name would be correct. A separate genus for the third specimen is no doubt justified, and probably three genera are involved. The skull capacity of these forms seems to be the same as that of gorillas, while the palate shows definite resemblances to that of man and the dentition is essentially human. In the chimpanzee and gorilla the upper premolars usually have three roots, whereas in man they usually have only one root, but in Kaffirs they have two. In Plesianthropus there are two roots, the outer flattened near the tip as if the ancestors had three.

Shaw (1940) described from Sterkfontein another tooth. It was a third molar, much bigger than in Sinanthropus but with no "wrinkles." It was markedly taurodont, resembling teeth of Plesianthropus and Paranthropus but with no cingulum and a simple occlusal pattern. Shaw thinks it may belong to a human contemporary of the Australopithecinae at Sterkfontein, but this is unlikely on the basis of all the evidence. Hrdlička (1924) shows how closely the teeth of *Dryopithecus rhenanus* resemble human teeth, and Hellman (1928) shows the racial distribution of the Dryopithecus pattern in the molars of modern man. Without going into details, he finds the most advanced stages reached in Caucasians, the most primitive features retained in the west African Negro, with intermediate conditions in Mongoloids.

In Paranthropus the incisors and canines are relatively small, but the first premolar has three roots, like the chimpanzee, gorilla, and Dryopithecus. The molars also resemble those of Dryopithecus. Broom considers that reduction of the canines was the first stage in human evolution, the next being reduction in size of all the teeth, especially the premolars. Broom believes that man was derived from Paranthropus in Middle or Upper Pliocene by a rapid increase in the size of the brain, the resemblance being too close for parallel development to have taken place. There is thus the possibility that the African Australopithecinae were directly ancestral to Pithecanthropus. In Paranthropus the face is already flat above the canines and incisors and the canines are small, whereas in the Sterkfontein ape they are large. The group probably arose in the Pliocene even if it survived into the Pleistocene.

Dart (1929), who regards the age as at least Pliocene, points out that the Taungs skull was found among thousands of bone fragments, like the cavern lair of a carnivore. Australopithecus had preyed upon baboons, antelopes (bok), tortoises, rodents, bats, and birds, forming a kitchen midden like those of modern human tribes. There were also egg-shells and even crab-shells. Two of the baboons were contemporaneous and far more primitive than the modern baboons in South Africa. The Taungs child (Plate I, a, b) is reckoned, from its dentition, to be six years old. The group were cursorial apes living in caverns on a cliff overlooking a broad dry plain on the fringe of the desert and having a diet more varied than that of modern baboons. Broom (1924) considers that they used methods for trapping game, and they broke open brain-cases and the marrow cavities of long bones, probably with unfashioned sticks and stones. This was before the days even of eoliths.

These man-apes are "missing links" in everything except perhaps in geological age. They are the most generalized of all apes, having characters of the chimpanzee, the orangutan, and even the gorilla. They were plantigrade, probably walking erect on the soles of their feet, with the skull poised, as in man, on the top of the spinal column; but in *Paranthropus robustus* the foramen magnum is in a somewhat rearward position, as in the chimpanzee. Their neck was longer and more slender than in the apes, the face was upright, as in man, with no muzzle, so that the eyes could concentrate. The dentition and jaws were human but even the milk-teeth were much worn by mastication, and the nature of the wear shows that the movement was rotary, as in man.

The human brain differs from that of the apes in that three areas of the brain, the prefrontal, parietal, and inferior temporal regions, are greatly developed. These areas are the last to develop in childhood, as the child learns to stand erect and to run. According to Elliot Smith, the prefrontal area just above the eye was as well developed in Australopithecus as in Pithecanthropus. The temporal areas, associated with hearing and speech, were less developed than in Pithecanthropus but more so than in the modern apes. There was probably no speech then, but only cries, gestures, and facial expressions, the features being much more refined than in the apes. The hands, freed from functions of locomotion, became sensitive organs of touch, guided by closer vision. These proto-human man-apes in South Africa had achieved human status as regards environment, habits, posture, features, and dentition, but Dart excludes them from the Hominidae on the ground that they were subhuman in the temporal area of the brain and therefore lacked speech. They could have been the ancestors of Pithecanthropus if they lived in the Pliocene. Otherwise they must have become extinct, leaving no descendants. If so, they show remarkable parallelisms to man.

Dart suggests that the Australopithecinae are nearest to the common ancestry of man and apes, but it is difficult to believe that apes which had acquired the erect plantigrade method of progression on the ground would later develop the various adaptations of hands and feet necessary for re-adaptation to arboreal life, as seen in the living anthropoids. However, many now believe that man's ancestors had only an early arboreal phase. Koenigswald (1942) includes the group in the Hominidae but excludes them from human ancestry on the ground that they are geologically too young and their teeth are too large. They would then be a side branch, man-like but without civilization. It will be seen that much turns upon how we define man. Weidenreich (1939) excludes them from the Hominidae because of the ape-like premaxilla and small brain.[3] He believes that the hominid and anthropoid stocks must have separated in pre-Miocene times. Koenigswald compares the African man-apes with Pithecanthropus in Java. The heavy mandible is comparable with that of Paranthropus. He shows that as the brain increased in size in the series Pithecanthropus → Sinanthropus → Neanderthal → recent man,[4] the teeth diminished. The teeth of Pithecanthropus are smaller than in Australopithecus but the jaw has a diastema which is not present in the latter. Even if these African man-apes are really Pliocene in age it has been argued that they can probably be excluded from the direct ancestry of modern man, because Pithecanthropus has a much more anthropoid dentition and cannot therefore be a direct descendant.

Broom was evidently correct in his statement that "in South Africa there once lived apes which had almost become men, but the brain remained anthropoid." In Paranthropus the mandible was more massive than in *Sivapithecus indicus* and there was no diastema; but an interesting feature was the scoop-like upper lateral incisors, corresponding with the medium shovel-shape of Hrdlička (see Gates, 1946, p. 370). The molars of the Australopithecinae were intermediate between those of Sivapithecus and the gorilla, but much larger than in Sinanthropus or Homo. Also the third molars were the largest of the molars, as in Sivapithecus and the gorilla, whereas in Sinanthropus and

---

[3] Wood Jones (1947), in a recent note, also points out that man is distinguished from the anthropoids by the absence of the premaxilla as a separate bone, the sockets for the incisors in man being formed "from maxilla labially and premaxilla lingually." He finds that Australopithecus and Plesianthropus "show definite facial maxilla-premaxilla sutures in typical simian form" and Paranthropus fails to show them only because the individual is over the age at which it is obliterated, and not because the condition is the same as in man. He concludes, on this evidence, that the Australopithecines (or Dartians) are all "apes" and "throw no light whatever on the primate forms that were ancestral to *Homo*." This is rather a sweeping statement to base on a single character-difference when so many marked resemblances to man exist.

[4] The brain of historical man appears, however, to be no larger than in Neanderthal man.

Homo they have been reduced and are smaller than the other molars. According to Gregory and Hellman (1939) the teeth also show points of resemblance to those of the orangutan. In a restoration of the dental arches of Plesianthropus (1940) they emphasize that small, near-human canines and lateral incisors are combined with huge ape-like molars.

W. Abel (1931), in a study of *A. africanus* and a detailed comparison with the gorilla skull, concluded that Australopithecus is derived from an ancestry near to that of the gorilla but without its specializations such as the enlarged canines. The reduction of the face was compared with that of the chimpanzee as a convergent character, accompanied by a slight lengthening of the brain, and this anthropoid was regarded as not ancestral to man. In other words, it represents an evolution which is in some respects parallel to that of man, but the possibility is not excluded that it was derived from the same stock as the original ancestors of man in the Miocene. It was larger than chimpanzees and nearer the size of men; but the brow ridges of the skull were conspicuous and ape-like in the adults, the forehead however broader and the jaws less prognathous than in the living anthropoids but more so than in Sinanthropus.

In the same deposit two species of baboons were found, more primitive than the modern species and with smaller endocranial capacity. Broom (1940) has since examined a large collection of fossil baboons mainly from Sterkfontein and Kromdraai and described several new species of Parapapio, making eight in all. One giant baboon (*Dinopithecus ingens*) was nearly as big as a gorilla. While the man-apes hunted the smaller baboons, they may well have been preyed upon by the giant species. Probably their only refuge would be in the caves, which were inhabited for many thousands of years by jackals, hyenas, and saber-toothed tigers as well as baboons and Australopithecus.

The fact that these man-apes were already erect and living on the ground in unforested country is significant in connection with the early arboreal phase of man's ancestry. It is probable that the ancestors of the hominids never became structurally adapted to arboreal life to the extent that has happened in the anthropoids. While they may have developed the habit of brachiation, they probably never become specialized for climbing, with the great toe opposable, to the extent that it appears in the apes, although in Neanderthal and Chancelade man it was more opposable than in modern Homo. Straus (1927) has shown that the feet of foetal monkey and man are more alike than the adult feet. Nevertheless, the foetal foot and legs of man resemble those of the adult arboreal Primates, the simian characters *in utero* being a recapitulation. Straus shows that from the third to the seventh or eighth month of the human foetus the structure and proportions of the foot resemble a generalized type common in many lower Primates. The foot is supinated, has relatively long phalanges, a short tarsus, a short hallux highly divergent and more or less opposable. Such a foot is unfit for upright terrestrial progress. While in many respects this foot is like that of an adult gorilla, in some points it is more primitive. These primitive points are mostly lost before birth, but many persist in a vestigial condition. The fact that the feet of man and anthropoids are more alike in early stages points to a common ancestry. Since the relatives of our human ancestors were tree-dwellers back to the tarsioids and the tree-shrews, it is probable that our ancestors had a long arboreal period as small animals. There is practically no fossil evidence regarding the manner and the evolutionary stages by which the human foot took on the features necessary for successful rapid progression as an erect terrestrial biped. The giant birds and some of the Mesozoic reptiles did it by rising off their hind legs. Man's ancestors did it by a process of re-adaptation from tree-life.

In a detailed study of the dentition of all three forms, Gregory and Hellman (1939) agree that *A. transvaalensis* has essentially the skull of a gorilla with an almost human dentition. They concluded that Plesianthropus was intermediate between the ancestral Dryopithecine stock and primitive palaeanthropic men. It was also more primitive than most modern anthropoid apes. The edge-to-edge bite of the upper and lower incisors and canines in the Australopithecinae, in contrast to the somewhat projecting incisors and the interlocking canines of the modern living anthropoids is distinctly human, as is the absence of a diastema. Gregory and Hellman suggest that the Australopithecinae have such a mixture of characters because they were late Pleistocene survivors of a common Dryopithecus-Sivapithecus stock and were therefore related to its three derived branches, the chimpanzee-gorilla, orangutan, and human. This combination of apparently unrelated characters, an anthropoidal skull with human dentition, incidentally removes the difficulty from an understanding of Eoanthropus, with its human cranium and anthropoid teeth and jaw. The relatively close relationships of all these forms to the Dryopithecinae, to which they must all be related, clearly strengthens the view that man arose no earlier than the Miocene. The human-like dentition of the African Australopithecinae might then have developed independently, as a parallel to that of the hominid stock derived from the Asiatic Dryopithecinae, but it seems more probable that the more hominoid members of the Asiatic Dryopithecinae and the African Australopithecinae form a single group ancestral to man.

In a later reconstruction of the Plesianthropus skull (Plate I, d), Gregory and Hellman (1945) find that all their measurements fall between anthropoid and human limits, while Broom's lie chiefly within the anthropoids. The occipital region is very wide and the occipital torus merges at the sides into the supramastoid swelling, as in some modern apes and early man. Broom (1943) pictures the Australopithecinae as bipedal, living like the modern baboons among rocks on dry plains and hunting in packs. The cranial capacity he estimates at 450–650 cc., with a brain very much like that of Pithecanthropus but smaller, the brain of a female Pithecanthropus being estimated by Koenigswald at 750 cc.

Since this was written, the important monograph of Broom and Schepers (1946) has been received and some of its general features and conclusions can be added here. The three genera, Australopithecus from Taungs, Plesianthropus from Sterkfontein, and Paranthropus from Kromdraai, are probably justified, and Dart's original view that these are man-apes is fully confirmed. In the Sterkfontein cave a Pliocene hyena (Lycyaena) has been found with a fragment of baboon skull (*Parapapio broomi*). This baboon was a contemporary of Plesianthropus, and Lycyaena belongs to the Lower Pliocene of Europe and India. While it may have survived in South Africa to a later date, it is probable that Plesianthropus occurred at least as early as the Upper Pliocene.

The contemporary fauna described by Broom in the bone breccia at Taungs included fourteen species in about as many genera, all the species and five of the genera being extinct. None are forest forms but some are definitely of the desert. Broom points out that the many species of the succulent plant genera Mesembryanthemum, Stapelia, Aloe, and Euphorbia, which evolved in this region, shows that the central and western parts of South Africa adjoining Bechuanaland must have remained desert for many millions of years. The Taungs fauna is older than that of Sterkfontein or Kromdraai, having a larger percentage of extinct species, so it is probably Middle (or Upper) Pliocene. No species at Taungs is found at either of the other localities, but a number of the genera are the same.

At the Sterkfontein caves *Plesianthropus transvaalensis* was discovered by Broom in 1936. The bone breccia in the cave extends to a depth of over one hundred feet and may have been accumulating for 10,000 years or more. He found the base of the skull, the right maxilla with three teeth, most of the left maxilla with four teeth, parts of the face, and a wisdom tooth. Parts of at least three skulls have since been found, and a number of isolated teeth, so that a nearly complete account of the skull can be given. The maxilla resembles that of the chimpanzee more than man, but the canine is much smaller and the incisors less prominent, the front of the jaw less ape-like. The nose was probably more projecting and human. The teeth form a horseshoe-shaped arch as in man, the palate is nearer to that of Rhodesian man than to any anthropoid, the glenoid cavity is human in form, but the base of the skull is more like that of a chimpanzee and there is no trace of a chin. The symphysis is remarkably chimpanzee-like, as in the Piltdown mandible, but in size and structure the jaw is more like the Heidelberg mandible than like a chimpanzee. One incisor closely approaches the shovel-shape, which Broom regards as almost a Mongoloid character. Shovel incisors are therefore a pre-human character which was retained especially by Sinanthropus and the modern Mongoloids.

In a detailed account of the dentition of Plesianthropus, Broom finds that the incisors (of which only the alveoli are known) were probably rather large, the canine large and the premolars very large. The teeth are very like human teeth but a little larger, and Broom concludes that the third molar described by Middleton Shaw (1940) is more likely to belong to Plesianthropus than to be human. The canine is compared with a larger fossil canine described by Weidenreich from Yunnan, China, as belonging to a fossil orangutan. Broom makes the interesting suggestion that this is a canine of *Gigantopithecus blacki* and that this supposed giant is really an Asiatic member of the Australopithecinae. Pliopithecus is generally regarded as ancestral to the gibbons. Broom suggests that, since it and Plesianthropus both retain the cercopithecid second cusp in the lower canine, Pliopithecus may be nearer the ancestor of the Australopithecines.

The various remains of Plesianthropus are regarded as probably belonging to seven individuals. These remains include a first left lower deciduous premolar, remarkably human but in a fragment of a Plesianthropus jaw, and a jaw fragment with molars from an old male. The postcranial bones include only the distal end of a femur, part of a

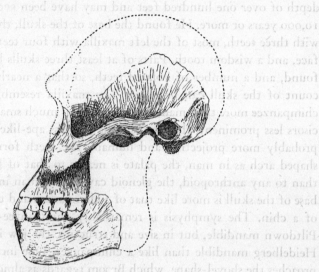

FIGURE 2a. Side view of skull of *Paranthropus robustus* Broom. The shaded portions are those preserved.

phalanx, and a carpal bone (os capitatum) which is intermediate between that of the orangutan and man in form but more nearly human. The femur is remarkably similar to that of man, and shows that Plesianthropus walked on its hind feet.

In a letter from Dr. Broom to Professor Hooton, dated August 8, 1947, which I am permitted to quote, Dr. Broom had recently found a pelvis of Plesianthropus. The brief description and accompanying sketch show that the pelvis was smaller even than that of the Bushmen, and quite unlike the chimpanzee or gorilla. The ilium is quite short and there is a very large tuberosity for the rectus muscle. This humanoid pelvis offers final confirmation that Plesianthropus walked erect and was in fact in the human line of descent.

Mammal bones have been obtained from some twenty workings in this area. Those associated with Plesianthropus include several species of baboons (*P. whitei, P. broomi,* and *P. jonesi*) and saber-tooth cats. Broom describes a large saber-tooth, *Meganthereon barlowi,* found near the type skull of Plesianthropus, and a small species, *M. gracilis,* also at Sterkfontein. A small hyena allied to the modern species was found in one cave, as well as *Lycyaena silberbergi* which helps in dating the deposits, although the latter extend over a considerable period. Many other mammals, large and small, were described, the age being apparently Pliocene, as already mentioned.

The finding of *Paranthropus robustus* is a romantic story. In 1938 a schoolboy broke out a portion of the skull from a weathered outcrop of bone breccia at Kromdraai and carried around some of the teeth in his pocket. Broom, hearing of this, was able to bring together most of the palate, the left side of the face, zygomatic arch and skull base, part of the parietal, the right half of the mandible and most of the teeth, the right side of the skull having been weathered away (Figure 2a). This skull is better preserved than any previous fossil anthropoid. The premaxilla is fused with the maxilla, as in man; the face is less prognathous than the chimpanzee; the nasal opening is more like that of a chimpanzee, but approaches the condition in some Australians. A very peculiar feature is found in the upper part of the maxilla, which is convex in Plesianthropus, concave in Paranthropus (Figure 2a), the malar process projecting forwards unlike the condition in modern anthropoids or man. The orbital part of the malar is large, as in the orangutan, but the ascending facial portion is human. The zygomatic process is almost human in structure but twice as large as in an Australian aboriginal, and the glenoid fossa resembles that of Eoanthropus. The arch of the teeth is of the same horseshoe shape as in man, but the posterior portion of the palate is large, as in baboons and gorillas. A palatine torus is present. The mandible is very massive, but much more like Heidelberg man than like the apes; and the chin, though rounded, is nearer that of the Heidelberg jaw. The canine is smaller than in Sinanthropus or the Rhodesian skull. The first premolar is also human, but much larger than in any human race.

An imperfect mandible of a Paranthropus child three or four years old was found in 1941 within a few feet of the type skull. The milk teeth and a partly erupted first molar make possible a comparison with the Australopithecus child. The milk sets show striking differences, both being very unlike the chimpanzee, but those of Australopithecus are "amazingly like" those of a Bushman child. The lower incisors of all three genera are not greatly larger than in man but very different from the chimpanzee. The first lower molar of Paranthropus is smaller than in Australopithecus, but the cusps are more pointed and other differences are noted. In Paranthropus we have grinding molars with low flattened cusps, whereas the anthropoids such as Ramapithecus and Sivapithecus have crushing molars with fairly sharp cusps. The molars are also worn as they are in man and not as they are in the living anthropoids.

The postcranial bones of Paranthropus include the lower end of the humerus, upper end of the ulna, a hand bone, two toe bones, and a talus, all probably from the same individual as the type skull. The Paranthropus humerus, like the Plesianthropus femur, is nearly but not quite human, and the same is true of the ulnar fragment and the metacarpal. The structure of the talus favors the view that Paranthropus was bipedal and slightly larger than a chimpanzee. The toe bones are very like those of man and very unlike the chimpanzee, but the toes are longer and more finger-like than in man today. Broom concludes that Paranthropus, although less human than Australopithecus in its milk molars and face, had a much larger brain (about 650 cc.).

The associated fauna at Kromdraai includes a large baboon, *Parapapio major* and a small one, *P. angusticeps,* a hyrax, also not found at Sterkfontein, and two extinct species of Equus. The saber-tooths and pigs at Sterkfontein are not found at Kromdraai, and the two deposits probably differ in age by many thousands of years.

In a detailed study of the endocranial casts, Schepers finds the convolutional pattern of sulci and gyri remarkably clear, so that they could virtually all be identified. This is better than has usually been thought possible even from endocasts of a living species. He finds marked differences between the two individuals of Plesianthropus. While Australopithecus is dolichencephalic, Plesianthropus I is brachencephalic and II is intermediate. In all three genera the parietal lobe is greatly expanded and almost human, while the occipital lobe is reduced by the enlargement of the parietal. Dart was correct in his identification of the lunate sulcus. In all three it is 2 or 3 cm. from the parallel sulcus, whereas in anthropoids this distance is seldom more than 1 cm. The well-preserved sulci of the frontal lobe show a relatively simple pattern intermediate between anthropoids and man. Detailed diagrams are given of the cortical areas of the brain. The premotor area 6 is fully developed as in man (for walking erect and associated postures), the temporal lobe cortex is nearly as complex as in man, and the parietal areas are fairly expanded. Schepers concludes that the adult Australopithecus probably had a larger frontal lobe than the other two, nearer to Pithecanthropus. The Plesianthropus I cast gives a probable volume of 435 cc., Plesianthropus II 560 cc., Paranthropus 650 cc., and Australopithecus 500 cc. (in accordance with Dart). Schepers believes that these man-apes were not only bipedal with emancipated hands, but that they had articulate speech. The brains, while simian in character, show distinct advances in shape and proportion. The auditory area is much expanded, as well as the primary auditory association area. The visual association areas are also well developed, especially in Australopithecus. "They must have been virtually true human beings, no matter how simian their external appearance may have remained."

Since this account was written, Broom (1947a, b, c) has unearthed at Sterkfontein, and close to the original site, remains belonging to at least five more individuals of Plesianthropus. Two upper milk molars were found to be closely like those of the Taungs ape or a modern Bushman child. The Carabelli cusp on the second molar is a human character, not known in the apes. The other finds were, (1) the crushed snout of an adolescent Plesianthropus, (2) a fragment of the snout of a child, (3) an upper canine of a male, (4) a lower molar, (5) a perfect skull but without mandible or teeth. The skull is *ca.* 150 mm. long, 100 mm. broad (C. I. 66), with a probable capacity of 500 cc. Two features of this skull are regarded as relating it to man and distinct from the anthropoids: (a) the structure of the anterior fossa, (b) in the orbit, the ethmoid has a long articulation with the lacrimal bone.

In comparing Paranthropus with Plesianthropus, we find a striking difference in the shape of the face. The former is nearly flat in the infraorbital region but the cheeks are as far forward as the nose. In Plesianthropus this condition is absent and there is no bony ridge from the infra-orbital foramen to the canine root. Plesianthropus is thus more human, but the foramen magnum is farther back, so that Paranthropus was probably even more erect. It also had a larger brain, much smaller incisors and canines and much larger premolars. All three genera are nearly related and very near the ancestry of man, but possibly Plesianthropus may prove to be only specifically different from *Australopithecus africanus*.

Broom formerly agreed with Gregory and others that man was derived from some Miocene Dryopithecine ancestor. He now tends to derive man from an earlier stem, and agrees with the view of Weidenreich that this stem divided into two branches; one having more homomorphic canines and leading to man; the other having heteromorphic canines and leading to Dryopithecus and the anthropoids. Broom regards Bramapithecus, Sivapithecus, and Sugrivapithecus as genera all of which are close to the chimpanzee and orangutan, belonging in the same family. He supposes that the Australopithecinae separated from these anthropoids as early as the Lower Oligocene and that the hominids arose from them in the Pliocene. Broom concludes (p. 140) that "neither man nor the Australopithecines can be at all nearly related to the living anthropoids," which implies much parallel evolution. He derives the human line from a Lower Oligocene primate allied to Propliopithecus, perhaps as early as the Eocene. The Australopithecines, whether human or subhuman, probably gave rise to man in the Pliocene.

This primitive type of man, ancestral to Pithecanthropus, was of slight build, ran on their hind legs, had delicate hands, a small brain of more advanced character than the anthropoids, a large face and well-developed supra-orbitals but no definite brows, teeth and jaws essentially human but larger than in Homo, incisors and canines human— an astonishing assemblage of human and simian characters. Broom thinks Neanderthal man was not an ancestor of Homo but a mutation from a type at the level of the Wadjak skull.

To the present writer it seems probable that such Dryopithecine genera as Ramapithecus of Pliocene age from the Siwaliks may be classed with the Australopithecinae when more is known of their skulls, and that the human stem may have taken its rise from the plexus of Australopithecine ancestors in the Miocene. It seems unnecessary to go further back when, even in the Pliocene species and genera, so many human characters were present mingled with the anthropoid.

Having now considered the pre-hominid ancestors of man and the immediate source in the vicinity of the Dryopithecine and the Australopithecine groups from which he sprang, it is possible to conclude that the Hominidae took their origin probably not earlier than the Lower Miocene, at about the same time that the lines of descent leading to the modern orangutan, gorilla, and chimpanzee were being defined. The idea of an Eocene man from some tarsioid ancestor vanishes into thin air. Yet the variety of Dryopithecines from the Middle and Upper Miocene of the Siwalik Hills alone shows that there is plenty of time left for all the human evolution we know, and more.

The time elapsed since man took his rise in the Pliocene was formerly estimated at two million years. More accurate estimates, based on the radioactivity of the rocks, extend this time to thirteen million years, and from the Upper Miocene eighteen million, the Pleistocene having begun a million years ago. If it is true, as some writers hold (for instance, Zeuner, 1943), that the rate of species-formation is essentially the same in insects, birds, and mammals, the time required for the formation of a good species being five hundred thousand to a million years, then we may expect to find many species and genera of fossil Hominidae developed since Miocene times. But Broom (1943) suggests that man arose from some pre-Dryopithecine type of the Oligocene.

It has been suggested (Bourne, 1935) that the adrenal glands are intimately associated with the origin of the Primates and man. The adrenals are believed to play a part in gestation, and would therefore be involved in the change from oviparous to viviparous reproduction, the origin of the apes and man depending on an efficient organization of these glands. Bourne finds also a close relation between the adrenals and the brain, the adrenal cortical hormone controlling lipoid metabolism and myelinization of the brain cells. In support of this hypothesis, the adrenals of foetal Primates and man have a hypertrophied cortex, while anencephalous monsters have practically no foetal cortex to the adrenals. Changes in the adrenal cortex may then have made primate evolution possible.

Views regarding the relations of the great apes to man have differed widely. The more recent work in comparative anatomy supports the view that the Hominidae developed independently of the great apes and the Hylobatidae (gibbons). In a study of the ilium in Primates, Straus (1929) shows that this bone is much broader in man than in any other primate. The human ilium is in some respects more specialized and in others more primitive than in the great apes. From a similar comparative study of the thoracic and abdominal viscera of Primates, Straus (1936) finds a number of features which are shared by and peculiar to the five genera, orangutan, gibbon, chimpanzee, gorilla, and man. The chimpanzee, gorilla, and man are most primitive in certain points, whereas man, orangutan, and gibbon are most advanced in certain others. The general conclusion is that in visceral characters man is as near the gibbon as he is to the chimpanzee-gorilla stock. In a general discussion of the Primates, Schultz (1936) points out that the six genera of higher Primates show much greater variability than the sixteen recent genera of Cercopithecidae. He concludes that the morphological differences between the three great apes are smaller and less significant than between any of them and man. The Hominidae and Hylobatidae had already begun before the orangutan had diverged from the common stem of the great apes, and man separated from the common stock after the gibbons but before the great apes. These views appear to represent our present knowledge of primate morphology and they imply a large measure of parallel development. Gregory (1935), on the other hand, adheres to the more usual view that man had an arboreal brachiating ancestry. This he bases on the structure of the foot and the pelvis. Although the opposability of the great toe in man is much less than in the apes, it is difficult to see how it could have developed except under arboreal conditions.

Straus (1947) has recently emphasized that man has many monkey-like features which are not present in the anthropoids. He believes that the human line separated from the anthropoids prior to the differentiation of the gibbons. These pre-hominids (paleopithecoids) would be essentially monkeys with an expanding brain, a reduced tail and generalized extremities. They never brachiated but became terrestrial bipeds, developing parallel to the anthropoid apes. The Dartians would seem to represent such a condition in part,—erect and presumably without a tail, the brain enlarged in relation to their body size and the teeth already humanoid. Straus thus sees four lines— the monkeys, gibbons, great apes, and man—evolving independently and to a considerable extent parallel, and possessing many of the same genes with different frequencies.

In the following chapters we propose to consider briefly the known human fossil types in their relation to the evolution of the Hominidae and the living types of man.

REFERENCES

Abel, O. 1934. Das Verwandtschaftverhältnis zwischen dem Menschen und der höheren fossilen Primaten. *Zeits. f. Morph. u. Anthrop.* 34:1–14.

Abel, W. 1931. Kritische Untersuchungen über *Australopithecus africanus*. *Morph. Jarhb.* 65:539–640. Pl. 1. Figs. 33.

Beattie, J. 1927. The anatomy of the common marmoset (*Hapale jacchus*). *Proc. Zool. Soc. London.* 593–718. Pls. 2. Figs. 39.

Bourne, G. 1935. The phylogeny of the adrenal gland. *Am. Nat.* 70:159–178.

Broom, R. 1924. On the fossil remains associated with *Australopithecus africanus*. *S. Afr. J. Sci.* 31:471–480. Figs. 7.

———. 1939a. The dentition of the Transvaal Pleistocene Anthropoids, *Plesianthropus* and *Paranthropus*. *Ann. Transv. Mus.* 19:303–314. Figs. 4.

———. 1939b. A restoration of the Kromdraai skull. *Ann. Transv. Mus.* 19:327–329. Figs. 3.

———. 1940. The South African Pleistocene cercopithecoid apes. *Ann. Transv. Mus.* 20:89–100. Figs. 6.

———. 1943. South Africa's part in the solution of the problem of the origin of man. *S. Afr. J. Sci.* 40:68–80. Figs. 2.

———. 1945. Age of the South African ape-men. *Nature.* 155:389–390.

———. 1947a. The upper milk molars of the ape-man, Plesianthropus. *Nature.* 159:602. Fig. 1.

———. 1947b. Discovery of a new skull of the South African ape-man, Plesianthropus. *Nature.* 159:672. Figs. 2.

Broom, R., and J. T. Robinson. 1947c. Two features of the Plesianthropus skull. *Nature.* 159:809–810. Figs. 2.

Broom, R., and G. W. H. Schepers. 1946. The South African fossil ape-men: The Australopithecinae. *Transv. Mus. Mem.* No. 2. Pretoria. pp. 272. Pls. 18. Figs. 57.

Clark, W. E. LeGros. 1925. On the skull of Tupaia. *Proc. Zool. Soc. London.* 559–567. Figs. 6.

———. 1930. The thalamus of *Tarsius*. *J. Anat.* 64:371–414. Pls. 2. Figs. 22.

———. 1932. The brain of the Insectivora. *Proc. Zool. Soc. London.* 975–1013. Figs. 16.

———. 1935. Man's place among the Primates. *Man.* 35:1–6.

———. 1936. Evolutionary parallelism and human phylogeny. *Man.* 36:4–8.

———. 1939. The scope and limitations of physical anthropology. *Advancement of Sci.* pp. 52–75.

Clark, W. E. LeGros. 1940. Paleontological evidence bearing on human evolution. *Biol. Revs.* 15:202–230. Figs. 12.

Colbert, E. H. 1935. The correlation of the Siwaliks of India as inferred by the migrations of *Hipparion* and *Equus*. *Am. Mus. Novit.* No. 797. pp. 15.

———. 1937. A new primate from the Upper Eocene Pondaung formation of Burma. *Am. Mus. Novit.* No. 951. pp. 18. Figs. 3.

———. 1938. Fossil mammals from Burma in the American Museum of Natural History. *Bull. Am. Mus. Nat. Hist.* 74:255–434. Figs. 64.

Dart, R. A. 1925. *Australopithecus africanus*: the man-ape of South Africa. *Nature.* 115:195–199. Figs. 6.

———. 1926. Taungs and its significance. *Nat. Hist.* 26:315–327. Figs. 10.

———. 1929. A note on the Taungs skull. *S. Afr. J. Sci.* 26:648–658. Pl. 1.

Gates, R. R. 1946. *Human Genetics*. New York and London: Macmillan. 2 vols. pp. 1518.

Gregory, W. K. 1916. Phylogeny of recent and extinct anthropoids. *Bull. Am. Mus. Nat. Hist.* 35:258–355. Pl. 1. Figs. 37.

———. 1920–21. The origin and evolution of the human dentition. *J. Dent. Res.* 2:89–183, 215–282, 357–426, 607–717, 3:87–288. Figs. 353. Pls. 15.

———. 1920a. On the structure and relations of Notharctus. *Mem. Am. Mus.* 3:53–243. Figs. 84. Pls. 27.

———. 1927. How near is the relationship of man to the chimpanzee-gorilla stock? *Quart. Rev. Biol.* 2:549–560. Figs. 10.

———. 1935. The roles of undeviating evolution and transformation in the origin of man. *Am. Nat.* 69:385–404. Figs. 12.

Gregory, W. K., and Milo Hellman. 1926. The dentition of *Dryopithecus* and the origin of man. *Anthrop. Papers Am. Mus.* 28:1–123. Pls. 25. Figs. 32.

———. 1926a. The crown patterns of fossil and recent human molar teeth and their meaning. *Nat. Hist.* 26:300–309. Figs. 9.

———. 1939. The dentition of the extinct South African man-ape *Australopithecus* (*Plesianthropus*) *transvaalensis* Broom. *Ann. Transv. Mus.* 19:339–373. Figs. 14.

———. 1940. The upper dental arch of *Plesianthropus transvaalensis* Broom, and its relations to other parts of the skull. *Am. J. Phys. Anthrop.* 26:211–224. Pls. 4. Figs. 1.

———. 1945. Revised reconstruction of the skull of *Plesianthropus transvaalensis* Broom. *Am. J. Phys. Anthrop.* N. S. 3:267–275. Pls. 3.

Gregory, W. K., M. Hellman, and G. E. Lewis. 1938. Fossil anthropoids of the Yale-Cambridge India Expedition of 1935. *Carnegie Publ.* No. 495. pp. 28. Pls. 8.

Hellman, M. 1928. Racial characters in human dentition. *Proc. Am. Phil. Soc.* 67:157–174. Figs. 7.

Hopwood, A. T. 1933. Miocene Primates from Kenya. *J. Linn. Soc. Zool.* 38:437–464. Pl. 1.

Hrdlička, A. 1924. New data on the teeth of early man and certain fossil European apes. *Am. J. Phys. Anthrop.* 7:109–132. Pl. 1.

Koenigswald, G. H. R. von. 1942. The South African man-apes and Pithecanthropus. *Carnegie Inst. Publ.* No. 530. pp. 205–222. Pls. 10. Figs. 6.
Lewis, G. E. 1934. Preliminary notice of new man-like apes from India. *Am. J. Sci.* 27:161–179. Pls. 2.
———. 1937. Taxonomic syllabus of Siwalik fossil anthropoids. *Am. J. Sci.* 34:139–147.
MacInnes, D. G. 1943. Notes on the East African Miocene Primates. *J. East Afr. and Uganda Nat. Hist. Soc.* 17:141–181. Pls. 6. Figs 2.
Matthew, W. D. 1904. The arboreal ancestry of the Mammalia. *Am. Nat.* 38:811–818.
———. 1927. The evolution of the mammals in the Eocene. *Proc. Zool. Soc. London.* 947–985. Figs. 16.
Pocock, R. I. 1927. The gibbons of the genus Hylobates. *Proc. Zool. Soc.* 719–741. Figs. 5.
Schultz, A. H. 1936. Characters common to higher Primates and characters specific for man. *Quart. Rev. Biol.* 11:259–283, 425–455. Figs. 21.
Senyürek, M. S. 1941. The dentition of Plesianthropus and Paranthropus. *Ann. Transv. Mus.* 20:293–302.
Shaw, J. C. M. 1940. Concerning some remains of a new Sterkfontein Primate. *Ann. Transv. Mus.* 20:145–156. Fig. 1.
Simpson, G. G. 1935. The first mammals. *Quart. Rev. Biol.* 10:154–180. Figs. 19.
———. 1937. The Fort Union of the Crazy Mountain Field, Montana, and its mammalian faunas. *Smithson. Inst. U. S. Nat. Mus. Bull.* 169. pp. 287. Pl. 10.
———. 1945. The principles of classification and a classification of Mammals. *Bull. Am. Mus. Nat. Hist.* 85:1–350.
Straus, W. L., Jr. 1927. Growth of the human foot and its evolutionary significance. *Carnegie Publ.* No. 380. pp. 93–134. Pl. 1. Figs. 6.
———. 1929. Studies on Primate ilia. *Am. J. Anat.* 43:403–460. Figs. 8.
———. 1936. The thoracic and abdominal viscera of Primates, with special reference to the orang-utan. *Proc. Am. Phil. Soc.* 76:1–85. Figs. 16.
———. 1940. The posture of the great ape hand in locomotion, and its phylogenetic implications. *Am. J. Phys. Anthrop.* 27:199–207. Fig. 1.
———. 1947. The riddle of man's ancestry. *Am. J. Phys. Anthrop.* N.S. 5:243 (abstr.).
Werth, E. 1919. *Parapithecus*, ein primitive Menschenaffe. *Sitz-ber. Ges. Naturforsch. Fr. Berlin.* 1918:pp. 327–345. Figs. 7.
Wood, Jones F. 1947. The premaxilla and the ancestry of man. *Nature.* 159:439.
Woollard, H. H. 1925. The anatomy of *Tarsius spectrum*. *Proc. Zool. Soc. London.* 1071–1184. Figs. 53.
Zeuner, F. E. 1943. Studies in the systematics of Troides Hübner (Lepidoptera Papilionidae) and its allies: distribution and phylogeny in relation to the geological history of the Australasian Archipelago. *Trans. Zool. Soc. London.* 25:107–184. Fgs. 115.

4

Evolution of the Hominidae

During the past forty years there has been an increasingly rapid accumulation of fossil forms belonging to the Anthropoideae and the Hominidae. Many gaps have been filled and the known forms have become so numerous that authorities frequently differ regarding their exact relationship, especially when the geological age is uncertain. There are a number of skulls or jaws which, on present knowledge, might be put in either group. Indeed, Elliot Smith considered that, but for human vanity, the two groups should be combined into one family. This is essentially what Simpson has done in creating his group Hominoidea. We have seen that the Siwalik Hills in northern India were a center of radiation of higher primate types over a long period. From that plexus of forms many at least of the later Primates appear to have been derived, although man may be more immediately descended from the Australopithecines of Africa.

Viewing all the known fossil men, beginning with the Upper Pliocene or Lower Pleistocene, they seem to fall quite clearly into three stages or evolutionary steps, representing three genera which are partly successive in time; but since evolution proceeds at varying rates, they are also partly contemporary. These stages are now fairly generally recognized as (1) the Pithecanthropus level, (2) the Paleoanthropus or Neanderthal level of Sergi, (3) the level of Homo or modern man. We have already considered the possibility that Pithecanthropus arose from an ancestry in the South African Australopithecinae, but it is still possible that these man-apes terminated in a dead end. If so, Pithecanthropus may have arisen from some Asiatic derivative of the Dryopithecinae having more or less similar characters. The latest concensus is that the Australopithecines were at any rate very near to the line of human ancestry.

Keith (1947) has recently accepted the view that Dart's original interpretation of Australopithecus as essentially a direct human ancestor was correct. He, therefore, suggests replacing the long name Australopithecines by the more colloquial term Dartians for this group of "ground-living anthropoids, human in posture, gait and dentition, but still anthropoid in facial physiognomy and in size of brain." The Dartian phase of human evolution would thus be intercalated between the Dryopithecine level and Pithecanthropus.

In any case, the Pithecanthropus level of human evolution is now represented not only by the famous skull cap and femur found by Dubois in Java in 1891, but by *Sinanthropus pekinensis*, discovered by Davidson Black in northern China, Africanthropus, and perhaps Rhodesian man [1] in Africa, and certain other less well-established types. It will be necessary to say something about each of these very important discoveries in the ancestry of man.

After his discovery by Dubois at Trinil in central Java on the Solo River, in 1891, *Pithecanthropus erectus* figured for many years as the "missing link" between man and apes, but is now recognized as a primitive human type. It may be pointed out that Cunningham (1895) recognized the true position of Pithecanthropus when he wrote that it has many Neanderthaloid characters "but stands very nearly as much below the Neanderthal skull as the latter does below the ordinary European skull." He also recognized that the accompanying femur was of ordinary human type.

The shape of the human femur has generally been regarded as a result or a sign of the erect posture, and Dubois gave the specific name *erectus* on the assumption that this shape was necessarily accompanied by the erect posture. As shown by Walmsley (1933), the lateral condyle (at the distal end of the femur) is the weight-bearing condyle in modern man, and in Pithecanthropus too the load line passes from the head of the femur through the lateral condyle. In Neanderthal man the load is distributed equally on both, whereas in the gorilla the medial condyle is larger than the lateral and the load line passes through it. A similar shift of the weight axis from the anthropoid condition to that in modern man occurs in children be-

[1] Rhodesian man is, however, more advanced except in the tremendous brow ridges.

tween birth and ten years of age. The Neanderthal femur is also bowed, but this may not exceed the variations in some modern men. Mivart (1874) states that in the baboons and spider monkeys, as in modern man, the inner condyle projects further downwards than the outer, so that here also the weight comes on the lateral condyle. If that is the case, then the Pithecanthropus type of femur is not necessarily accompanied by erect posture. The position of the load line will also depend on the length of the femur neck and the angle the head makes with the shaft of the femur.

Dubois' find consisted only of a skull cap and a femur, the latter apparently of modern type but not certainly associated with the skull.[2] Abundant remains of other vertebrates were found with them, which help to date the skull as Middle to Lower Pleistocene. For many years there was nothing to fill the great gaps between modern man, Pithecanthropus, and Dryopithecus. Dubois (1924) described a mandibular fragment from the Trinil beds of Kedung Brubus, 40 km. from the original site, which he kept secret for thirty-three years and then announced that *P. erectus* was not a man but a giant gibbon!

The more recent work of Koenigswald (1940) in Java has unearthed portions of three more Pithecanthropus skulls. He gives a full account of the Java discoveries and of the contemporary fauna and the geological conditions. He concludes that the Trinil beds are of mid-Pleistocene age, contemporary with the Mauer jaw, the Swanscombe cranium, the Cromer Forest Beds, and the Choukoukien Sinanthropus, the Djetis fauna being Lower Pleistocene. The skull which has been placed in various genera under the specific name *soloensis* Koenigswald regards as developed from Pithecanthropus towards modern man, but less advanced, belonging to the Neanderthal stage but more primitive.[3] The Pithecanthropus teeth from Trinil fall,

---

[2] Drennan (1936) found in South Africa a Hottentot femur almost identical with the one from Trinil. The Hottentot skeleton was excavated in Cape Province in 1928, two and one-half feet below the surface. It had been buried in a sitting position over a century ago. If the original femur belonged to Pithecanthropus, then this Hottentot had a pithecanthropoid femur.

[3] Oppenoorth (1937), in a careful comparison of the Ngandong (soloensis) remains, which include parts of eleven skulls and a tibia, shows that they are quite a different type from Neanderthal man. While conforming generally with the Rhodesian skull, they differ from Neanderthal man in the shape of the supra-orbital torus, the orbits (which must be quadrangular rather than round), and the flat occiput. The cranial capacity is much less (1160–1300 cc. as against 1350–1600 cc.), while the mandibular articulation, the tympanic

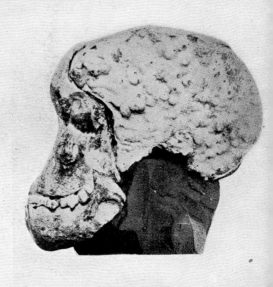
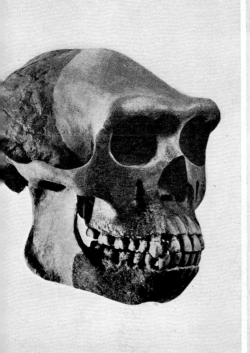
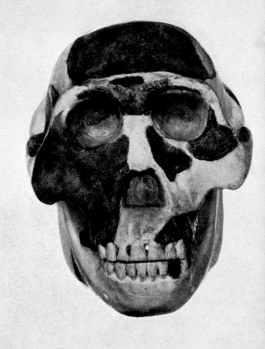

a. (upper left): Restoration of the head of Australopithecus by Forestier (by permission of the *Illustrated London News*)
b. (upper right): The Taungs skull, *Australopithecus africanus*
c. (lower left): Reconstruction of *Pithecanthropus robustus* (skull IV) with the mandible B of *P. erectus*
d. (lower right): Reconstruction of *Plesianthropus transvaalensis* Broom

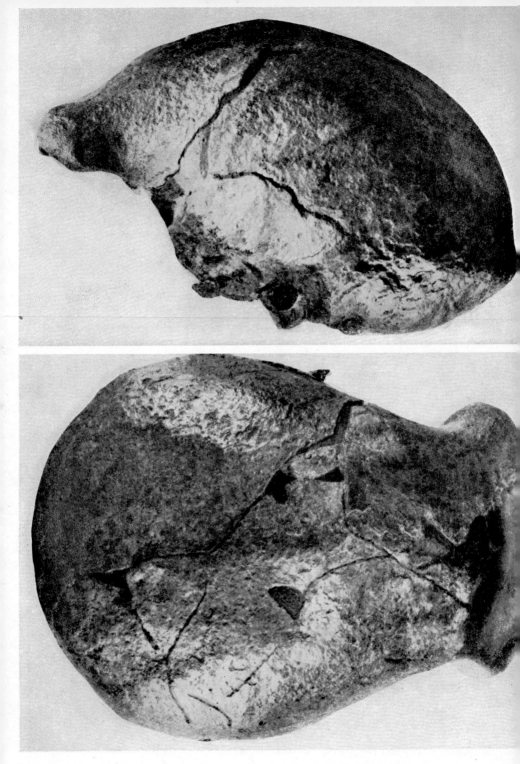

Side view and vertical view of *Pithecanthropus erectus*, skull II

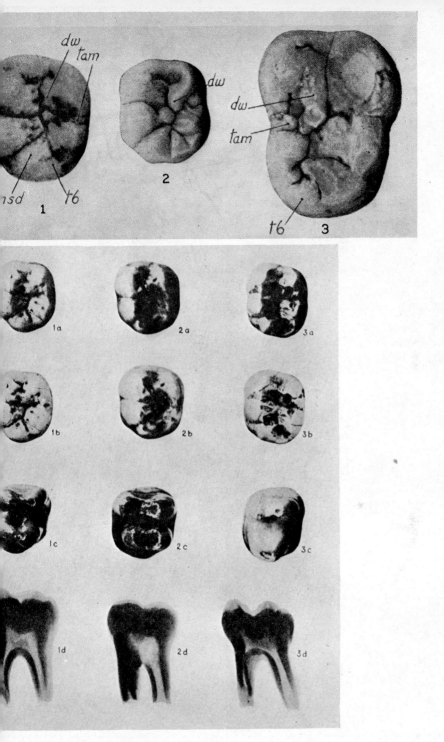

a. (upper): (1) left front molar of Sinanthropus pekinensis, (2) left second molar of a modern Amerind, (3) right third molar of Gigantopithecus, showing relative sizes

b. (lower): Four views of (1) molar tooth of a chimpanzee, (2) the original molar of Sinanthropus, (3) molar tooth of a Chinese child

a. Skull of adolescent Sinanthropus from Locus D in preparation, showing the natural endocast and the great thickness of the bones

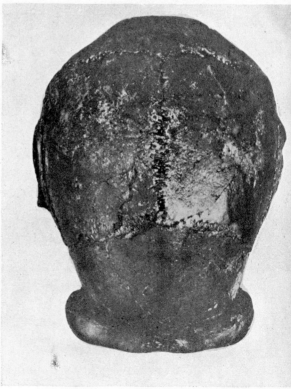

b. Vertical view of the same skull, showing heavy occipital torus

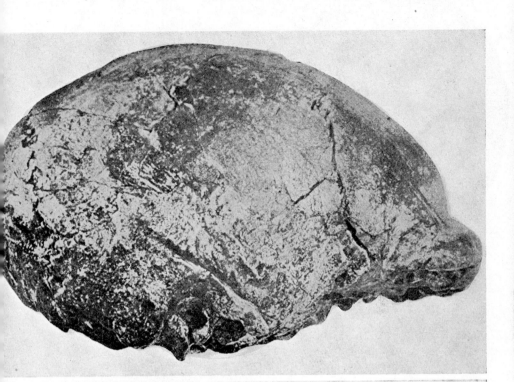

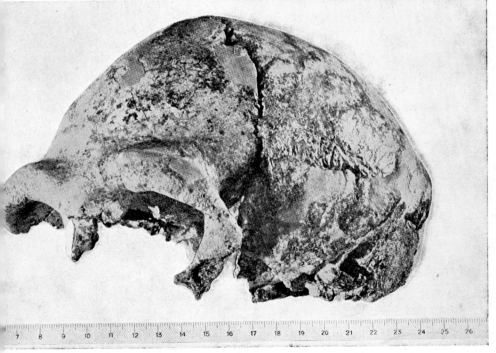

a. (upper): Sinanthropus skull in side view
b. (lower): Sinanthropus skull III of Locus L, showing frontal and occipital tori

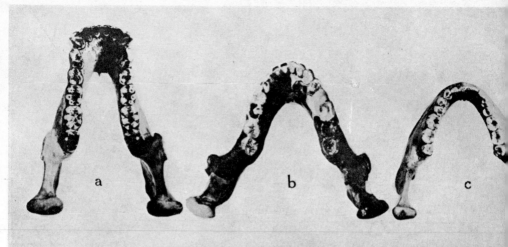

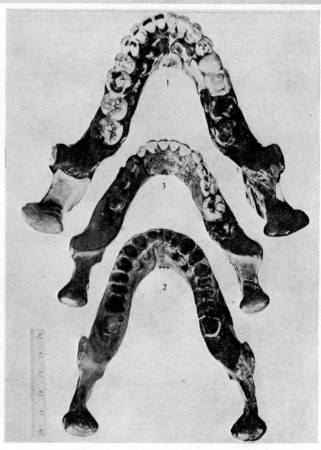

a. (upper): Mandibles of (a) female gorilla, (b) Sinanthropus male, (c) modern northern Chinese male
b. (lower): Sinanthropus mandibles, (1) an adult male, (2) an adult female, (3) a female child

a. The Gibraltar (Neanderthal) skull

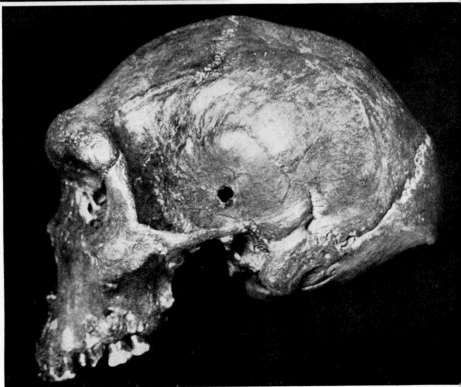

b. The Rhodesian skull in side view

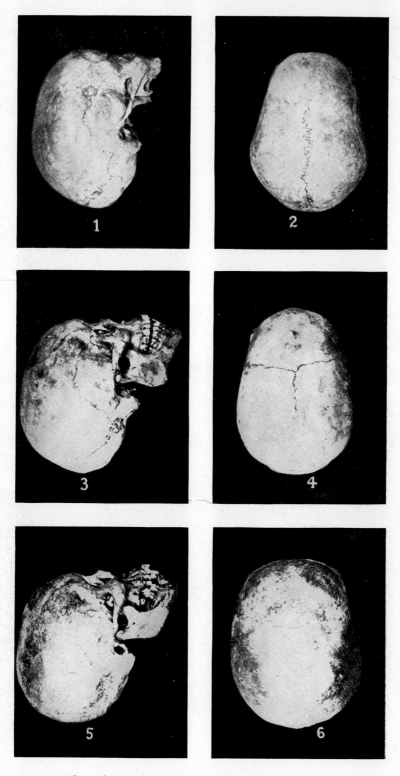

Three shapes of skull in vertical and lateral aspect

in size, within the range of living orangutans, the molars of modern man being reduced 20–25 per cent in size from those of Pithecanthropus. In the sum of its characters Pithecanthropus is found to be more primitive than Sinanthropus but nearly related, both deriving from the same stem. The femur from Trinil, if found alone, would undoubtedly be placed with modern man. The Sinanthropus femur differs in several respects, including its much thicker walls, but the differences are less than Weidenreich thinks.

The No. IV skull of Pithecanthropus, being nearly complete, confirms and greatly extends our knowledge of this type. There are also teeth, jaws, and limb bones. They were listed as seven or more distinct finds by Koenigswald and Weidenreich (1939). In 1936 Koenigswald described a small Lower Pleistocene skull under the name *Homo modjokertensis* which has been regarded as belonging to a baby of *Pithecanthropus erectus,* but Weidenreich (1945) suggests with good reason that it belongs to *P. robustus* or Meganthropus. Its tympanic plate is purely of anthropoid type. Another mandible fragment (Koenigswald, 1937) shows that *Pithecanthropus erectus* survived into the Middle Pleistocene as a relict species. Of skull IV (Weidenreich, 1940; Koenigswald, 1942) the jaw was found first, coated with matrix. When Koenigswald returned with the collector to the spot he found the skull from which the jaw had been freshly broken. The whole, as restored by Weidenreich (1940) is shown in Plate I (c). It consists of three-quarters of the brain-case and portions of the upper jaw. The frontal bone and face are missing but the upper face can be restored with comparative certainty from the other skulls. A side view and a vertical view of skull II are shown in Plate II, the great supra-orbital torus with its lateral extensions being conspicuous. This skull was found in the deepest Trinil beds at Sangiran in 1937 (Koenigswald, 1938) and is an almost exact duplicate of the skull cap found forty-six years earlier by Dubois. The bones of the skull are extremely thick, the forehead flat, the brain-case low, the eyebrow ridges very heavy. Skull IV is "the most primitive type of fossil man ever found" and because of its larger size was at first regarded as a

---

plate, and the mastoid process are much more modern. The foramen magnum is also more forward, the tibia more slender and straight, and the estimated height greater (166 cm. for males). Instead of stone implements, they used bone antlers (as did Sinanthropus) but they had the stone balls (bolas) which are also found in the Mousterian cultures of Europe and South Africa.

male of *Pithecanthropus erectus*, whereas the others are smaller and were believed to be female. It has since, however, been regarded as a distinct species, *P. robustus* (Weidenreich, 1944). Another distinct feature in this skull is a crest extending along the sagittal line. The cranial capacity is estimated at 900 cc., whereas 600 cc., represents about the maximum in the living apes.

The upper jaw of this skull is wider than any known human or simian jaw and it shows alveolar prognathism. Another remarkable feature is a diastema or space between the upper canine and the first premolar. This feature is present in the apes, where the large, projecting canines require a space to interlock. It is found in no other human and is only present in the upper jaw of *P. robustus*. The mandible has no chin, the teeth are large and robust, the canines projecting but slightly (as in Sinanthropus). The molars are not only large but they show increasing size from first to third, whereas in modern man the third molar (wisdom tooth) is smaller than the others.

Another mandible found in Trinil beds at Sangiran in central Java is of such extraordinary size that Koenigswald gives it another generic name, *Meganthropus palaeojavanicus*. It is larger than the massive Heidelberg jaw and the teeth, unlike those of the latter, are also gigantic, and the dental arch is intermediate between anthropoid and human. Thus a really gigantic human type, at least of jaw, has been found buried in the volcanic ash and sands of the Trinil formation. Weidenreich (1944) regards *P. robustus* as intermediate between Meganthropus and the more normal sized *P. erectus*. The Meganthropus mandible is fully described by Weidenreich (1945) and shown to be entirely hominid in character. In average thickness it surpasses all known hominid or anthropoid mandibles. The greatest thickness is not, as in anthropoids, at the alveolar level, but lower down. As in Heidelberg man, the symphysis has a basal arch or submental incisura, but on the lingual aspect of this arch is a depression which Weidenreich calls the digastric recess. This is present in no other hominid or living anthropoid, but somewhat similar conditions are found in *Dryopithecus pilgrimi* and *D. fontani*. There is, of course, no chin. Koenigswald regards the mandible fragment found in 1939 at Sangiran as belonging to a female of the same species, but Weidenreich, from a study of the cast, thinks it probably belongs to an orangutan-like anthropoid approaching the hominid pattern.

All these fossils have been listed and their status fully explained in another publication (Weidenreich, 1945c).

There are thus four hominid maxillae from the Trinil formation, all of different sizes and showing other differences as well corresponding to the size sequence. But the story of human gigantism in southeast Asia is not yet complete. It is well known that Chinese apothecaries collect "dragon's bones" for grinding and distribution to their patients as medicine. Knowing this practice, Dr. von Koenigswald was able to rescue from chemists' shops in Hong Kong between the years 1934 and 1939 (besides various animal teeth, including a big orangutan), three molars—two of them wisdom teeth—which probably came from at least two individuals. One of these giant molars he named *Gigantopithecus blacki* (Koenigswald, 1935), considering it to belong to an anthropoid. Weidenreich (1944), however, believes it to be definitely human—Gigantanthropus—from the pattern of the occlusal surface, but other anthropologists who have expressed an opinion regard it as the tooth of a giant orangutan or some other anthropoid. The volume of the crown is about six times that in modern man, or twice the size of a gorilla third molar. Broom (1939) points out that the Gigantopithecus tooth is very much like that of Paranthropus in South Africa.

Weidenreich estimated that Meganthropus must have been of the size and strength of a big male gorilla, and Gigantopithecus must have been correspondingly larger. In a detailed study of these teeth, Weidenreich (1945a) shows their hominid but primitive character. There is no cingulum. The third molar, unlike Sinanthropus and modern man, is as large as the others. The mass of the crown is 4170 cu. mm., compared with an average of 926 cu. mm. and an extreme of 1526 cu. mm. in modern man. It is then four times the average size in modern man (Plate III, a). Weidenreich estimates the length of the Gigantopithecus mandible as 180 mm. compared with 135 mm. in Meganthropus, 109 mm. in Heidelberg man, and 85–100 mm. in modern man. The Trinil femurs, found by Dubois, indicate a tall race, and from homologous measurements Weidenreich believes that the femur of Gigantopithecus would be only a little longer and stouter than a tall race of modern man. The small teeth in the heavy Heidelberg jaw show the danger of drawing conclusions regarding gigantism from tooth size alone.

These teeth of giant hominids come from the "yellow deposits," of Lower or Middle Pleistocene age, found in caves all over China south of the Yangtze River. The fauna of these caves is also characteristic of the Trinil beds in Java (which are believed to be also of Lower or Middle Pleistocene age), and is known as the Sino-Malayan fauna. The South Chinese Gigantopithecus, Weidenreich thinks, is nearly related to the Javan Meganthropus and Pithecanthropus, and a contemporary of Sinanthropus. The frequency of pig bones and shells in caves of Kwangsi suggests human activity, although no early human bones have yet been found there. The volcanic eruptions of this period in Java were often accompanied by mud streams and torrents, disturbances which increase the difficulty of determining the exact age of the contained fossils.

It is well recognized that Java emerged from the sea in Upper Pliocene times and was linked with Sumatra, Borneo, and Malaya to form a vast extension of southeast Asia known as Sundaland. The fauna entering this region from southern China, Burma, and India was of Upper Siwalik age. According to DeTerra (1943), the greatest emergence of Sundaland was in Early and Middle Pleistocene, during the second Glacial stage, when the ocean level was lowered by the accumulation of land ice and at the same time there was uplift as well. Vulcanism was active in Java during this period and new drainage (river) lines resulted from the land uplift. The Sino-Malayan fauna formed a unit over this great area. Koenigswald (1935) has described this fauna in Java. He points out that the orangutan, tapir, and gavial occurred in Borneo and Sumatra as well, showing that these islands were formerly united, apparently several times, during the Pleistocene. In a further discussion of Sundaland, Zeuner (1943) points out that Wallace's faunal line (between Bali and Lombok and northwards between Borneo and Celebes) forms its eastern border and that at some time in the Pleistocene sea-level was seventy to ninety meters below the present level, at another time one hundred meters higher than now.[4] It is thus natural that the human types in this area

---

[4] For a discussion of the geology of this region see Molengraaff (1921), and for evidence from the flora and fauna Merrill (1946). Sundaland included Sumatra, Java, Bali, and Borneo connected with Asia, and adjacent island chains (forming temporary isthmuses) extended northeastward to the Philippines. Australia and New Guinea with adjacent islands formed Papualand (Merrill), which has been separated from Asia (Sundaland) by deep water since the Cretaceous. The region between, including Lombok, Sumba, Timor, and Celebes, has

at that time should have been nearly related, and we shall find that Sinanthropus of northern China was also related to Pithecanthropus of Java, just as Meganthropus and Gigantopithecus may be allied.

Weidenreich (1945) suggests that these gigantic hominids were the source from which the smaller genera and species of man evolved in later ages. This is impossible to believe, for two reasons. First, if they were the ancestral hominids, where are the equally gigantic anthropoids from which they would presumably have been derived? The second and more serious difficulty arises from the general law, applicable apparently to all vertebrates and to many of the invertebrate groups, that each new group is at first of small or very small size, gradually increasing to a maximum size and then dying out leaving no descendants.[5] There is no doubt that the theory of side lines has been much overdone in human phylogeny, but this appears to be a case where forms like Gigantopithecus and Meganthropus are more likely to be lateral dead ends than ancestors. If these giants were ancestors to modern man they would form a marked exception to a very general rule. But Weidenreich reads the series, Gigantopithecus ⟶ Meganthropus ⟶ *Pithecanthropus robustus* ⟶ *P. erectus*. He believes (1945) that Pithecanthropus in Java and Sinanthropus in northern China are both descended from Meganthropus or even Gigantopithecus, one species having moved south and the other northwards from an origin in southern China which might be in touch with the anthropoids of the Siwalik Hills. In an elaboration of these views, Weidenreich (1946) suggests that Gigantopithecus was twice the size of a male gorilla. This might be so, but anthropological opinion regards him as a giant orangutan and not human—a sane view compared with the speculations regarding man's ancestry in Weidenreich's little book.

In favor of the thesis that in the higher Primates the evolution of size has sometimes proceeded from gigantism to smaller sizes is the presence of *Paedopithex rhenanus*, a giant gibbon, in the Lower Pliocene of Germany, and the fact that some of the Dryopithecinae were apparently of gigantic stature, as were some early baboons.

---

always remained archipelagic, receiving flora and fauna from east, west, and south and transmitting it northwards through island chains to the Philippines. Man's early migrations in this area from Java to Australia would be controlled partly by the numerous local rises and falls which took place in land and sea levels during the Pleistocene.

[5] However, there are exceptions to this rule. See chapter 9.

There are also early giant lemurs, but there is no evidence that the Primates have reversed the usual course of events and it is probable that among Primates, as in other groups, the giants were terminal dead ends. In any case, the paleontological evidence is that each phylum began with small forms which later progressed in size but, at least during the Pleistocene, sometimes produced smaller descendant species. Man, with his many generalized and primitive anatomical features, might be the starting point for a new phylum, though it is difficult to conceive just what form it would take, since the practicable limit of increase in brain size appears to have been reached.

Koenigswald (1937a) lists the Pleistocene zoning of Java as follows:
Recent—
Sampoeng (Neolithic)
Ngandong (Upper Pleistocene)
Trinil (Middle Pleistocene)
Djetis (Lower Pleistocene fauna)
Kali Glagah (Mastodon)
Tji Djoelang (Siwalik fauna)
Tji Sande.

In the lowest (Tji Sande) zone only western Java was above water. The stone implements found at the Trinil level are too advanced to have been produced by Pithecanthropus. They include hand-axes of Abbevillian (Chellean) type (rostrocarinates), scrapers, and cores, showing that the *coup-de-poing* culture spread all the way from western Europe to South Africa and through India to Java in Middle and Upper Pleistocene. The affinities of Ngandong man (soloensis) to Rhodesian man are emphasized. The skulls had all been broken at the base to remove the brain.

SINANTHROPUS

Sinanthropus from the cave of Choukoutien, near Peking, has proved to be the richest find in all human paleontology. Dr. O. Zdansky first found here two molar teeth which were recognized as human. Then Dr. Davidson Black organized excavations on a large scale and on October 16, 1926 Dr. Bohlin discovered a unique tooth which Black (1927) with great scientific courage and prevision described as *Sinanthropus pekinensis*, a new type of primitive man. The later discovery of abundant skulls and other bones shows how fully his predictions regarding this remarkable hominid were justified. The tooth was recognized as a "slightly worn left lower permanent molar tooth, probably the first." This tooth showed moderate taurodontism and was compared with the corresponding tooth of a chimpanzee and a Chinese child. The differences, which mean much to an expert, are indicated in Plate III(b). Taurodont molar teeth have relatively large and capacious crowns with reduced roots. They are frequently found in all early hominids and are characteristic of the teeth of cattle, as Keith's name indicates (see Gates, 1946, p. 369). Black (1925) had previously published a full account of skeletons representing about forty-five Chinese individuals from another cave deposit of Aeneolithic age, so he was well prepared to cope with the great discoveries which followed. These skeletons proved to be similar to the present Chinese.

Soon more or less complete skulls of Sinanthropus began to be found, and a series of monographs followed. The first adult skull was described (Black, 1930) and later an adolescent. Some of the remarkable features of Sinanthropus, as well as its obvious relationship to Pithecanthropus, can be seen from Plates IV and V.

The work of Movius (1943) leads to the conclusion, first expressed by Hooton (1940), that the Pithecanthropus-Sinanthropus stage of evolution in man, occupying the eastern half of Asia from Peking to Java, and quite independent of contemporary man in Europe, had also a characteristic and independent culture development. From a detailed study of stone implements in Burma, Movius describes the Early and Late Anyathian culture, contemporary with the Lower and Upper Paleolithic of Europe, but a separate culture province. Its development is shown in relation to the terraces of the Irrawaddy. The material used consisted of fossil wood, silicified tufa, and quartzite, and it passes finally into the Neolithic without the succession of developmental types which is so characteristic of western Europe. The Anyathian begins with coarse choppers—core tools which he calls hand-adzes, as common as the hand-axes of the European Lower Paleolithic. There are other types of cutting and scraping implements, but the tools are mainly heavy choppers, partially flaked into shape. They continued with little change during the whole period of the Paleolithic, when a succession of European types was being produced. There is no blade industry in Burma until post-Paleolithic times. An analogous development is found in northwest India, but the Late Soan of this area begins earlier than the Late Anyathian. Contact of Late Soan with the west is indicated by the presence, in addition, of a Levalloisian technique.

At Choukoutien, associated with Sinanthropus, are chopping tools of the same basic type as the Soan and Anyathian. There is also a small flake industry with quartz. At Patjitan, Java, Koenigswald found a Lower Paleolithic site with choppers of the Anyathian type, as well as cores and flakes. There were a few hand-axes which seem to represent a pointed bifacial chopper with longitudinal flaking—not a true Acheulian *coup-de-poing*. The Levalloisian technique is absent from Java, Burma, and China. Thus the Choukoutienian of China, the Anyathian of Burma, the Tampanian of Malaya, and the Patjitanian of Java are local cultures of the Lower Paleolithic, belonging to one general culture province which extended from southeastern Asia to northern India and China. This was the culture of Pithecanthropus in Java and Sinanthropus in China in the Middle Pleistocene. It was distinct from the classical culture development in the great triangle bounded by western Europe, South Africa, and India, but the two made contact in northwestern India.

A general account of Peking man was published by Black, Teilhard de Chardin, Young, and Pei (1933), dealing not only with the skulls and other bones but making a comprehensive study of the abundant crude Paleolithic implements and the accompanying fauna. The latter included species of Equus, Machairodus, Hyena, a cave bear, a rhinoceros near *R. mercki*, and a fossil dog, *Canis* (*Nyctereutes*) *sinensis*. *Pliopithecus postumus*, an anthropoid from southern Mongolia probably of Pliocene age, was a probable successor of the European *P. antiquus*. It was concluded that the remains were older than Upper Pleistocene but younger than Late Pliocene. Ashes in the cave deposits showed that Sinanthropus not only hunted the animals for food but probably cooked them. A further account of the Choukoutien caves and their inhabitants was given by Black (1934) in a Croonian Lecture to the Royal Society.

Shellshear and Elliot Smith (1934) made a study of the endocranial cast of Sinanthropus and compared it with living humans. Among the significant features they found the occipital region of the brain identical with that of the apes, while other features were typically human. The primitive human brain shows a degree of symmetry of the cerebral hemispheres which is quite exceptional in modern man. The attainment of human rank is associated with precocious expansion of the posterior end of the second temporal convolution and the orbital margin of the frontal territory—both probably connected with the acquisition of speech. The expansion of the mid-temporal area is probably connected with greater skill in movement and locomotion. Hirschler (1942) has recently made a comparative study of anthropoid and human endocranial casts, including those of thirty-two Chinese from Hongkong. He gives measurements and indices and relates the endocasts to features of the brain. In anthropoids the sulcus lunatus is always in front of the lambda, and in man it is always far behind that point.

After Black's unfortunately early death in 1934, Weidenreich took up the work and in 1943 produced his great monograph, which may be considered a summary of a series of earlier monographs by the same author. Twelve skulls, some quite incomplete, were then available, as well as a number of maxillae, mandibles, teeth, and other bones.

The most striking features of the Sinanthropus skull may be mentioned as (I) a heavy and continuous occipital torus, forming an attachment for the heavy neck muscles, (II) an extremely heavy and continuous supra-orbital torus. Sinanthropus was compared in detail with Pithecanthropus. It was found that the skulls agreed in fifty-seven out of seventy-four characters but differ in five main characters: (1) Cranial capacity. The average capacity of Sinanthropus according to the latest estimate (Weidenreich, 1946) is 1040 cc., although the smallest is only 915 cc., the largest 1225 cc. Pithecanthropus is about 20 per cent smaller, the skulls available showing an average of 775 cc. This is due partly to small dimensions of the brain-case except in height, and partly to greater thickness of the bone. There is about the same amount of difference in size of the brain-case of Eskimos and Australian aborigines. (2) There are differences in the form of the vault in the frontal, obelion, and occipital regions. In horizontal outline Pithecanthropus belongs to Sergi's spheroides and Sinanthropus to the ellipsoideus form, the former having a broad and rounded occiput while in the latter it is narrow and elongated. (3) Sinanthropus has a characteristic tendency to develop an Inca bone in the occiput of the head as well as exostoses (the torus mandibularis, maxillary, and ear exostoses). These are not found in Pithecanthropus. (4) There are various differences in mandible and teeth. The frontal part of the mandible is thicker in Pithecanthropus and the presence of a diastema in the upper jaw is an extraordinarily anthropoid condition. (5) In Sinanthropus the frontal sinuses are small and contracted, but in Pithecanthropus they are very large and laterally expanded.

Weidenreich regards these differences as only of racial value, but there is no doubt that if they were found in any animals but humans they would be regarded as specific. We will return to this aspect in a later chapter. Several authorities have pointed out that the differences are not great enough to be generic, and since *Pithecanthropus erectus* was described first, Sinanthropus then becomes *Pithecanthropus pekinensis*. While, on the whole, *Pithecanthropus erectus* and *P. robustus* of Java are more primitive than *P. pekinensis* of China, especially when we consider the smaller brain-case and the diastema, yet there is obviously no need for one to be more advanced than the other in order to regard them as separate species. Weidenreich points out various characters in which Sinanthropus is more primitive.

Although from a nomenclatorial point of view the Peking man should now be known as *P. pekinensis*, indicating its relationship, the name Sinanthropus will probably continue in frequent use as a sort of nickname. It is little short of astonishing that Weidenreich, after establishing *P. robustus* as a species distant from *P. erectus* in the same area (Java) almost entirely on the basis of its greater size, is unwilling to accord the same specific rank to *P. pekinensis*, whose differences are multiform and much greater, in addition to the fact of its different geographical distribution. This is but one of the many distortions in biological thought occasioned by adherence to the statement of Linnaeus that all the living types of man belong to one species, *Homo sapiens*. Biologists the world over will only begin to think straight about racial questions when this eighteenth-century dogma is repudiated.

Peking man, as we have seen, has a larger brain than Pithecanthropus, but its cranial capacity is much less than that of Neanderthal man, as Weidenreich (1936b) has pointed out. Regarding all three species as a unit, their cranial capacity is about 400 cc. less than that of the European Pleistocene Neanderthals, while on the other hand their brains are about an equal amount greater than those of the anthropoid apes. Brain enlargement is thus the most conspicuous and significant feature in human evolution, and this enlargement, according to Weidenreich, has taken place chiefly in the parietal region (especially in height) rather than in the frontal areas. The cerebral hemispheres have increased, he finds, almost equally in all directions but especially in height.

In another study, Weidenreich (1938) compared the course and ramifications of the middle meningeal artery in early and modern man. The grooves on the inner face of the skull in the parietal and temporal regions show the course of arteries and veins in an endocast. The change in ramification he ascribes to increase in size, especially in the parietal region. Two types of branching were recognized by Giuffrida-Ruggeri in man. In type I the anterior (frontoparietal) ramus is by far the largest. In type II the trunk divides into two branches, both of which may occasionally be of the same size. This is the more primitive condition. In Sinanthropus the trunk divides into two or even three branches, with fewer ramifications than in modern man. This resembles the condition in the apes, especially the gorilla. In Pithecanthropus the condition is like the more advanced Sinanthropus types, whereas in Neanderthal man it is close to that in recent man.

In another monograph, on the mandibles of Peking man (*Pithecanthropus pekinensis*) Weidenreich (1936a) made a comparative study of eleven jaws, five of them adult and six juvenile, pointing out many significant details (see Plate VI). Among these may be mentioned the torus mandibularis and the torus palatinus. The former is a variable bulge on the inner surface of the jaw, generally opposite the premolar teeth, dense and compact like ivory. It was present in half the Peking mandibles but is not found in Pithecanthropus or in Neanderthal man. These structures were formerly believed to be functional in origin, arising as a strengthening response to heavy chewing (Hooton, 1918), but it may be that they are simply inherited structural elements of no functional value. They are found not only in Sinanthropus but have a high frequency in Chinese, Neolithic Japanese (62 per cent), Eskimos (up to 97 per cent), Ainus (24 per cent, perhaps from crossing with Japanese), Ostiaks (31 per cent), and Lapps (32 per cent). They are thus characteristic of the Mongolian races, which are linked with Sinanthropus in this and other characters. This is not surprising, since Sinanthropus developed in the heart of what is now the Mongolian region. Their presence in Iceland, however, favors the view of their functional development in Arctic peoples.

In most other races, such as American Indians and Italians, these structures are sporadic (*ca.* 4 per cent), but in Scandinavia they reach about 17 per cent and in Iceland 68 per cent. The high frequency, especially in modern Iceland, may perhaps be one of the results in distribution of a population originally containing by chance a high frequency of the genes for these two more or less knob-like processes. But there is some evidence (Hooton, 1918) that the early Icelanders, who were of Nordic origin and who colonized Greenland in A.D. 986, may have subsequently received from that colony elements which had already mixed with the Eskimos and might thus increase the frequency of these tori still further.

The torus palatinus, which is a knob or ridge in the midline of the palate, was shown by Hooton (1918) to be frequently associated with the torus mandibularis. The frequency is shown in Table 1.

TABLE 1

FREQUENCY OF TORUS MANDIBULARIS AND TORUS PALATINUS

	Torus mandibularis	With torus palatinus	Without torus palatinus
Eskimo (Hooton)	81%	56.6%	43.4%
Lapps (Schreiner)	32.5%	82%	18%

There is every reason to believe that the torus palatinus is a useless feature transmitted by heredity from the Sinanthropus ancestor. Neither of these structures has been found in Pithecanthropus, but the torus palatinus is not uncommon in the Indian temple monkey, *Macacus rhesus*. Weidenreich (1936a) records the torus mandibularis in two American sisters, and its inheritance should be studied in modern families.

Other Mongoloid features which Weidenreich (1943) emphasizes in the Sinanthropus skulls are the Inca bone (os epactale), shovel-shaped incisors, and a pronounced sagittal crest or ridge. The last feature, found also in *P. robustus* and in the Rhodesian skull, is characteristic of Eskimos and American Indian crania, frequent in modern

Chinese, Melanesians, and Australians, but lost in Negroes and in Caucasians. Weidenreich (1945) shows that in *P. robustus* the sagittal crest is very marked, with a depression on either side and that it continues from the vertex to the occipital torus in the form of a series of knob-like thickened areas, whereas the *P. erectus* skull is smooth in this region. Above the occipital torus is also a deep sulcus in *P. robustus*, which is not present in *P. erectus*, again showing the more primitive character of the former. Shovel- or scoop-shaped incisors (see Gates, 1946, p. 370) are less frequent in Eskimo but occur in some 95 per cent of the Mongolian peoples.

The Inca bone is typically triangular, at the point in the occiput where the sagittal suture meets the lambdoid suture. It has a low frequency in many races (Gates, 1946, p. 478) but is believed to be more frequent in the Incas of Peru and some other Indian tribes, again suggesting dispersal of the gene from Sinanthropus to American Indian descendants. It has been found in 15 per cent of Amerindians and 6 per cent of Caucasians. Some have claimed that in the Red Indians it is associated with artificial skull deformation. Similarly, platymeria of the femur and a strong deltoid tuberosity, which are characteristic of Sinanthropus, are also found in the prehistoric population of Kansu and in the modern Fuegians (Weidenreich, 1943, p. 252). There is true evidence of "racial" differentiation towards the Mongolian type at this early Sinanthropus stage, or perhaps it would be better to say that these Sinanthropus features have been handed down to many of their Mongoloid and Amerind descendants. We shall find even clearer evidence that the Pithecanthropus species of Java were ancestral to the modern Australoid type of man.

The dentition of Sinanthropus is the subject of another monograph by Weidenreich (1937a). In this study 147 teeth, 83 of them found in or near their jaws, were carefully analyzed. There were large teeth in large mandibles and small teeth in smaller, weaker mandibles. The latter were recognized as belonging to female and the former to male skulls, a sexual dimorphism which is more marked than in any modern race of man. These teeth are larger in root and crown than Neanderthal teeth, the crowns (except the incisors) being higher than in other hominids. One characteristic feature is the abundant accessory ridges (wrinkles) on the chewing (occlusal) surface, another is the general persistence of a cingulum (a U-shaped ridge especially at the base of the canines or upper incisors), the upper canines being relatively very large. The canines of Sinanthropus show a striking resemblance to those of a fossil orangutan in Yunnan. Dryopithecus, like the anthropoids, has heteromorphic canines, while in Sinanthropus, Australopithecus, and modern man they are homomorphic. The very pronounced shovel-shaped incisors have already been referred to.

The taurodontism turns out to be present not only in Sinanthropus and typically in the Heidelberg molars, but in the lower molars of some Neanderthal men. It also survives in the orangutan and chimpanzee and in certain modern types of man, such as the Eskimo and the Bushman. Weidenreich (1943, p. 244) concludes that the hominids were derived from an anthropoid type with taurodont molars. The third molar is generally smaller than the second, as in modern man. In Pithecanthropus the canines lack a cingulum. The molars are larger but less primitive in form. In Neanderthal man there are only traces of the cingulum and the wrinkles. Sinanthropus is thus clearly more primitive in these respects and nearest to the anthropoids, but differs markedly in having reduced canines. It is probable that in the hominid stock the canines never were large. Weidenreich suggests that the Hominidae branched off from the Anthropoideae before the Dryopithecus level was reached, but in Ramapithecus and other genera of this group, the teeth already showed some striking hominid characters.

The dentition of Sinanthropus is closest of all to that of the Australopithecinae. Weidenreich reads the series Sinanthropus ⟶ Neanderthal ⟶ Homo as a reduction series, in teeth, jaws, and skull, associated with the progressive organization of the brain. The primary increase is in brain size, however it can be accounted for, and the other numerous skull changes are refinements resulting from the evolutionary progress of the brain. Another striking feature of Sinanthropus dentition is that it combines lower incisors like those of modern man (Homo) with premolars which are the most ape-like of any hominid. This seems to show that separate genes are concerned not only with different teeth but even with parts of teeth, as in the case of the cingulum and the Carabelli cusp or tuberculum anomalum on the molars, which is believed to be derived from it (see Gates, 1946, p. 365).

Another feature of the Pithecanthropus level of evolution is the absence of a chin. Weidenreich (1934, 1936a) has discussed the technical features involved in the development of a chin. Much has been written on the subject. We may only say that in the development of the human mandible the "simian shelf" as an inner support for the symphysis is done away with and its place is taken functionally by the mentum osseum which, on the outer aspect of the chin, helps to hold the two halves of the jaw together. The jutting of the chin is mainly a result of the thinning of the alveolar region in man. In Sinanthropus the mentum osseum is absent and the trigonum mentale, which is a special development of it at the chin triangle, has only begun to develop. In Neanderthal man the mentum osseum is more advanced.

In yet another monograph, Weidenreich (1941) has considered such other skeletal bones, mainly of the arm and leg, of Sinanthropus as are available. Seven more or less fragmentary femora were found as well as two imperfect humeri, a clavicle, and an os lunatum (from the wrist). The femora were relatively short and slightly bent forward but the ends are missing, only the shaft being present. Among other features they show hyperplatymeria and an absence of muscular lines, as in the apes and Neanderthal man. The femur is less stout and less bent than in Neanderthal man; the medullary canal is very narrow and the walls very thick. In modern man this canal represents half the thickness of the shaft, whereas in Sinanthropus it was only one-third of the whole diameter. By contrast, the six femora associated with Pithecanthropus are of modern type and Weidenreich believes they belong to modern man, but the possibility remains that here, as in other proven cases, some parts of the skeleton remained primitive while others advanced rapidly. It would be strange indeed if, in the same formation in Java, only primitive skulls and only modern femora survived. If these femora do belong to Pithecanthropus, then the difference between it and Sinanthropus is greater than has been supposed. In an X-ray study of the osteone arrangements in the ends of the six femora, Dubois (1937) showed their structure to be the same as in modern man.

The humerus of Sinanthropus also had a narrow medullary canal, it was peculiar in having a triangular form in cross section, a slender distal half, and a remarkable development of the deltoid tuberosity. These features are found as variations in modern man, or in certain racial groups. The Neanderthal short and heavy femur and the humerus were not intermediate, leading towards modern man, but were specialized, at any rate in some members. The femur of Sinanthropus was longer than the humerus. From its length his height is estimated at 156 cm. or 5 ft. 1½ in., comparable with such short races as the Eskimos, Ainu, and Japanese. The female is estimated to be only 144 cm. or 4 ft. 8½ in.

There is much evidence that the long bones were fed upon in the cave by hyenas, bears, and other carnivores. Hyenas certainly inhabited the cave at one time. They cracked off the ends and ate them. That these men were cannibals there can be little doubt, since the long bones were split open to obtain the marrow and only man can do this. Every one of the human skulls had been extensively broken at the base around the foramen magnum, apparently to feed upon the brains. Some of the skulls of Sinanthropus show scars on top caused by cutting instruments or a blow. Evidently man at this time hunted his own kind as he did the animals, for food. Skeletons of baboons and macaques have also been found here, like the modern species but larger.

Coming now to other human ancestors in Java, eleven skulls and skull fragments from Ngandong are recognized as a distinct type, *Homo soloensis,* and have been placed in a separate genus, Javanthropus,[6] by Oppenoorth in 1932. They have not, however, been fully described. Weidenreich (1937b) recognizes *Javanthropus soloensis* as probably transitional between Pithecanthropus and Sinanthropus, on the one hand, and recent Mongolians on the other. Hence it would correspond roughly in Asia with the Neanderthal level of evolution in Europe, but it is in some respects more primitive than Neanderthal and bears a near resemblance to Rhodesian man in Africa. It is essentially like a Pithecanthropus with enlarged cranial capacity (1100 cc.). The Rhodesian skull is much higher and has a much more conspicuous superciliary torus. One of the striking differences between Sinanthropus and Pithecanthropus is that in the former the superciliary ridges form an extremely heavy ledge running continuously across the nasal base as well as the orbits and separated from the forehead by a relatively deep and broad furrow (see Plate v, b). In the other three genera these brow ridges are less heavy and they form a gradual con-

[6] It is now recognized that Javanthropus is not a happy generic name for the soloensis skulls in Java and the Rhodesian skull in Africa, representing nearly the same evolutionary level as Neanderthal man. Probably Palaeoanthropus would be the best generic designation for all of them.

tinuation of the brow, without a furrow (Plate I, c), except in the Rhodesian skull, which has a depression behind the torus (Plate VII, b). Moreover, in Javanthropus and Pithecanthropus there are large frontal air sinuses over the nose and the orbits, whereas in Sinanthropus, although the jutting ledge is conspicuously larger and more continuous, there is only a small air sinus in the interorbital region. The Australian aborigines have more marked superciliary ridges than any other living race, but 30 per cent of them have no frontal sinus.

The relatively flat forehead of Javanthropus, sloping directly into the torus, is decidedly like the Australian aborigines, but in the latter the skull is much higher (see Plate IX, c). *J. soloensis* also has an occipital torus which, as already described, is even more marked in the Rhodesian skull. It thus appears that Javanthropus and the Rhodesian skull are both more primitive than Neanderthal man, each in his own continent tending to fill the gap between the prehominids and the paleohominids (Neanderthal). From this point of view the Rhodesian man might be put in the genus Javanthropus, but this seems an inappropriate name to apply to a South African type of man. It seems much more reasonable to connect its evolution with that of the Njarasa skull in the same continent and call it *Africanthropus rhodesiensis*. From Table 2, arranged from Weidenreich's (1943) data, it

TABLE 2

MEASUREMENTS OF PITHECANTHROPUS AND RELATED SKULLS

	Max. length of skull	Max. breadth of skull	Auricular height	Inner skull length	Thickness *
Pithecanthropus	170.5–183 mm.	135 mm.	89 mm.	148 mm.	22 mm.
Sinanthropus	188–199 mm.	141 mm.	98.4 mm.	166 mm.	28 mm.
Javanthropus	193–219 mm.	146 mm.	107.4 mm.	161 mm.	48 mm.
Rhodesian	208–210 mm.	144.5–148 mm.	106–107 mm.	—	—

* Average thickness of walls at the level of maximum length.

will be seen that the first three genera form a graded size series; but the extreme length of Javanthropus skulls is due mainly to the massive superstructures and the cranial capacity is not correspondingly increased. It is only 1035–1255 cc. (av. 1100 cc.). The conspicuous tori of Sinanthropus, in their later evolution towards modern man, break into three parts, a central and two lateral. There are signs of this already happening in the occipital torus of Javanthropus. We can then agree with Weidenreich, (1) that it is largely a matter of convention where the line between prehominids and paleohominids is to be drawn, and (2) that Javanthropus represents a stage of evolution directly derived from Pithecanthropus in Java. Weidenreich (1945) suggests that the six femurs found by Dubois in Java, being of essentially modern type, may belong to Javanthropus rather than Pithecanthropus. All the evidence goes to show that man acquired his erect posture early in the evolution of the Hominidae. Judging from Australopithecus, he was already an erect biped when his brain was no larger than that of a gorilla.

We can probably never know anything about the epidermal characters of these fossil men. There are many reasons for believing that early man had dark skin, eyes, and hair. When his body became relatively hairless we can only conjecture, but it seems quite possible that Sinanthropus still retained a considerable hairy covering, even more than the most hairy Ainu, which would serve him well in the winter climate. It would be curious if Neanderthal man lost his hairy covering beyond recovery during an interglacial period. It seems more likely that early man lost his hairy coat in the warmer regions of Africa and Asia. As Neanderthal man in Europe survived into the Würm I glaciation he must have had either a natural or an artificial covering. The elephants developed (or retained) a hairy coat as an adaptation to glacial conditions in northern Eurasia. There is every indication, however, that Neanderthal man in Europe had lost this protection and was forced to rely upon the skins of animals for clothing—a necessity which would increase as the last interglacial climate passed into the fourth glacial advance. In the Upper Paleolithic, needles, which imply sewing and clothing (of skins), are already known. The use of fire might be a real danger to man with a hairy coat, but it has not eliminated the more hairy Australians. It is evident that something else, perhaps man's endocrine development, led to the almost completely hairless condition of the modern Negro and Mongolian. Australopithecus may well have been as hairy as any ape. Leakey has recently found that Acheulian man in Kenya had no fire, but he lived in a relatively warm climate.

Africanthropus njarasensis was the name given to fragments of three skulls found by the German Kohn-Larsen Expedition at the northeast end of Lake Njarasa (Eyassi) in Kenya in 1935. Leakey (1936) gave a general account of them. Weinert (1939-1940), who described these finds, regards them as representing the Pithecanthropus stage in Africa. The main skull was in nearly two-hundred pieces, so that accurate reconstruction is difficult, but the main features of the occipital, parietal, and temporal bones could be made out, as well as a heavy supra-occipital torus. The foramen magnum inclined backwards as in the apes, and the tympanic plate was like that of a chimpanzee. Later, in 1938, twenty more skull fragments including two molars were found in three sites, but they could not be pieced together. Weinert believes that the Rhodesian skull could be derived from Africanthropus by enlargement and certain changes. He also compared it with Javanthropus. While probably more advanced than this later type, it is perhaps nearest the Rhodesian level. The dentition showed no resemblance to the South African Australopithecinae but was more like that of Sinanthropus. The age is apparently a pluvial period of Middle or Late Pleistocene and implements were found which Leakey says are Lavelloisian-Mousterian, as with other similar types of man. There were also abundant animal bones, including a three-toed Hipparion and some recent species, but the accompanying fauna does not clear up the age.

RHODESIAN MAN

The Rhodesian skull, found in mining operations at Broken Hill in northern Rhodesia, is one of the most remarkable and one of the most complete fossil human skulls ever recovered. Practically only the mandible is missing. The front view is almost gorilla-like because of the very heavy supra-orbital torus, the very wide face, and large orbits. The skull is very long (210 mm.; Plate VII, b) and the cranial index is only 69.4. The frontal torus closely resembles those of Pithecanthropus and Javanthropus, it even exceeds them; but there is no marked groove behind it, as in Sinanthropus. The occipital region is also like that of Javanthropus. The condition of the skull suggests that it belongs to a primitive type which survived to relatively recent times. The Heidelberg jaw almost fits this skull.

The skull was unearthed by a miner on June 17, 1921, in a kopje or small hill originally fifty or sixty feet high and 250 feet long, which was being mined for its lead, zinc, and vanadium, and has now been entirely put through the smelters. The zinc impregnation was in the upper part of the kopje and the lead in the lower levels. Leakey (1935) obtained in 1929, from dumps of this zinc ore, a sacrum of Bushman type as well as mammalian remains and worked flakes. Mining had begun in 1895 and eventually a cave was found, when a tunnel was driven thirty-four feet into the hill. It had no apparent entrance but contained large quantities of mineralized animal bones with some rude, chipped implements of quartzite and chert. An account of these bones was published (Mennell and Chubb, 1907; Mennell, 1907; White, 1908), showing that the accumulation extended over a long period but included almost entirely living species, except a rhinoceros, *R. whitei*. This cave and another were apparently occupied over a long period alternately by man and hyenas. There were wet and dry seasons in which the water level changed greatly, and the "bone cave" was gradually filled with debris mingled with sand and soil, many of the bones being impregnated with mineral salts and lime. With further mining operations, a crevice from the cave was found to slope down below the soil level and here, on an inclined plane sixty feet below the surrounding country, the skull was found. There was no mandible and no skeleton at this spot, but according to some statements it was in rather loose earth surrounded by "bat bones." Many of these details are from Hrdlička (1926), who visited the mine and made full enquiries in 1925. Nearby was found a human tibia and at a lower level a few feet away were parts of a smashed lion's skull.

The Rhodesian skull and several other human bones were deposited in the British Museum, but Hrdlička contends that the upper jaw, sacrum, and femur later found have no connection with the Rhodesian skull but belong to modern Africans. There would seem to be a reasonable chance that the tibia, at least, belongs with the skull. Bonin (1930) believes that the other bones go with the skull. The great quantities of animal bones and some human bones had been broken and split to obtain the marrow. How the Rhodesian skull, which was in a part of the cave isolated from the bone debris, got there, remains a mystery. Smith Woodward (1921) found the skull remarkably fresh in preservation, having lost its animal matter but not become mineralized. In a further statement, Smith Woodward

(1922) takes the view that the enormous frontal torus of Rhodesian man was a late specialization and that he therefore represented a *later* descendant of the Neanderthal type, which became extinct. From his extensive paleontological knowledge he says, "When a race of animals begins to develop skeletal excrescences, it has reached the end of its course and will not give rise to any higher race." He anticipates that when the ancestors of the gorilla are found they will have smaller, not larger, brow ridges. The question turns upon whether these great brow ridges are to be regarded as skeletal "excrescences." Later discoveries favor the view that the human skull in this line of descent has experienced a softening down from the almost gorilla-like Sinanthropus and Rhodesian type to modern man.

There is no clear evidence of the age of this skull, but it appears to be a late survival of a very primitive type of man. Incidentally, the teeth in the upper jaw showed extensive caries and alveolar abscesses which must have been very painful. There is evidence in the bone that the man probably died from a mastoid abscess. Keith (1931), in the *Antiquity of Man*, gives a very clear account of the skull, which we have classed with Palaeoanthropus, but it could equally well be classed with Africanthropus or Javanthropus.

The British Museum published a full description of the skull and other bones as well as the accompanying artifacts and animal bones, by Pycraft (1928) and several other authors. Bather took the view that the entire skeleton was originally present, covered with stalagmite which may have been a cast of the body. The teeth were of great size and were worn nearly flat, the third molar was smaller than the second. The orbits were 49 mm. wide and 37 mm. high, larger than those of Chapelle-aux-Saints. The huge frontal torus was 139 mm. long, of unique thickness (21 mm.), bigger than that of a gorilla. Pycraft claimed from the character of the pelvis, the form of the acetabular cavity, the nuchal plate of the skull, and the form of the proximal condyles of the tibia, that this man walked with a stoop, and gave him the name Cyphanthropus; but this view has not been accepted. The perfect endocranial cast led Elliot Smith to recognize five depressed areas in the brain, especially in the prefrontal, Sylvian fissure, and parieto-occipital areas, and deficiency of the temporal development, as well as a precocious expansion of the inferior parietal area which is the most characteristic feature. The cranial capacity is

1280 cc., decidedly smaller than in Neanderthal man, except the Gibraltar skull (Plate VII, a). The brain was also more primitive than Neanderthal, but bigger than Piltdown. The associated implements were Mousterian in character, as were the stone balls (bolas), the largest of which was 3.4 in. in diameter. A clinker showed the presence of fire. In addition to the rhinoceros already mentioned, Hopwood found a new species of extinct Serval cat. All the other animals are still found in the locality. As the climate was tropical and apparently not subject to the fluctuations of the Ice Age, there is no evidence that the skull is anything but recent in age.

NEANDERTHAL MAN

Having considered the prehominids, now represented by so many skulls, teeth, and other fragments, we may next pay attention to the paleohominid (Neanderthal) type. They represent the next marked stage of human evolution and were placed by Sergi in the genus Palaeoanthropus, having a number of different species. Morant (1927) measured nine European Paleolithic skulls or casts, including LaChapelle, LaQuina, Spy I and II, Neanderthal, LeMoustier, and Gibraltar, and found them metrically a very uniform type, "hardly more variable than a single race of modern man." He thinks they might all belong to one contemporary population. Morant (1928) subjected the Rhodesian skull to the same set of measurements. He found the differences between Rhodesian and Neanderthal man all associated with skull breadth, and that both were about equally related to modern man. But the tremendous frontal torus of Rhodesian man is markedly heavier than in any Neanderthalian and surpasses even that of *Pithecanthropus erectus*. Morant's conclusions seem to show that measurements have their limitations and that skulls can differ in important observational features of a non-metrical character. Such anatomical observations indicate considerable variety in these skulls, which extend over a geological period during which the type was undergoing evolution in various features. Morant finds the Neanderthal faces longer than in any modern race, but the orbits relatively small (see Plate VII, a). In modern man (Homo) he finds four chief differences: (1) the frontal bone is more vertical; (2) the facial skeleton is relatively smaller; (3) the foramen magnum slopes slightly forward instead of backward; (4) the face is almost vertical—there is no projecting muzzle.

An analytical study of the Gibraltar skull, which was excavated in 1868 from the talus under the north front of the Rock of Gibraltar, was made by Sollas (1908). It is a typical Neanderthal skull (Plate VII, a) with a cranial capacity of about 1260 cc., round orbits, a long face, and broad nasal aperture (nasal index 95.65). This is extraordinarily high, the Australian aborigines having, by comparison, a nasal index of 60.5, and one wonders if this may indicate African influence. With a cephalic index of 76.3, the skull is mesocephalic. There is very little prognathism and the eyes are wide apart. In his comparison with Australian skulls, Sollas points out that in the latter the supra-orbital tori have been further reduced, as shown by a shallow oblique ophryonic groove which separates the outer temporal from the inner medial portion. He suggests that the Neanderthal and Australian races had diverged from a common ancestry, but later (*Ancient Hunters*) concluded that the Australian race is a Mousterian survival, not related to European races.

Work in many fields, including glacial geology, the study of varves by DeGeer and Antevs and of pollen by Erdtman and others in Pleistocene and post-glacial deposits, has led to the division of the glacial period of Europe into many stages, with much more definite estimates of the length of each phase in years. These results are of great value in relation to the various discoveries of Neanderthal man in Europe, as they help fill in a much more complete picture of the conditions, including the development of different types of forest following the changes in climate. It only need be said here that four main glacial advances, named successively Gunz, Mindel, Riss, and Würm, are recognized, with three interglacial periods between them. In the great Mindel-Riss interglacial—longest of the interglacial periods—the climate of Europe was warmer than now. The glacial periods have themselves been subdivided into two or (in the case of the Würm) even three episodes of advance separated by an interstadial period representing partial retreat of the ice. Zeuner (1940) has published a correlation of the later glacial and climatic phases with the corresponding Paleolithic industries in Europe (Table 3). For a much more detailed account of the Pleistocene in relation to man, see Zeuner, 1945. For a study of the Pleistocene geology of Cashmir in relation to Soan and other human cultures in India, see DeTerra and Paterson (1939).

In a more extensive table, Zeuner indicates the approximate ages, in

TABLE 3
DIVISIONS OF THE ICE AGE IN EUROPE

Climatic phase	Climate in central and western Europe	Industries in central Europe
Postglacial	Temperate	
Würm III	Periglacial more humid	Mesolithic Magdalenian
Würm II/Würm III	Cool	Magdalenian
Würm II	Periglacial	Magdalenian Solutrean Aurignacian
Würm I/Würm II	Mild	? Mousterian
Würm I	Periglacial	Mousterian
Interglacial Riss/Würm	Hotter summers than now	Mousterian

years, of the various climatic phases, with the relation of these phases to the various known Neanderthal skulls. These are reproduced, with modifications, in Table 4. The relative dates in Würm I are from Obermaier's chronology. The exact place of certain skulls in this chronology is still doubtful. The oldest Neanderthal skulls are of Riss-Würm interglacial age, but the Steinheim skull may belong to the Riss I/Riss II interstadial. The Swanscombe cranium is from the (older) Mindel-Riss interglacial. While most Neanderthal skulls are from the Würm I or earliest phase of the last glacial, Neanderthal man (Palaeoanthropus) may have survived in France and Italy to a time when Cro-Magnon man (Homo) occupied much of Europe and the Mediterranean area. The Ehringsdorf skull antedates the last glaciation but postdates the cool pre-Würm period. The Taubach skull, near Weimar, belongs to the last half of the last interglacial. The Monte Circeo skull was very late interglacial or Würm I. In the large cave Cotte de St. Brelade, in Jersey, extensive excavations, especially by Marrett, unearthed thirteen Neanderthal teeth, a rich Lavellosian and Mousterian industry, and mammals including the reindeer, woolly rhinoceros, Bos, and horse. At Carmel the implements are Lavellosian-Mousterian, the Tabun cave showing three levels, A, B, and C, while the Skhul cave implements belong between levels B and C. The Galilee skull may be of the same age as Carmel. At Krapina the floor of the

Evolution of the Hominidae 105

TABLE 4

NEANDERTHAL REMAINS IN RELATION TO GLACIAL PERIODS

Approximate age in years	Climatic phase	Approximate geological age of Neanderthal remains	Exact relative date (in Würm I according to Obermaier)
22000 72000	Postglacial Würm III Interstadial W II/W III Würm II Interstadial W I/W II	Homo	
115000	Würm I	Gibraltar II, Jersey, La-Quina, La Ferrassie, Poch de l'Azé, La Chapelle-aux-Saints, Spy, La Naulette, ?Monte Circeo, Mount Carmel in part	Late Mousterian: La-Quina (upper), Spy, Jersey Middle Mousterian: La Ferrasie, La Chapelle-aux-Saints, LaQuina (base), Sipka Early Mousterian: ?Le Moustier
	Interglacial Riss-Würm or Last Interglacial	Late ?Monte Circeo, Mt. Carmel (in part), ?Galilee, Krapina, Saccopastore, ?Steinheim, Taubach, Ehringsdorf	PW/WI-Taubach, Ehringsdorf Pre-Würm (145,000)—Saccopastore R II/PW
187000	Riss II		
230000	Interstadial RI/RII Riss I	?Steinheim	
	Interglacial Mindel-Riss (Great Interglacial)		
435000	Mindel II		
476000	Interstadial MI/MII Mindel I	Heidelberg jaw	
	Interglacial Gunz-Mindel Gunz Glaciation		

cave is twenty-five feet above the river level and the presence of *Rhinoceros Merckii* indicates the last interglacial, the implements being Mousterian.

Paterson (1940) attempts a world correlation of the Pleistocene, using geological and faunistic evidence as well as human implements (treating them as zonal fossils), river terraces, and sedimentation. The terraces show sedimentation cycles, arising through variations in the power-volume of rivers, which result from combined orogenic and precipitation changes. These are found to be coincident over the whole of Europe, Asia, and Africa. The three great cycles of sedimentation are divisible into seven phases. Instead of the Gunz-Mindel nomenclature, which is applicable to local conditions in Europe, he divided the Pleistocene into Lower, Middle, and Upper, the seven phases being termed l_1, l_2, m_1, m_2, u_1, u_2, u_3. The first, second, and third glacial episodes coincide with the beginning of each cycle of sedimentation, the fourth coincides with the third phase of the last cycle and the fifth glacial episode marks the end of the Pleistocene. The types of human artifacts or industries are placed in these seven phases. The river terraces are believed to represent not a cold phase but an increase in precipitation.

In a more ambitious study, Zeuner (1946) has recently attempted a world-wide correlation of the Pleistocene phases back to the beginning of the Ice Age some 600,000 years ago. Tree-ring analysis covers the last 3,000 years in certain areas, varved-clay analysis some 15,000 years, the solar radiation cycles are applied to the last million years, and radio-activity methods of dating to the 1500 million years of geological time. This very useful book gives an immense amount of detail regarding the Pleistocene climatic phases, especially in Europe and America, in relation to the changing flora, fauna, and man, particularly as regards the types of human implements. However, while much progress has been made in dating past prehistoric events, yet there are still gaps, for instance in the varved clays, which necessitate some estimates even in the late Pleistocene. Flint (1945), in a critique of all such methods of absolute chronology, finds the radium concentration of marine sediments the most accurate method of dating against time, but this relation is only constant for the last 300,000 to 400,000 years, while he concludes that the Pleistocene lasted at least 1,000,000 and possibly as much as 5,000,000.

For many years Neanderthal man was believed to be found only in Europe, but more recently he has been discovered further afield. The Galilee skull was found in Palestine and later a dozen skeletons in a cave on Mount Carmel. McCown and Keith (1939), in their important monograph on the Carmel skeletons, established the genus Palaeoanthropus for the Neanderthal type, thus recognizing its generic distinction from the prehominids, as a higher evolutionary level.[7] They also recognized the following species of Palaeoanthropus:

1. *P. heidelbergensis* (the Heidelberg or Mauer jaw) attributed to the second (Mindel) glaciation or more precisely the Mindel I/Mindel II interstadial, much older than the true Neanderthal and probably as old as Peking man.
2. *P. ehringsdorfensis* (the Ehringsdorf skull).
3. *P. neanderthalensis* (Düsseldorf and La Chapelle-aux-Saints).
4. *P. krapinensis* (Krapina).
5. *P. palestinensis* (Mount Carmel and Galilee). To these I have suggested (Gates, 1944) the addition of
6. *P. rhodesiensis* (the Rhodesian skull, although it has some more primitive characters.
7. *P. soloensis* (*Javanthropus soloensis*), although it too is transitional in some respects.

To these must be added the Steinheim skull, belonging to the third (Riss) glaciation or the Riss I/Riss II interstadial. It is thus much earlier than typical Neanderthal but much more modern in type, especially in the temporal and occipital regions. Such cases afford the clearest evidence that some skull characters have advanced rapidly while others remained relatively stationary. There is much to be said for the view that in western Europe Neanderthal man became extinct, as a dead end, transmitting but little of his inheritance by hybridization with Cro-Magnon man, whereas farther east the gradual transition from Neanderthal to modern man took place. We may suppose that the Cro-Magnon invaders from northern Africa largely exterminated the Neanderthal type as they advanced in western Europe, probably generally killing the men and sometimes retaining the women, as frequently happens when one primitive tribe invades the territory of

---

[7] Boule (1923) places Neanderthal man in the genus Homo, but he admits (p. 239) that "it would probably be a different matter if we were dealing with a feline, a ruminant, or a monkey!"

another. The idea that "sexual aversion" is a criterion of specific or even generic difference is wholly inapplicable to man. The conception of species has to be placed on a wider and more stable basis.

The recognition of Palaeoanthropus as a distinct genus representing the Neanderthal level of evolution appears to me to be an important advance in the understanding of human evolution. Weidenreich (1943) is not prepared to accept this, and he even goes so far as to suppose that Neanderthal and "modern" man are one species because (on his hypothesis, which I am inclined to accept in part) the former somewhere gave rise to the latter by continuous breeding! The biological aspects of the species question will be discussed in a later chapter. Here we may be content to point out that Weidenreich himself (p. 237) recognizes four "groups" of Neanderthalians: (1) the Rhodesian group (one skull); (2) the Spy group (Morant's Neanderthalians); (3) the Ehringsdorf group (Ehringsdorf, Steinheim, the Tabun skull of Carmel, probably Krapina, and another from Judaea); (4) the Skhul skulls from Mount Carmel and the Galilee skull (found by Turville-Petre in a cave with Mousterian culture in 1925), this group being intermediate between Neanderthal and modern man. It will be seen that this grouping agrees with the one given above except in classing the skulls from Krapina (Croatia) somewhat doubtfully with the Carmel skeletons and in regarding the two types at Carmel as distinct, whereas McCown and Keith (1939) regard them as one variable population.

On Mount Carmel Miss Garrod explored the strata of the fifty-two foot cave deposit. McCown and Movius found fifteen skeletons, six of them nearly complete. McCown and Keith, in their study of these skeletons of Neanderthal man in Palestine, recognize a series of contemporaneous forms, with Neanderthal man at one extreme and Cro-Magnon at the other. The Tabun cave type is nearest that of Krapina. The Krapina skeletons represent a small people with low-vaulted skulls, but strong jaws and taurodont teeth. They bridge the gap between the Carmel type and the Neanderthals of western Europe, but their chief affinity is with the latter. The Ehringsdorf group also belongs to the last (Riss-Würm) interglacial period. It consists of a skull of Neanderthal type and two mandibles, but the skull has a relatively high vault, a neanthropic mastoid process, and an incipient external occipital protuberance—three conditions found in Carmel skulls.

The Steinheim skull, found in western Germany (Weinert, 1936) seven meters below the surface, is the earliest Neanderthaloid in Europe, belonging to the early Riss-Würm interglacial. It is thus probably 200,000 years old. It was accompanied by a fauna including *Elephas antiquus* and *Rhinoceros Merckii*. Though small and delicate (♀) this skull combines an eyebrow region like Pithecanthropus with other features more like Homo, having little of the occipital torus. The skull length and breadth agree with those of Pithecanthropus. This skull thus combines a prehominid feature with others more advanced than Neanderthal. It should be called *Palaeoanthropus steinheimensis*. This type could have transformed into modern man through elimination of the superciliary torus, so Steinheim man may be directly ancestral to one strain of Homo. McCown and Keith compare this skull with the Tabun woman. It is thus clear that the massive frontal and occipital tori of the prehominids could be lost independently in their later descendants.

To these Neanderthal remains should perhaps be added the taurodont teeth described by Keith (1924) from Malta. They were found in a cave with bones of two extinct species of hippopotamus, an extinct elephant, and Neolithic pottery. Neanderthal man may have survived here in isolation beyond his time in Europe.

A fragment of a child's maxilla and an adult molar tooth excavated by Coon in one of the caves of Hercules at Tangier on the Atlantic coast of Morocco, appear to belong to Neanderthal man. They are described by Senyürek (1940). They were associated with animal bones, apparently of Pleistocene age, and flint artifacts, in the ninth layer from the top. The maxillary fragment was large and thick, the simian nasal groove was present but the canine fossa was absent, the zygomatic process sloping outward and backward as in Neanderthal man. This Tangier child has a large palate with large teeth and very thick bone. The teeth are only mildly taurodont and the adult molar is larger than in Neanderthal man but should probably be classed with that species. The associated industry resembles the Upper Mousterian of Gibraltar, but with additional elements from northern Africa. The find is tentatively dated as Late Pleistocene with a Middle Pleistocene culture.

Neanderthal man (Palaeoanthropus) has also been found in Italy. On the coast south of Rome, where Ulysses landed to be the guest of

Circe, the limestone mountains come down to the sea and over thirty caves have been found. Many contained Paleolithic implements and the bones of extinct animals. In one cave which was practically closed and had not been entered for at least 70,000 years, the floor was littered with fossil bones, antlers, and animal skeletons covered with pearl-like concretion. The bones included the elephant, hyena, wild ox, and deer (Blanc, 1940). In the third crypt a Neanderthal skull rested on the ground. The man had been killed by a blow which fractured the right temple, then beheaded outside the cave. No skeleton was found, but a circle of stones suggested a ceremonial "burial" after the base of the skull had been broken open to remove the brain. We may conclude that ceremonial cannibalism had survived from the time of Sinanthropus, as it has in some tribes to the present day. Probably a belief in survival after death already existed. Sergi found this skull to have a cranial capacity of 1550 cc., larger than the average for modern man, but the forehead was very low, the face, eyes, and nose very large. This skull and the two discovered by Sergi in a gravel pit at Saccopastore, a suburb of Rome, appear to be typical Neanderthalians.

Recently a skeleton (the oldest known burial) was found at the rock-shelter of Kiik-Koba near Kerch Straits in the Crimea (Keith, 1944). From the tibia the height is estimated at 5 ft. 2½ in. The hands were massive and rugged. The stone culture was late Achulean as in the Tabun cave on Mount Carmel, and the age was Middle Pleistocene, showing that he was earlier than Neanderthal man of western Europe. Still further east, in Usbekistan (central Asia) near Baisun in 1938 the skull and lower jaw of a child was discovered, buried in a cave (Weidenreich, 1945). The implements were Mousterian. The child was eight or nine years old, its superciliary ridges were already well developed, the chin was not receding as in Neanderthal man, the permanent upper lateral incisors were extremely shovel-shaped, and the brain-case was as large as in a modern adult man (1490 cc.). In the adult it would probably have reached 1600 cc. Weidenreich concludes that this Usbek child was like the Skhul population in Palestine, more advanced than the ordinary Neanderthalian, perhaps intermediate between Neanderthal and modern man, also showing some Mongolian features in the face and dentition. He thinks that the Podkumok skull, found north of the Caucasus, between the Black and Caspian seas, may also be of similar intermediate type. Further illustrations of the Teshik-Tash skull and its restoration by a sculptor are given in *American Journal of Physical Anthropology* (N. S. 4:121; 1946) from photographs brought from Moscow by Dr. Henry Field. V. V. Bunak and A. P. Okladnikov have prepared a monograph, to be published by the Moscow University Press, on this skull and the associated implements and fauna.

It is now clear that Neanderthal man was not a uniform type remaining constant for 100,000 years, but that a series of evolutionary stages are represented which give an essential continuity with small gaps all the way from *Pithecanthropus erectus* through Neanderthal to the modern Homo. Whereas there is a distinct hiatus between the Neanderthal and modern types in western Europe, farther east in Palestine and central Asia fossil men are found which had advanced nearer to the modern level. Whether the Carmel population represented a center of evolutionary advance beyond the Neanderthal level, as McCown and Keith (1939) maintain, or whether they represent a hybrid population at a point where the two species met, is a very difficult question, particularly as there are no adequate studies of genetic segregation in the skeletal characters of human racial hybrids. One must ask first, are the character-combinations in these people such as might arise through crossing? The authors say (p. 52), "Great as is the individual variation, the group [Skhul-Tabun] is made homogeneous by the possession of a series of common characters." The pelvis (p. 71) agrees with the Cro-Magnon (modern) rather than the Neanderthal type, yet there are more Neanderthal features than in any modern race. The Tabun woman has a unique pelvis, nearest to that of the gorilla but nevertheless neanthropic. They suggest that the Tabun woman may belong to an "older strain," which would imply crossing. They also admit that if there were no transitional forms in the population she would be regarded as a true Neanderthalian, and they say (p. 144), "We are still in doubt concerning her true status, whether she is really an extreme variant or the representative of a group which has strayed into a Skhul community." The latter would of course imply crossing.

The Palestinians numbered a dozen skeletons, four of whom died in childhood and only one or two who reached the age of fifty years. They were perfectly erect but the vertebral column was remarkably short, even shorter than in Neanderthal. The vertebrae show certain peculiarities which are found also in the Australian aborigines and the Bushmen. The ribs range in form from Neanderthaloid (Skhul v) to nearly modern (Skhul iv). If the Skhul men belong to the Cro-Magnon race, then this race appeared in Palestine before it appeared in Europe, but the brow ridges, most of the teeth, the vertebrae and ribs are Neanderthaloid. McCown and Keith regard the Skhul type as proto-Caucasian or proto-Cro-Magnon. The hands form a series ranging from near the Neanderthal end to Skhul hands near the modern end. The forearms and humeri were essentially of two types corresponding to Tabun and Skhul. The Skhul skeleton iv was tall like Cro-Magnon, but some of this type were short, with small, slender humeri, as in Krapina. Modern races are full of people of markedly different stature, and these differences are inherited. In dentition again there is the near-modern and the near-Krapina type, the Tabun series showing shovel-shaped incisors. The mandibles also fall into two types, but Skhul iv and v resemble Tabun ii in many points, while Skhul vii is Neanderthaloid in mandible, teeth, and forearms. The conclusion is drawn that the Tabun type is Neanderthalian with a few modern features, while the Skhul type is modern with a few Neanderthal features. Obviously in such a mixed population crossing between types must have been taking place. These are clearly not first hybrids, and if the results are from crossing it must have been going on for several generations.

The Carmel people are summed up as having three Neanderthal characters (the frontal torus, the form of the malar, and the pattern of the molar teeth). In eight characters they agree with Cro-Magnon man and in eleven they are intermediate, while all agree in three characters (dolichocephaly, a wide ascending mandibular ramus, and a moderate to large cranial capacity). Every grade in the development of the chin is found in different mandibles.

In a critical summing up of the main features of the Tabun and Skhul skeletons, which are remarkably complete, Keith and McCown (1937) point out that the age is the latter half of the Riss-Würm interglacial, the moist climate of the early Acheulean being gradually replaced by drier and cooler conditions. The Tabun woman, in addition to her Neanderthalian features, showed primitive anthropoid characters in her pelvis and forearm. The feet of the Carmel skeletons were essentially modern rather than Neanderthaloid, but they retained "traces of their anthropoid heritage." None of the teeth were taurodont like those of Neanderthal man in Europe. These skeletons are in some respects a mosaic of characters derived from two different species of man.

The Upper Paleolithic Alfalou in North Africa differ from Cro-Magnon in having wider noses (Coon, 1939; Gates, 1946, p. 1370). It appears probable that this tall type originated in the Sahara region during a pluvial period, when it was a fertile plain supporting a rich fauna. From thence the Cro-Magnon could have crossed the pillars of Hercules into Europe, leaving the Alfalou behind to develop a broad nose, while others passed eastwards into Palestine where they met and mingled with the more primitive Neanderthals of Asia Minor. This seems more likely than the hypothesis that in the Mount Carmel Palestinians we are witnessing a process of racial differentiation or progressive development from near the Neanderthal to near the Cro-Magnon level. Paleontology and genetics both seem to teach that species differentiation only occurs under conditions of relative isolation, and the Mount Carmel population seems too mixed to be regarded as a single isolated community.

In the upper cave at Choukoutien, China, of Late Magdalenian age, the three best-preserved skulls, all of modern type, are regarded by Weidenreich (1943, p. 251) as respectively proto-Mongoloid, Melanesoid, and Eskimoid. But the "proto-Mongoloid" also resembles the Australian type in having rather marked supra-orbital ridges and in the conformation of the occipital bone. Some anthropologists at least regard them as all belonging to one race, and consider that Weidenreich has much exaggerated the differences between them. At best, the differences he sees are only incipient and not fully developed racial characters in the three skulls. However these differences may have arisen, through variation and inheritance of particular skull characters, the three types must surely have undergone crossing in the northern Chinese area. They are less widely separated than the Tabun and Skhul types at Carmel, but Weidenrich (1946, p. 86) suggests, on the basis of this meager evidence, that isolation has played no part in human speciation!

We have already seen that various characters of Sinanthropus, such as the Inca bone, shovel-shaped incisors, platymeria, and a strong deltoid tuberosity of the femur, have been perpetuated in the American

Indians, even as far as the natives of Tierra del Fuego. This confirms other evidence that the Amerinds received at least a part of their inheritance from their Sinanthropus ancestry.

In another study, Weidenreich (1939) emphasizes the remarkably low brain case of Sinanthropus. The ratio length: height as taken from the nasion-opisthion base-line to the vertex (Weidenreich, 1943, p. 192) is 64.2 for Pithecanthropus skull II, 69.4 for Sinanthropus, 77.7 for Neanderthal, 91 for modern man, but only 54 for anthropoids.

In this chapter we have traced the genera and species of Hominidae from their earliest known forms in Java, northern China, and intermediate regions (the prehominids), in the first (Gunz-Mindel) interglacial period of the Lower Pleistocene to the paleohominids (the Neanderthal series) in western Europe, Asia Minor, and eastwards to Java in Upper Pleistocene (Würm interglacial) times, when Neanderthal man in Europe disappeared. Many types of man arising in different parts of the world are disclosed, and in later chapters we shall see even more clearly local evolution taking place independently on different continents in remote areas. That these types would all have been interfertile there is little doubt, yet no paleontologist would dream of placing them all in one species, or even in one genus. When Weidenreich (1943, p. 246) proposes calling Sinanthropus *Homo sapiens erectus pekinensis*, stating that even specific distinction between Peking man and modern man "remains doubtful, to say the least," he is only reducing to absurdity an outlook which arises from endeavoring to fit his views into the frame which regards all living men as belonging to one species. This dogma, that *Homo sapiens* is one species representing all living mankind, dates from Linnaeus in the eighteenth century. Before the end of that century it had been enunciated in the political form that "all men are born free and equal." This statement might have been more credible had it not been made by men who kept large numbers of Negro slaves. Every student of heredity knows that no two men are equal either in physique or mentality, unless they happen to be identical twins. Considered literally, this eighteenth century political doctrine is hopelessly at variance with the facts of science, and has been the cause of much obscure thinking. But in the form that all men are men (though widely diverse in potentialities and achievements) and entitled to respect and fair treatment as such, we have a statement which every scientific man can fully endorse.

REFERENCES

Black, Davidson. 1925. The human skeletal remains from the Sha Kuo T'un Cave deposit in comparison with those from Yang Shao Tsun and with recent North China skeletal material. *Palaeont. Sin.* Ser. D, vol. 1, fasc. 3. pp. 148. Pls. 14. Tables 86.

———. 1927. The lower molar hominid tooth from the Chou Kou Tien deposit. *Palaeont. Sin.* Ser. D, vol. 7, fasc. 1. pp. 28. Pls. 2. Figs. 8.

———. 1930. Interim report on the skull of Sinanthropus. *Bull. Geol. Soc. China.* 9:1–22. Pls. 6.

———. 1931. On the adolescent skull of *Sinanthropus pekinensis* in comparison with an adult skull of the same species and with other hominid skulls, recent and fossil. *Palaeont. Sin.* Ser. D, vol. 7, fasc. 2. pp. 144.

———. 1934. On the discovery, morphology and environment of *Sinanthropus pekinensis*. *Phil. Trans. Roy. Soc.* B223:57–120. Pls. 10.

Black, D., Teilhard de Chardin, C. C. Young, and W. C. Pei. 1933. Fossil man in China. *Geol. Mem. China.* Ser. A, No. 11. pp. 166. Figs. 82.

Blanc, A. C. 1940. The fossil man of Circe's mountain. *Nat. Hist. Mag.* 45: 280–287. Figs. 18.

Bonin, G. von. 1930. Studien zum *Homo rhodesiensis*. *Zeits. f. Anthrop. u. Morph.* 27:347–381. Pl. 1.

Cunningham, D. J. 1895. Dr. Dubois' so-called missing link. *Nature.* 51:428–429. Figs. 2.

DeTerra, H. 1943. Pleistocene geology and early man in Java. *Trans. Am. Phil. Soc.* 32:437–464. Pls. 2.

DeTerra, H., and T. T. Paterson. 1939. Studies on the Ice Age in India and associated human cultures. *Carneg. Publ.* No. 493. pp. 354. Pls. 54.

Drennan, M. R. 1936. Report on a Hottentot femur resembling that of Pithecanthropus. *Am. J. Phys. Anthrop.* 21:205–216. Pl. 1. Fig. 1.

Dubois, E. 1924. On the principal characters of the cranium and the brain, the mandible and the teeth of *Pithecanthropus erectus*. *Proc. K. Akad. Wetens. Amsterdam.* 27:265–278.

———. 1937. The osteone arrangement of the thigh-bone compacta of man identical with that, first found, of Pithecanthropus. *Proc. K. Akad. Wetens. Amsterdam.* 40:864–870. Pls. 2. Figs. 1.

Flint, R. F. 1945. Chronology of the Pleistocene epoch. *Quart. J. Fla. Acad. Sci.* 8:1–34.

Harris, W. E. and A. Keith. 1921. The finding of the Broken Hill skull. *Illustr. London News,* Nov. 19. p. 680.

Hirschler, P. 1942. Anthropoid and human endocranial casts. Dissertation. pp. 150. Figs. 32. N. V. Noord-Hollandsche Uitgevers Maatschappij.

Hooton, E. A. 1918. On certain Eskimoid characters in Icelandic skulls. *Am. J. Phys. Anthrop.* 1:53–76. Pls. 3.

———. 1940. *Why Men behave like Apes and vice versa*. Princeton: Princeton University Press.

Hrdlička, A. 1926. The Rhodesian man. *Am. J. Phys. Anthrop.* 9:173–204. Figs. 2.
Keith, Sir A. 1924. Neanderthal man in Malta. *J. Roy. Anthrop. Inst.* 54:251–275. Pls. 2.
———. 1931. *New Discoveries Relating to the Antiquity of Man.* 2nd Edition, London and New York. pp. 512.
———. 1944. Pre-Neanderthal man in the Crimea. *Nature.* 153:515–517.
———. 1947. Australopithecinae or Dartians. *Nature.* 159:377.
Keith, Sir A., and T. D. McCown. 1937. Mount Carmel man: his bearing on the ancestry of modern races. In *Early Man* (ed. G. G. McCurdy). pp. 41–52. Pl. 1.
Koenigswald, G. H. R. von. 1935a. Eine fossile saügetierfauna mit Simia aus Südchina. *Proc. K. Akad. Wetens. Amsterdam.* 38:827–879. Pl. 1.
———. 1935b. Die fossilen Saügetierfaunen Javas. *Proc. K. Akad. Wetens. Amsterdam.* 38:188–198.
———. 1936. Erste Mitteilung über einen fossilen Hominiden aus dem Altpleistocän Ostjavas. *Proc. K. Akad. Wetens. Amsterdam.* 39:1000–1009. Pl. 1.
———. 1937. Ein Unterkieferfragment des Pithecanthropus aus dem Trinilschichten Mitteljavas. *Proc. K. Akad. Wetens. Amsterdam.* 40:883–893. Pl. 1.
———. 1937a. A review of the stratigraphy of Java and its relations to early man. In *Early Man* (ed. G. G. McCurdy). p. 362.
———. 1938. Ein neuer Pithecanthropus-Schädel. *Proc. K. Akad. Wetens. Amsterdam.* 41:185–192. Pl. 1.
———. 1940. Neue Pithecanthropus-Funde 1936–1938. *Wetens. Mededeelingen.* No. 28. pp. 232. Pls. 14. Figs. 40.
Koenigswald, G. H. R. von, and F. Weidenreich. 1938. Discovery of an additional Pithecanthropus skull. *Nature.* 142:715.
———. 1939. The relationship between Pithecanthropus and Sinanthropus. *Nature.* 144:926–929. Figs. 3.
Leakey, L. S. B. 1936. A new fossil skull from Eyassi, East Africa. *Nature.* 138:1082–1084. Figs. 3.
McCown, T. D., and Sir A. Keith. 1939. *The Stone Age of Mount Carmel.* Oxford University Press. Vol. II. pp. 390. Pls. 28. Figs. 247.
Mennell, F. P., and E. C. Chubb, 1907. On an African occurrence of fossil Mammalia associated with stone implements. *Geol. Mag.* 4:443–448.
Merrill, E. D. 1946. Correlation of the indicated biologic alliances of the Philippines with the geologic history of Malaya. *Chron. Botan.* 10:216–236.
Mivart, St. George. 1874. *Man and Ape.* New York. pp. 200. Figs. 61.
Molengraaff, G. A. F., 1921. Modern deep-sea research in the East Indian Archipelago. *Geogr. J.* 57:95–121.
Morant, G. M. 1927. Studies of Palaeolithic man, II. A biometric study of Neanderthaloid skulls and of their relationships to modern racial types. *Ann. Eugen.* 3:318–381. Pls. 12. Figs. 23.

———. 1928. III. The Rhodesian skull and its relations to Neanderthaloid and modern types. *Ann. Eugen.* 3:337–360. Pls. 6. Figs. 3.

———. 1930. IV. A biometric study of the Upper Palaeolithic skulls of Europe and of their relationships to earlier and later types. *Ann. Eugen.* 4:109–214. Pls. 12. Figs. 63.

Movius, H. L., Jr. 1943. The Stone Age of Burma. *Trans. Am. Phil. Soc. N. S.* 32:341–393. Pls. 6. Figs. 23.

———. 1944. Early man and Pleistocene stratigraphy in Southern and Eastern Asia. *Papers Peabody Mus. Am. Arch. and Ethnol., Harvard* 19:No. 3. pp. 125. Figs. 47.

Oppenoorth, W. F. F. 1937. The place of *Homo soloensis* among fossil men. In *Early Man*, pp. 349–360. Pls. 2.

Paterson, T. T. 1941. On a world correlation of the Pleistocene. *Trans. Roy. Soc. Edinb.* 60:373–425. Figs. 23.

Pycraft, W. P. *et al.* 1928. Rhodesian man and associated remains. *Brit. Mus. (Nat. Hist.).* pp. 75. Pls. 5. Figs. 23.

Senyüsek, M. S. 1940. Fossil man in Tangier. *Papers Peabody Mus., Harvard* 16:No. 3. pp. 27. Figs. 3.

Shellshear, J. E., and G. Elliot-Smith. 1934. A comparative study of the endocranial cast of Sinanthropus. *Phil. Trans. Roy. Soc.* B223:469–487. Pls. 2. Figs. 14.

Smith Woodward, A. 1921. A new cave man from Rhodesia, South Africa. *Nature.* 108:371–372. Fig. 1.

———. 1922. The problem of the Rhodesian fossil man. *Sci. Prog.* 16:574–579.

Sollas, W. J. 1908. On the cranial and facial characters of the Neanderthal race. *Phil. Trans. Roy. Soc.* 199B:281–339. Pl. 1. Figs. 25.

Walmsley, T. 1933. The vertical axes of the femur and their relations. A contribution to the study of the erect position. *J. Anat.* 67:284–300. Figs. 16.

Weidenreich, F. 1934. Das Menschenkinn und seine Entstehung. *Ergeb. d. Anat. u. Entwick.* 31:1–124. Figs. 89.

———. 1936a. The mandibles of *Sinanthropus pekinensis*: a comparative study. *Palaeont. Sin.* Ser. D., vol. 7, fasc. 3. pp. 162. Pls. 15. Figs. 100.

———. 1936b. Observations on the form and proportions of the endocranial casts of *Sinanthropus pekinensis*, other hominids and the great apes: a comparative study of brain size. *Palaeont. Sin.* Ser. D., vol. 7, fac. 4. pp. 50. Figs. 21.

———. 1937a. The dentition of *Sinanthropus pekinensis*: a comparative odontography of the hominids. *Palaeont. Sin.* No. 101. pp. 180. Pls. 36.

———. 1937b. The relation of *Sinanthropus pekinensis* to Pithecanthropus, Javanthropus and Rhodesian man. *J. Roy. Anthrop. Inst.* 67:51–65. Pls. 4. Figs. 4.

———. 1938. The ramification of the middle meningeal artery in fossil hominids and its bearing upon phylogenetic problems. *Palaeont. Sin.* No. 110. pp. 16. Pls. 6.

Weidenreich, F. 1939. Six lectures on *Sinanthropus pekinensis* and related problems. *Bull. Geol. Soc. China.* 19:1–110. Pls. 9.
———. 1940a. The torus occipitalis and related structures and their transformations in the course of human evolution. *Bull. Geol. Soc. China.* 19:479–558. Pls. 6.
———. 1940b. Man or Ape. *Nat. History* 45:32–37. Figs. 11.
———. 1941. The extremity bones of *Sinanthropus pekinensis*. *Palaeont. Sin.* No. 116. pp. 150. Pls. 34.
———. 1943. The skull of *Sinanthropus sinensis*. *Palaeont. Sin.* No. 127. pp. 484. Pls. 93.
———. 1944. Giant early man from Java and South China. *Science.* 99:479–482.
———. 1945a. Giant early man from Java and South China. *Anthrop. Papers, Am. Mus. Nat. Hist.* Vol. 40, part 1. pp. 134. Pls. 12. Figs. 28.
———. 1945b. The Palaeolithic child from the Teshik-Tash cave in Southern Usbekistan (Central Asia). *Am. J. Phys. Anthrop.* N. S. 3:151–162. Pls. 1. Figs. 2.
———. 1945c. The puzzle of Pithecanthropus. *Science and Scientists in the Netherlands Indies.* New York. pp. 380–390. Figs. 4.
Weinert, H. 1936. Der Urmenschen schädel von Steinheim. *Zeits. f. Morph. u. Anthrop.* 35:463–518. Pls. 6. Figs. 18.
———. 1939. Africanthropus. *Zeits. f. Morph. u. Anthrop.* 38:18–24. Figs. 4.
Weinert, H., W. Bauermeister, and A. Remane. 1940. Africanthropus njarasensis. *Zeits. f. Morph. u. Anthrop.* 38:252–308.
White, F. 1908. *Proc. Rhodesian Sci. Assoc. Bulawayo.* 7:13–23.
Zeuner, F. E. 1940. The age of Neanderthal man, with notes on the Cotte de St. Brelade, Jersey, C. I. *Occas. Paper No. 3, Inst. of Archaeology, Univ. of London.* pp. 20.
———. 1943. Studies in the systematics of Troides Hübner (Lepidoptera Papilionidae) and its allies: distribution and phylogeny in relation to the geological history of the Australasian Archipelago. *Trans. Zool. Soc. London.* 25:107–184. Figs. 115.
———. 1945. The Pleistocene period: its climate, chronology and faunal successions. London: B. Quaritch. pp. 322. Figs. 76.
———. 1946. *Dating the Past.* An Introduction to Geochronology. London: Methuen. pp. 444. Pls. 24. Figs. 103.

5

Head Shapes and Their Inheritance

Before proceeding further with the study of human evolution it is necessary to consider certain features of head shape from a genetic point of view. The genetics of many anatomical features of the head are as yet insufficiently known. Some of these have been referred to in a previous work (Gates, 1946), and the inheritance of dolichocephaly and brachycephaly, as well as the racial development of the latter, were considered at length (pp. 1361–1388), including possible effects of the cradle and of migration on head shape. Some of the conclusions reached may be briefly stated here. Cradle effects are generally temporary, but forms of hard cradle used by certain races may flatten the occiput. This appears to apply especially to the Dinaric peoples. The alleged effects of migration from parts of Europe to America have not been substantiated. In the migration of Japanese to Hawaii the head in the Hawaiian-born generation became somewhat shorter and broader, with narrower face and nose.

As a result of the claim of Boas that the head shape of Jews was largely affected by environment, Pearson and Tippett (1924) pointed out the "close equality of cephalic index in English, German and Russian Jews." They remarked that "centuries have failed in Europe to strip the Jew of those racial characters, which a voyage across the Atlantic accomplishes in the American born Jew." They made a statistical study of some 4500 English school children of ages 4–19 years, using cephalic index (100 B/L) and height index (100 H/L). From these and some other very consistent results they conclude that the "shape of the head of the professional classes in England remains substantially of the same type whether measured at 7 or 14 or 21 years of age." There was no significant change in C.I. from 8 to 20 years of age, hence no correlation of head shape with age. In different classes, ranging from students to criminals and artisans, the C.I. only varied from 78.87–78.15. They also found no association of dolichocephaly with light pigmentation or of brachycephaly with dark pigmentation, which would seem to indicate that the population is now fairly thoroughly hybridized.

Coon (1939) has traced the racial history of the Jews in some detail. He shows (p. 442) that "the Jews form an ethnic group" and that like all ethnic groups they have their own racial elements distributed in their own proportions. They were a group of Mediterranean Semites who early absorbed the population of Palestine, which was also mainly of Mediterranean race. They have developed "a special racial sub-type and a special pattern of facial and bodily expression easy to identify but difficult to define." The Jews of the Mediterranean region from Morocco to Persia are partly of Sephardic origin (descended from the Jewish expulsion from Spain in 1492) and are "remarkably constant in their racial unity," representing "a stable combination of several brunet Mediterranean sub-races."

This basic Jewish racial entity has been modified partly by crossing in the case of the Ashkenazic Jews in Europe and the Oriental Jews in parts of Asia. Coon shows (p. 638) how these racial alterations have come about, especially as regards brachycephalization. In most parts of central and eastern Europe the Jews are less brachycephalic than the Gentiles, and less than 10 per cent are blond. The original Mediterranean blend of Palestine remains most important, and brachycephalization has taken place in the Jews parallel to its development in central Europe, but to a lesser extent.

All early species of man were dolichocephalic. In Paleolithic times, and later, brachycephaly has arisen and spread in all races, presumably through repeated mutations. In central Europe the transformation from dolichocephaly to brachycephaly appears to have taken place partly *in situ* and perhaps partly (but this is less certain) as the result of invasion by broad-headed peoples from the east. Weidenreich, Bunak, and others hold that there is no certain evidence that these invasions within historical times contained any broad-headed peoples. Weidenreich (1945) has given good reasons for concluding that broad-headedness is a further natural step in the evolution of mankind. In any case, it is a relatively recent development in the evolution of the human skull.

Brachycephaly is generally dominant to dolichocephaly in Europeans, but in certain families it appears to be recessive. These two forms of brachycephaly are not the same, but differ in certain other elements of head-shape. In dominant brachycephaly the frontal breadth is large, the post-auricular part of the skull relatively short, whereas in recessive brachycephaly in Europeans the frontal breadth is small and the post-auricular length great. The same cephalic index may include several forms of skull, and there are probably several forms of brachycephaly. The Lapps are broad-headed rather than short-headed, whereas in most races brachycephaly represents a head in which, as compared with dolichocephaly, the shortening is more marked than the broadening. Kappers and others regard the broad brain as resulting from a process of racial foetalization, to be discussed later in connection with the Bushman race.

It has been clearly demonstrated that genetic segregation of dolichocephaly and brachycephaly occurs in crosses. This has been shown in families. That it occurs in populations too, follows from the bimodal and trimodal graphs obtained of the cephalic index of the males in many populations. The females of a race are always somewhat more brachycephalic than the males and they show similar bimodal curves. The genetic analysis of Frets shows that shape factors (affecting two or even three dimensions of the head) as well as size factors (affecting single dimensions), are involved. A whole series of shape factors is thus concerned as between ultra-dolichocephaly and hyper-brachycephaly. These factors or genes presumably all segregate independently to give a series of cephalic indices between one extreme and the other. A factor for hypsicephaly (high vault above the ears) increases not only head-height but head-size in other dimensions.

Dart (1939) finds that in African races such as the Bushman, Boskop, and Negro, brachycephaly is recessive to dolichocephaly. He also finds the hypsicephaly of races in Europe and Asia recessive to the African orthocephaly (medium height) and even to the primitive low vault (chamaecephaly) of the Bushman and Boskop races. A definite beginning has thus been made in analysis of the genes for head-shape in so far as they affect the cephalic index.

The evidence indicates, as already pointed out, that brachycephaly has arisen and spread in many races of the genus Homo, the genera Palaeoanthropus, Pithecanthropus, and Javanthropus ancestral to Homo being almost, if not entirely, dolichocephalic. It is sometimes assumed that brachycephaly has spread because it is dominant to dolichocephaly. This, however, is a fallacy. A dominant character will not spread in a population unless, (a) it has some advantage over the corresponding recessive character and is therefore selected, or (b) it appears repeatedly as a mutation from the recessive character. In the latter case, in the absence of any selection pressure either way, the rate of spread will depend on the rate of mutation. This subject has been discussed elsewhere (Gates, 1946) particularly in relation to the blood groups. From what is known of mutation rates in other organisms it is unlikely that any mutation in man would have a higher frequency than 1 in 50,000 or 1 in 100,000. The mutation rates for many characters may be more like 1 in 1,000,000. That being the case, they are very difficult to detect and prove. Nevertheless, many abnormal inherited conditions in man are known to have arisen as mutations from the normal. In some cases the records show that this has happened repeatedly, but many conditions make it difficult to determine the mutation frequency in any population.

There is evidence to show that mutations occur from one subtype of color-blindness to another (Gates, 1946, p. 142). As these conditions are only slightly abnormal, there is every reason to suppose that normal conditions such as dolichocephaly are also subject to the law of mutation. There is in fact a record (Gates, 1946, p. 699) of a mutation in blood groups, from the A_2 to the A_1 subgroup.

A mutation which can occur once as a germinal change can naturally occur again. And in the absence of natural selection the rate of spread of any particular mutation will depend upon the frequency with which it occurs. It can be shown mathematically (Gates, 1946, p. 717) that if a mutation from, for example, dolichocephaly to brachycephaly occurred with a frequency of 1 in 100,000, then in the absence of any selective advantage, 40 per cent of the population would be brachycephalic at the end of a million years (counting four generations to a century). Brachycephaly was already present in Cro-Magnon man, although he was mainly dolichocephalic, and we may assume about 100,000 years since Cro-Magnon man appeared in Europe. Assuming four generations to a century, what mutation rate would be necessary to accumulate 10 per cent of brachycephali? We have the equation $(1 - \mu)^{4000} = 0.9$, or $4000 \log (1 - \mu) = \log 0.9$. Dr. R. B. Montgomery has kindly looked up these log values for me: $\mu = 0.000025$. In other words, a mutation rate of 1 in 40,000 would be necessary in order to accumulate 10 per cent of brachycephaly in a dolichocephalic population in 100,000 years. Such a mutation rate, while rather high, is of the right order. One can only say that the spread of brachycephaly through repeated mutations is not excluded.

There is no evidence that any change in the size or weight of the brain is associated with change in head shape. The shorter length and greater width essentially compensate for each other in the change from dolichocephaly through mesocephaly to brachycephaly. Parsons (1924) found that brachycephalic skulls in England were both shorter and broader, but mainly shorter, than dolichocephalic skulls. The latter were longer by increased growth in the posterior part, the former being wider chiefly by growth in the temporal region. He also found that brachycephalic skulls have broader noses, compared with the width of the face, and that the orbital openings were not so high as in the Nordic skull.

From measurements of twenty-nine dissecting room crania and brains, Pickering (1930) finds a positive correlation between them, which was highest between the brain and the internal cranial measurements. He suggests that the brain controls the head shape to the extent of 28 ± 10 per cent, other factors, such as size of face and muscular development, determining the remaining 72 per cent. He believes that the cranial capacity is greatest in the brachycephalic head.

In a general study of cephalization (proportion of brain weight to body weight) in relation to body size in mammals, Kappers (1927) shows that the more highly cephalized species of an order tend to brachencephaly—the forebrain is rounder rather than longer. Larger body size, on the other hand, does not produce brachencephaly, the larger species in some groups having longer skulls. In man a certain correlation between greater stature and dolichocephaly is well recognized. Kappers concludes that there is a different brachencephaly of primitive species in a family, which results from shortness of brain (lack of development), whereas in highly cephalized animals it is due to greater width accompanying increase in size of brain. He thus finds that the forebrain of primitive living mammals is usually strongly brachencephalic, while higher cephalization produces brachencephaly of a different type.

In a previous study dealing with the indices of human brains after hardening treatment, Kappers (1927) compared the brains of Dutch and Chinese. The latter are greater in occipital, temporal, and frontal height than brachycephalic Dutch, and still greater than dolichocephalic. Only part of the shortening of the Dutch brachycephalic brain is compensated for by increase in height. Dutch foetuses and newborn had more hypsicephalic crania and higher brain height indices than adults. Kappers naturally applies Bolk's retardation or foetalization theory in explanation, the adult Chinese brain being like the Dutch foetal brain. He shows that many other racial characters of the Chinese can be regarded as resulting from a process of foetalization.

In the Upper Paleolithic of Europe brachycephaly began to appear. Five skulls of Solutrean age measured by Morant (1930) ranged from 76 to 85.2 in cephalic index, the average being 80.2. The culturally Mesolithic people from Alfalou, northern Africa, include both mesocephalic and brachycephalic skulls. According to Weidenreich (1945), of the eleven Mesolithic skulls from Ofnet and Kaufertsberg in Bavaria, three were dolichocephalic, five mesocephalic, and three brachycephalic, one having an index of 86.7. In the Neolithic, the percentage of brachycephaly in Europe increases, and even in Sweden 9 per cent of seventy skulls measured were brachycephalic, while in British Neolithic long barrows 25 per cent were brachycephalic and 21.5 per cent mesocephalic. There appears to be no basis for the suggestion of Retzius and others that mesocephaly arose as a cross between a long-headed and a short-headed race. Rather, dolicho-, meso-, and brachycephaly represent a progressive series of mutational changes which has been going on steadily in Europe since Upper Paleolithic times. Crosses between dolichocephali and brachycephali may, however, be expected to increase the numbers of mesocephali (who would be heterozygous) unless dominance is complete, which is unlikely. Sibs in the same family can differ widely in cephalic index. Some mesocephali can thus represent a heterozygous condition while others are probably determined by intermediate factors. The spread of brachycephaly has been discussed at length by Coon (1939) and Gates (1946).[1]

[1] It may be pointed out that brachycephaly and dolichocephaly in the Titanotheres (an extinct large group of Eocene and Oligocene mammals) is a very different and much more

Weidenreich gives many other figures showing the steady increase in brachycephaly all over Europe. In the Bronze Age the British round barrow skulls increased in brachycephaly from 25 per cent to 39.7 per cent while the mesocephali remained practically unchanged. In Sweden, dolichocephaly has decreased from 65 per cent in the Bronze Age to 30.2 per cent today, mesocephaly having increased a corresponding amount, while brachycephaly (C.I. 80 or more) has remained stationary. In Sweden since Neolithic times the skulls have become slightly longer (186.1 to 188.6 mm.) but considerably broader (138.4 to 143.4 mm). Hug (1940) shows that in Switzerland and southern Germany the C.I. has changed steadily from 76.1 in the Neolithic to 84.4 in the sixteenth and seventeenth centuries, the corresponding increase in brachycephaly being from 22.2 per cent in the Neolithic to 85.7 per cent in the Roman Period, followed by a steady decline through the Middle Ages to 59.7 per cent in the seventeenth century. This probably indicates some shifting of populations as well as the autochthonous change in head shape. Hug found that while the male skulls in the Iron Age ranged from 167 mm. to 211 mm. in length, in modern times the range is only from 160 mm. to 200 mm. In breadth, however, the early range was from 132 mm. to 165 mm. and the modern from 130 mm. to 174 mm.

Dart (1939) has traced the brachycephaly in Egypt, from the protohistoric Badarians some 7000 years ago, with no brachycephali among 79 skulls, to 1 per cent in the El Amrah (101 skulls measured), 1.9 per cent in Nagada (314 skulls), and 11.6 per cent in Dynasty 9 (69 skulls). A decline follows. The brachycephali fail to maintain themselves, but in Dynasties 18–21 (New Empire) brachycephaly returned, about 3,000 years ago, to 11.6 per cent (250 skulls). This is followed by a second decline, to 6 per cent, in the late dynastic and

---

complicated condition than in man. Osborn (1929) shows that from primitive dolichocephalic ancestors (C.I. 50) some genera became broad-headed (C.I. 74–77) while others became extremely long-headed (C.I. 46–49). Head shape was a part of the adaptation to diverse feeding habits, in which the character of the grinding teeth was also involved. The very brachycephalic Palaeosyops had heavy canine tusks, long outer incisors, a heavy zygomatic arch and lower jaw, a high thin crest, and large areas for the attachment of the temporal and masseter muscles. These were all adaptations for browsing and tearing up underground roots and bulbs, whereas in the extreme dolichocephalic genera elongation was notably between the orbits and the post-glenoid processes, an approach to the grazing type with finer trituration of the food. But "every bony element of the skull has a different rate of elongation" (p. 257) and some Titanotheres show abbreviation of the face combined with a long cranium.

Ptolemaic period, followed by 10 per cent in Roman times, then increasing to 24.1 per cent (141 skulls) in modern times, the last increase being chiefly through Moslem invasion. This appears to show an influx of brachycephali in the 9th and again in the 18th Dynasty, which failed to maintain themselves. There seems no other reasonable way to account for these fluctuations, and some kind of selective process must be involved. In the third era, since Roman times, the brachycephali have been more successful, until now they represent a quarter of the population. These results are based on measurements of 2,861 skulls, 55.3 per cent of which were dolichocephalic, 38 per cent mesaticephalic, and only 6.7 per cent brachycephalic. Dart's further analysis of the Egyptian population in his masterly paper will be considered later.

Other surveys, which we need not enter into here, lead to the conclusion that in many cases, including the Chinese, the cephalic index has increased since Neolithic times. The only genetical explanation appears to be the repeated occurrence of more genes for broader heads in many autochthonous populations in different parts of the world, but not in all parts. In the same way the B blood group seems to have increased rapidly in many races, particularly in eastern and southern Asia, Melanesia, and central Africa, while in races like the Australians and Bushmen the mutation has not appeared, but a small percentage has been acquired through crossing. The ultimate causes of both these mutational developments seem entirely without an explanation, but their very wide-spread nature in the eastern hemisphere and their probably simultaneous spread can hardly be without significance. As regards brachycephaly, it appears not only to have developed independently, for instance, in central Europe and in the Chinese, but in each case the head shape changes without materially affecting the face and other characters. This is a fact which needs to be impressed on anthropologists, who are not usually accustomed to thinking in terms of the independent variation of particular characters, which is familiar and indeed fundamental to geneticists. This is apparently what Hug means when he says, "A description and explanation of brachycephalization can undoubtedly be achieved without any reference to racial differences." The brachycephalic members of any two races are obviously no more (and no less) nearly related than are their dolichocephalic members. Brachycephaly has arisen in many races, so also has dwarfness.

Even blue eyes have probably arisen independently in Europe and Ceylon—as parallel mutations.

Regarding the American Indians the interpretation is less clear. The earlier immigrants from Asia were apparently dolichocephalic. The indications are clear that the brachycephali were a later development, but how many of them came directly from Asia and how many arose in the autochthonous American population cannot be determined at present. The subject will be referred to again, in Chapter 9. The fact that Amerinds for the most part lack the B blood group is, however, strong evidence that any brachycephali who came from Asia came before northeastern Asia, which is now high in B, had acquired that condition (see discussion in Gates and Darby, 1934).

Ariëns Kappers suggested that the brachycephalic mutation may result from some special environmental or endocrine circumstances, but there is no genetical evidence of the production of mutations in this way. Since brachycephaly has arisen contemporaneously in many races living under very different conditions it is highly improbable that they have been "touched off" by some common environmental factor. It appears necessary to look for the cause more deeply, in the germplasm itself. The statement of Weidenreich (1940) that "the heredity of the skull index does not follow the Mendelian rules" is a gratuitous assumption contrary to all the probabilities and facts. While the analysis is complex, as already pointed out, and still far from complete, there is no reason to exclude head shapes or any other characters, qualitative or quantitative, from having a genetic basis. The extensive studies of Frets on the inheritance of the cephalic index were certainly not in vain, as can be seen from my analysis of his and other results (Gates, 1946, pp. 1361–1388) on head shape. No one supposes that the index as such is inherited, but length and breadth measurements are among the data which have to be used in any analysis of head shape. Weidenreich relates the development of brachycephaly to reduction of the face and especially of the jaws. This may be so, but it does not exclude an analysis of the process in genetical terms, although the inheritance of shapes is one of the most difficult fields in genetics. The deflection and shortening of the skull base is an associated change which has happened in human phylogeny. It will be seen that many factors in head shape have already been subjected to considerable genetic analysis.

Notwithstanding the enormous number of measurements of the

length-breadth index of human skulls, the cephalic index remains a very partial method of determining skull shape. In addition to these metrical differences there are characteristic non-metrical differences. Skulls which are ellipsoid, ovoid, or pentagonoid when viewed from above (in norma verticalis) may all have the same cephalic index. Sergi (1899) instituted a system of naming skull shapes from this point of view. Easily recognized forms (see Fig. 3) are pentagonoid, ellipsoid, ovoid. But Sergi instituted too many classes of forms. Frassetto drew up a simpler classification, referring these shapes to particular bones, especially the parietals, according to their forms in the foetal, infantile, and adult stages of life, and adopting six basal forms of skull. It is clear that different head-forms are diagnostic of different races, but as yet very little is known of the inheritance of these skull forms and their genetic relations. The cranial shape in norma verticalis is best obtained when the skull is oriented in the glabella-metalambdal plane. Frassetto also introduced the third dimension, each shape of skull as seen in side view could be hypsi-, ortho-, or chamaecephalic, according to whether the vault was high, medium, or low.

Sergi's classification was into nine shapes: 1. ellipsoid; 2. pentagonoid; 3. rhomboid; 4. ovoid; 5. beloid; 6. sphenoid; 7. sphaeroid; 8. cuboid; 9. platycephalic, each of which was subdivided into secondary groups. Frassetto simplified these to six main classes: 1. pentagonoid; 2. ovoid; 3. ellipsoid; 4. eurypentagonoid; 5. sphenoid; 6. sphaeroid. The frontal, parietal, and occipital are of course the bones involved, but as the parietals make up the larger part of the cranial contour in norma verticalis the classification depends mainly on the condition of these bones in the adult. When these cranial bones are of the long type, the skull will be dolichocephalic and when of the short type brachycephalic, the former being characteristic of Eurafricans, the latter of Eurasians. Thus arise six typical forms of cranial configuration, as shown in Figure 3 and Table 5.

TABLE 5
SHAPES OF HEAD

Eurafrican	State of parietals	Eurasian
Ellipsoid	Foetal	Eurypentagonoid
Pentagonoid	Infantile	Sphenoid
Ovoid	Adult	Sphaeroid

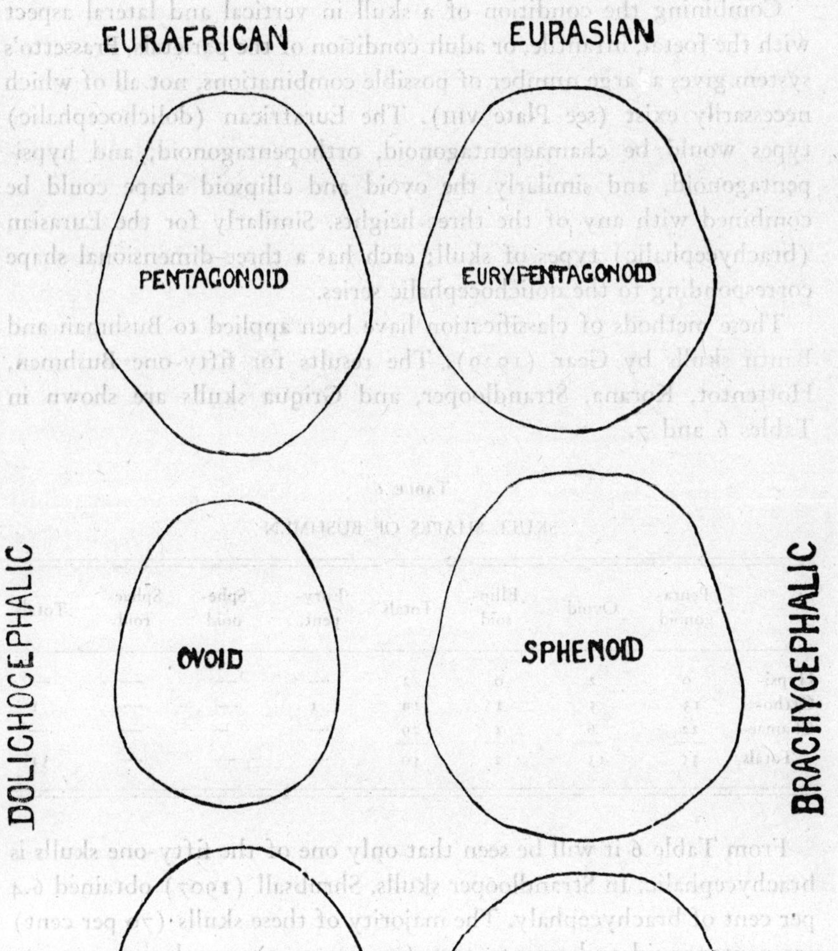

DOLICHOCEPHALIC

BRACHYCEPHALIC

EURAFRICAN | EURASIAN

PENTAGONOID | EURYPENTAGONOID

OVOID | SPHENOID

ELLIPSOID | SPHAEROID

FIGURE 3. Six forms of skull in vertical view.

Combining the condition of a skull in vertical and lateral aspect with the foetal, infantile, or adult condition of the parietals, Frassetto's system gives a large number of possible combinations, not all of which necessarily exist (see Plate VIII). The Eurafrican (dolichocephalic) types would be chamaepentagonoid, orthopentagonoid, and hypsipentagonoid, and similarly the ovoid and ellipsoid shape could be combined with any of the three heights. Similarly for the Eurasian (brachycephalic) types of skull; each has a three-dimensional shape corresponding to the dolichocephalic series.

These methods of classification have been applied to Bushman and Bantu skulls by Gear (1929). The results for fifty-one Bushmen, Hottentot, Korana, Strandlooper, and Griqua skulls are shown in Tables 6 and 7.

TABLE 6

SKULL SHAPES OF BUSHMEN

	Penta-gonoid	Ovoid	Ellip-soid	Totals	Eury-pent.	Sphe-noid	Sphae-roid	Totals
Hypsi-	0	2	0	2	—	—	—	—
Ortho-	13	5	1	19	1	—	—	1
Chamae-	22	6	1	29	—	—	—	—
Totals	35	13	2	50	—	—	—	51

From Table 6 it will be seen that only one of the fifty-one skulls is brachycephalic. In Strandlooper skulls, Shrubsall (1907) obtained 6.4 per cent of brachycephaly. The majority of these skulls (70 per cent) are pentagonoid and twenty-two (44 per cent) are chamaepentagonoid. The Bushman skull is thus characteristically low vaulted and pentagonoid, and this is accompanied by chamaeprosopy (a low, broad face). The few ovoid and very few ellipsoid skulls may well be determined by genes for these shapes; but genetic analysis of the relation between pentagonoid, ovoid, and ellipsoid shapes is necessary before any conclusions can be drawn. Every race no doubt carries many genes for different features of head shape, so the rarer forms of head shape may or may not have arisen from race crossing, according to the evidence in any particular case.

Table 7 shows that the preponderance of pentagonoids results from

Head Shapes and Their Inheritance

TABLE 7
DEVELOPMENT OF SKULL SHAPES OF BUSHMEN

	Foetal Occiput	Infantile Occiput	Adult Occiput		Total
Pentagonoid	6	2	0	Foetal frontal	8
	9	16	0	Infantile frontal	25
	2	0	0	Adult frontal	2
Total	17	18	0		35
Ovoid	3	3	0	Foetal frontal	6
	1	2	1	Infantile frontal	4
	0	3	0	Adult frontal	3
Total	4	8	1		13
Ellipsoid	0	0	0	Foetal frontal	0
	0	2	0	Infantile frontal	2
	0	0	0	Adult frontal	0
Total	0	2	0		2
Totals	21	28	1		50

the persistence of a foetal or infantile condition of the parietal bones into adult life—another evidence of foetalization in the Bushman race. Only one of the fifty skulls showed an adult condition of the occipital, but in five (two pentagonoid and three ovoid) the frontal was adult. Thus in the Bushman skull a foetal parietal is usually combined with a more or less foetal frontal or occipital. Three of these skull types are seen in Plate VIII.

By comparison, ninety-four skulls from various tribes of Bantus gave the results in Table 8. They contrast strongly with the Bushmen.

TABLE 8
SKULL SHAPES OF BANTU

	Pentagonoid	Ovoid	Ellipsoid	Totals	Eurypent.	Sphenoid	Sphaeroid	Totals
Hypsi-	0	8	4	12	—	—	1	55
Ortho-	10	25	19	54	—	—		
Chamae-	6	10	11	27	—	—		
Totals	16	43	34	93	—	—	1	94

Only 17 per cent are pentagonoid, whereas 45.7 per cent are ovoid, and 37.3 per cent ellipsoid. Moreover, 58.5 per cent are orthocephalic and only 28.7 per cent chamaecephalic. Ovoid or ellipsoid orthocephaly is the most common condition. The large intermixture of various other skull shapes indicates much greater hybridity among the Bantu than in the Bushmen. Pentagonoid chamaecephaly in the Bantu is thus an indication of Bushmen admixture. Gear states that in the modern Bantu population steatopygia and the "Hottentot apron" are also frequently met with. He divides the Bantu into three groups, a northeastern division including such tribes as the Barotse, Bavenda, Shangaan, and Bechuana; a southeastern division including the Swazi, Zulu, Basuto, Pondo, Xosa, Fingo, and Tembo; and a third area including the Kalahari, Bechuanaland, Namaqualand, Ovamboland, Damaraland, and southwest Africa. By this division the northeastern Bantu are predominantly ortho-ellipsoid, the southeastern largely ortho-ovoid; the third geographical area (containing the Bushmen) being characteristically chamae-pentagonoid. Among thirty Zulus and Basutos only three were hypsicephalic, which may indicate the presence of dominant genes for ortho- and chamaecephaly acquired from the Bushmen through crossing.

The Ovambo in southwest Africa (Windhoek), from a study by Galloway (1937), evidently have in their ancestry Bush, Negro, and Bantu blood. Galloway found in facial types one Boskopoid and one Mongoloid, as well as many Bush and Negro. His Armenoid and Mediterranean facial types were presumably derived through the Bantu. The axillary hair of the Ovambo could be straight, curly, or peppercorn, and one man had peppercorn hair on his chest. Many other observations were made on eye and skin color, facial, head, hair, and body measurements. Steatopygia was seen in three out of fifty males, it was slight in nine out of twenty females, well developed in four. In the females steatopygia can be associated with a Negro face, while two steatopygous males had a Bush face and one a Boskop. The cephalic index of the males ranged from 68.9 to 83.5, two males in fifty being brachycephalic. These people are thus remarkably hybrid and variable as a result of movements and resulting contacts within recent centuries.

Among the prehistoric Matjes River people near the south coast, Dreyer (1935) found bizarre and striking examples of hybrid skulls.

There were long Hottentot-like skulls, with Boskop resemblances, very large broad paedogenetic skulls, very small almost typical Bushmen skulls, an original early Mossel Bay type, and very long and narrow skulls of Boskop type, but the pygmy stature of the Bushmen does not appear. This may be because, being recessive, it would only be likely to appear in some of the hybrids after considerable inbreeding.

Dart (1939) has made great use of this three-dimensional analysis of head shape in his studies, particularly of racial history in northern and eastern Africa. On the basis of extensive anthropological and field studies he concluded that hypsicephaly is foreign to all three of the main native races of Africa. The Bushmen are chamaecephalic, the Brown (Mediterranean) and Negro races orthocephalic. Yet he finds that hypsicephaly now pervades Africa so as to show an increasing gradient from southeast to northwest of the continent. The immigrant elements which have produced this result are, (1) the dolichomorphic hypsicephaly of Europe, as shown by the Nordics, (2) the brachymorphic hypsicephaly of Asia as manifested in the Mongoloids and Armenoids. The increasing short-headed hypsicephaly in passing up the east coast is explained by the age-long traffic of the Indian Ocean between Asiatics and Africa (see Chapter 7).

The hypsicephaly of eastern Africa is small, however, compared with its occurrence combined with dolichocephaly in the western Bantu of the Congo and Angola and the Negroes of Guinea, Sudan, and Senegambia. Since the only living race which combines dolichocephaly with hypsicephaly is the Nordic, Dart (1937) boldly introduced the explanatory hypothesis that during congenial periods of the last Ice Age, when the Sahara was a favorable region for man, Nordics of the Mediterranean littoral migrated freely southwards and there intermingled with Negro stocks expanding northwards. This Negro-Nordic mixture in the western Sudan produced the hypsicephalized negroid stock which afterwards thrust itself across the headwaters of the Nile and still later poured down the eastern coast, so that some South African Bantu tribes are as strongly hypsicephalized as the "Nilo-Hamites" of the Sudan. Such a southern thrust of Nordics through the Sahara could have as important effects on the peoples, languages, and cultures of Africa as the eastward migration of the Aryans had on India, but there is no certain evidence of Nordics in the Mediterranean littoral during the last Ice Age.

The Egyptian population has a longer continuous history than any other. In a study of Egyptian craniology, Morant (1925) concluded that in pre-dynastic times a primitive dolichocephalic people lived in Egypt, consisting of two nearly related races in Upper and Lower Egypt respectively. The race in Lower Egypt remained unchanged until Ptolemaic times, while the Upper Egyptian type was gradually transformed into that of Lower Egypt. He found no evidence of Negro blood in the population from early pre-dynastic times. In a later study of the Badarian skulls excavated by Sir Flinders Petrie from early pre-dynastic sites near Badari, thirty miles south of Asyut, and which are believed to be ancestral to the early Egyptians, Morant (1935) concluded that the Badari type resembled most clearly the other pre-dynastic Egyptian series. They were a generalized type, close to the modern Sardinians, widely different from African Negro types and nearest a Dravidian series in India. They clearly belonged to the Mediterranean race. Through the whole dynastic Egyptian era Morant found the type gradually becoming less prognathous, with slightly lower nasal index and higher cephalic index, but remaining remarkably constant during this long period.

From this starting point Dart was led to examine the head shapes, in three dimensions, of the Egyptian populations, from the earliest Badarian pre-dynastic times (about 5000 B.C.), to the modern era. More measurements of skulls are available from Egypt than from any other country. In this study the cephalic and altitudinal indices of 2,861 skulls, grouped in eighteen series belonging to successive historical periods, were used. Dolicho-, mesati-, and brachycephaly combined with hypsi-, ortho-, and chamaecephaly gives nine possible skull types. The skulls were classified as shown in Table 9.

The dolichocephalic-hypsicephalic group of 280 skulls is recognized as Nordic,[2] the dolicho-orthocephalic as Brown, and the dolicho-chamaecephalic as composed of Bush and Boskop elements (see Chapter 7). The brachy-hypsicephalic group, making up 16.2 per cent of the skulls, are regarded as Armenoid, while three of the remaining squares (the mesaticephals) are hybrid, and the last two squares represent very small numbers in the population. This simple classification of heads as long, medium, or short on the one hand and high, medium,

---

[2] In the following discussion the term "Nordic" is used as applied to head alone. There is no evidence that these people had Nordic pigmentation.

TABLE 9
ANALYSIS OF EGYPTIAN SKULLS OF ALL PERIODS

	Hypsicephalic	Orthocephalic	Chamaecephalic	Total	Per Cent
Dolichocephalic	*Nordic* 280 35.5% / 17.7%	*Brown* 952 59.1% / 60.2%	*Bush & Boskop* 350 76.6% / 22.1%	1582 100%	55.3
Mesaticephalic	*Hybrid* 381 48.3% / 35.1%	*Hybrid* 598 37.2% / 55%	*Hybrid* 108 23.3% / 9.9%	1087 100%	38.0
Brachycephalic	*Armenoid* 128 16.2% / 66.7%	*?Mongol* 59 3.7% / 30.7%	*?Oriental Negrito* 5 1.1% / 2.6%	192 100%	6.7
Total	789 100%	1609 100%	463 100%	2861	
Per Cent	27.6	56.2	16.2		100

or low on the other, can thus be used in the analysis of head shapes in any population or race. The percentage of types will also furnish some evidence of the racial composition of any population.

Applying these methods to the population of Egypt during its 7,000 years of history, Dart finds some striking results. In the predynastic period, covering a millennium or more, among 858 "complete" skulls, the brachycephali never numbered as much as 2 per cent of the population. In early dynastic times this suddenly increased to 6 per cent or more but soon decreased again. At the period of the New Empire (18th Dynasty and later) the brachycephali return to their higher levels, but again prove incapable of establishing themselves. By the 30th Dynasty and the Ptolemaic period they are again down to 6 per cent, their levels a millennium and two millennia earlier. In the Roman period a brachycephalic element again enters the Nile Valley and their frequency rises to 10.1 per cent. In the modern Mohammedan era the further influx of brachycephali has increased their frequency to 24.1 per cent, but in the whole period of seven millennia only 6.7 per cent of the population are brachycephalic (Table 9). Two-thirds of these are also hypsicephalic, 30 per cent orthocephalic, and only 2.6 per cent chamaecephalic. It thus appears that in all the history of Egypt down to modern times the invading brachycephali have been repeatedly swallowed up by the autochthonous Mediterraneans.

The Egyptian population has thus always been mainly dolichocephalic, but the percentage has fluctuated between 74.3 per cent in pre-dynastic times and 41.4 per cent in the modern population. These fluctuations in dolichocephaly are independent of the brachycephalic element. Mesaticephaly has arisen only to a small extent from crosses between dolichocephali and brachycephali. The records show that a high incidence of mesaticephaly represents a temporary and highly unstable condition of the population, although an average of 38 per cent of all the populations was mesaticephalic (Table 9). Dart finds that the mesaticephaly has been produced mainly by inroads of dolichocephali (Nordics). This will be a case of crosses between two dolichocephalic races producing mesaticephaly, and requires further genetic analysis. The mesaticephali have generally represented about 30 per cent of the Egyptian population, but they rose to 58 per cent in one colony in the 9th Dynasty and to 62 per cent in the 18th, receding to 30 per cent in the 30th Dynasty. Fifty-five per cent of the mesaticephali were orthocephalic and 35 per cent hypsicephalic.

From a very recent study of 44 Egyptian skulls of the 1st Dynasty, from Sakkara, and 55 from the 11th Dynasty at Thebes, Batrawi and Morant (1947) draw various conclusions. Sakkara is 20 miles south of Giza, and this series contains about 60 bodies, all males and in the prime of life, dating from ca. 3400 B.C. The Thebes crania are of soldiers slain in a battle ca. 2000 B.C. They find that the Sakkara skulls differ in important features from the pre-dynastic people, and conclude that Egyptian culture was not derived from the pre-dynastic race but from a far more advanced race which had great tombs and statuary, paintings and writing. The pre-dynastic people made stone bowls and vases as well as pottery. They used copper, but were little removed from the Neolithic. Batrawi and Morant (p. 19) find "No evidence that there was any appreciable change in the

Head Shapes and Their Inheritance

variation exhibited by Egyptian populations during the long period from early predynastic to Roman times." But the Sakkara and Thebes skulls are clearly differentiated by their mean cranial measurements. Using the coefficient of racial likeness, they conclude that the Lower (Northern) Egyptian type prevailed from the 1st to the 30th dynasties, and probably earlier and later. Further south, around Thebes and Abydos, the population was of a second racial type (Upper or Southern Egyptians), derived from the Badarians. They underwent slow modification down to about the 18th dynasty, becoming more like the Lower Egyptian type but remaining distinct. There was then an abrupt change to nearly the Lower Egyptian type. From comparison of 24 series of crania, they find that 22 series show no suggestion of mixture with alien stock, but rather a steady transfer of population from the Delta to the Thebes region (Upper Egypt), the Upper and Lower Egyptian types becoming later "almost indistinguishable." Two series of crania (11th Dynasty Thebes and 9th Dynasty Delta) stand apart from the rest and are close to the modern Cretans, probably representing the ancient Egyptian type $\times$ a European or Asiatic type. They believe there is "no need to suppose that any people foreign to the country played a substantial part in modifying its population from predynastic to Roman times," contrary to the views of Elliot Smith. They believe that the Thebes and Delta populations mentioned may represent a mixture of Egyptians with Armenoids.

As regards head-height, it becomes clear that the fundamental skull type of the Egyptian population was the ortho-dolichocephalic type of the Brown or Mediterranean race. This has remained fundamentally the same throughout Egyptian history. It is a juvenile racial type, of medium height and slender (almost effeminate) physique, the body glabrous brunet, the hair jet-black and glossy, iris deep brown, conjunctiva white; skull pentagonoid, of moderate capacity; forehead erect, occiput bulged; nose straight, of moderate length and elevation; jaws small, chin pointed; face ovoid, short, of moderate width. Hypsicephaly has increased in the population quite independently of brachycephaly, showing that it is derived mainly from north African and Asiatic Nordics and not from Asiatic Armenoids. The latter do not antedate 3000 B.C. and are mostly much later. Thus in the early dynastic period when the brachycephali were only 6.1 per cent of the population, the hypsicephali were already 24.4 per cent. The brachycephali then decrease, followed by an increase to 11.6 per cent in the 9th Dynasty, whereas the hypsicephali go on steadily increasing to 47.8 per cent at this period. Hypsicephaly then rapidly decreased, showing that the Nordics, like the Asiatic brachycephali, were unable to hold their own in the Egyptian population. But in the whole of Egyptian history there have been four times as many hypsicephali as brachycephali, showing a much greater influx of Nordics. The high percentage of hypsicephaly in the pre-dynastic Badarians can only be accounted for by a Nordic (Capsian) invasion of Egypt at this time. Hypsicephaly remained high through the 9th Dynasty, then suffered a catastrophic fall from 50 to 19 per cent and remained low (10–20 per cent) until the modern period when it increased (largely through hypsi-brachycephalics) to 46 per cent.

That men of Nordic type can have come to Egypt from Asia Minor as well as from north Africa is shown by Dart's evidence that of 206 skulls, ranging in time from the fourth to the second millennium B.C. and in space from the Dardanelles to the Caspian Sea and the Persian Gulf, less than 2 per cent are brachycephalic. The long, high-vaulted Nordic type of head was prevalent in Mesopotamia in the early Egyptian period. The population of Asia Minor and Mesopotamia, like that of Egypt, was primarily Brown, secondarily Nordic, and only brachycephalic (Armenoid) at a later date.

In his further analysis of the Egyptian population, Dart shows that the Negro element has always been very small. There was no period of influx of true Negroids from the south until Roman times, and even then the number of Negroes appears to have remained relatively small. The Negro skull is recognized as orthocephalic, like the Mediterraneans, but ovoidal or egg-shaped in vertical view. The real Negro was thus of negligible importance in the complex population of ancient Egypt. On the other hand, the Bush and Boskop types, to be considered in detail in Chapter 7, formerly extended from the Cape of Good Hope to the Red Sea and constituted a significant element in the early Egyptian population. The dwarf Bush race is dolichocephalic, with small (microcephalic), low-vaulted (chamaecephalic), pentagonoid head, and their skulls are thus easily distinguishable from the Negro. The Boskop type are taller, with very large dolichocephalic skulls 190–200 mm. in length, acutely pentagonoid. Other differences between these types are considered in Chapter 7. Dart shows that these two chamaecephalic types, the macrocephalic Boskop and the microcephalic Bush, are present in the Egyptian population from the earliest times. In the pre-dynastic period (Badarians) they were only 3.8 per cent. Then they ranged from 12 to 20 per cent until the 1st Dynasty, when they reached their highest level of 27.3 per cent. In the first millennium of the dynastic period, as hypsicephaly waxed chamaecephaly waned almost to extinction (1.5 per cent), but with the decline of hypsicephaly in the 12th and later Dynasties chamaecephaly increases to 20 per cent in the 18th Dynasty and 25 per cent in the 30th. Since Roman times it has again dwindled to 10 per cent. Thus during 7000 years chamaecephaly has been down to 10 per cent or less at three different times, while in the Ptolemaic and Roman periods, it was up to 25 per cent of the population. Dart (1939) also suggests that the small, microcephalic Bush race reached into Europe, where it is represented not only in the Neolithic of Switzerland, the Iron Age Kurgans of southern Russia, and old graves in Sardinia, but in the modern population of Sicily and southern Italy. There is no direct evidence, however, that these semi-dwarf peoples belonged to the Bushman race.

This analysis of the history of Egyptian populations on the basis of skull shapes shows that four dolichocephalic races are represented from the earliest times—Brown or Mediterranean (autochthonous), Bush and Boskop from Africa, and Nordic coming in from Asia Minor and perhaps from north Africa. A fifth race, not represented in the earlier period, was the brachycephalic Armenoids. Intercrossing of these types through 7000 years has produced no blend, but instead the original racial types continue to reappear as the result of the genetic segregation and recombination of the numerous genes for skull shape involved.

Evidently this method of analysis of skull shapes in populations can be applied to all races. Dart applies it to Sergi's catalogue of 1567 skulls from the Kurgans of southern Russia, ranging in time from Neolithic to the Iron Age. The results are shown in Table 10.

Of these skulls, 598 were of Sergi's Eurafrican type and 469 of Eurasiatic type. Of the former, 60 per cent were ellipsoids, 17.2 per cent ovoids, and 22.8 per cent pentagonoids.

In the last chapter various references were made to the supra-orbital

TABLE 10

CLASSIFICATION OF SKULL SHAPES FROM RUSSIAN KURGANS

	Hypsi-	Ortho-	Chamae-	Total	Per Cent
Dolicho-	14	24	7	45	50.0
Mesati-	19	21	2	42	46.7
Brachy-	3	0	0	3	3.3
Total	36	45	9	90	—
Per Cent	40	50	10	—	100.0

and the occipital torus. The former, and probably also the latter, was wanting in the juvenile Australopithecus, but in Pithecanthropus and Sinanthropus they were massive ledges of bone extending across the front and rear of the skull. In Sinanthropus a marked groove set off the supra-orbital torus from the rest of the frontal bone, but despite the great size of the torus the frontal sinuses were small or even absent. In the Rhodesian skull the frontal torus was equally massive and continuous. In the genus of Neanderthal man this torus was already broken into three parts, the glabellar portion being separate from the supra-orbital ridges, which are much less massive than in the ancestral genus Pithecanthropus. The next definite step in this simplification is in the modern *Homo australicus,* in which the supra-orbital ridges contain an oblique ophryonic groove which separates each into an inner and an outer portion. Traces of this separation of the superciliary ridges into an inner and an outer portion are found in the Neanderthal skull of La Chapelle-aux-Saints, in which the glabellar torus is also less developed. The last stage in the series is the complete loss of eyebrow ridges, which has proceeded farthest in *Homo mongoloideus, H. capensis,* and *H. africanus.* It may be pointed out that a similar process has occurred in the Anthropoidea, the gorilla having retained a heavy torus which is entirely absent in the orangutan. It appears probable that in the line of descent ending in the orangutan it was never present.

A similar story can be briefly recapitulated as regards the occipital torus. In Sinanthropus it stretches across the whole posterior aspect of the skull and is separated from the occipital plane above by a supratoral sulcus or groove. The inferior margin of the torus is similarly demarcated in some skulls but not in others. This occipital band across the skull is of enormous thickness. It is, like the frontal torus, a superstructure, with no corresponding moulding of the endocranial surface but serving for the attachment of the heavy neck muscles. This occipital ledge was more developed in males than females. In addition, at the sides of the skull is the torus angularis, shown by Weidenreich (1940) to complete the posterior ring of thickening by linking the torus occipitalis with the supramastoid crest on either side. The torus angularis has completely disappeared in modern man, although the occipital torus can be seen in some modern primitive races. The loss of the torus angularis is thus the first stage in the modernization of this region of the skull. The condition in Pithecanthropus is very similar, but in *Javanthropus soloensis* the angular tori show signs of separating from the occipital torus. In Neanderthal man the disintegration has proceeded much further and the components are all much reduced. The mastoid portion is now independent, the angular torus having disappeared. The central portion around the inion shows signs of disintegration into three parts. In most species of Homo these have disappeared.

These two reduction series, in the frontal and occipital tori, are by no means synchronous. The reduction of one may proceed while the other remains stationary. This of course applies to many other features, such as the chin. One of the most striking results of the modern study of human fossil skulls is this fact that characters evolve independently in different parts of the skeleton. The erect posture, for instance, was attained in Australopithecus when the brain was no larger than in the gorilla. This independence in the variation and evolution of different skeletal characters is naturally in accord with the modern principles of genetics. It cannot be too strongly emphasized that both in variation and hybridization we are dealing with genetic units and their recombinations in the germ cells. To determine what these units are in relation to skeletal characters is one of the main problems of human paleontology.

It appears that the only way to reconcile the known facts of human paleontology as regards the superciliary and the occipital tori is by recognizing that there have been two more or less independent streams in human evolution. One of them began with heavy brow ridges and occipital tori, as in Pithecanthropus. These were gradually reduced to the condition seen in the Australian aborigines. This I propose to call the gorilloid line of evolution. The other line, beginning with Eoanthropus and Kanam man, evolved without these heavy skull ornamentations, but has hybridized from time to time with representatives of the other line, such as Neanderthal man. We may call this the orangoid line. Australopithecus similarly reached essentially the human level without developing heavy brow ridges, and the same is true of the eastern Asiatic orangutan in contrast to the western African gorilla. The terms gorilloid and orangoid do not of course imply relationship to the gorilla or the orangutan, but simply indicate parallel developments in the human strains, as regards the production or failure to produce superciliary and occipital tori.

Since this was written, it has been very interesting to find an article by Woollard (1938) in which, after concluding that the modern type of man (Homo) was already present in the early Pleistocene, he writes (p. 27), "One type of man might have evolved without being encumbered by overhanging brows, and without a flat head, and have proceeded straight along an evolutionary line that ended in the production of the modern type." He suggests that this development might have been saltatory, with a few large gradations or steps. The reasonableness of this view is enhanced by the work which he cites (see Chapter 3)—the conclusions of Straus that man is nearer the monkeys than the great apes in the structure of the pelvis, the musculature of hand and foot, and in the form and arrangement of the abdominal viscera. Straus therefore placed hominid emergence between the Catarrhine monkeys and the gibbons. As already pointed out, Schultz concludes that the great apes are nearer to each other than they are to man, and Le Gros Clark similarly found that the human stem branched off from the Primates before the apes, so that man and the great apes are in many respects parallel developments. Among other facts in support of such views, Woollard cites the supracondyloid foramen or process (see Gates, 1946, p. 452), which occurs as an inherited condition in perhaps one in fifty of the modern population although it is otherwise found only in the basal primate stock. The parietal bones of the skull and the front part of its floor also present a primitive primate arrangement. The baboon-like habits of the Australopithecinae also support the view that man never passed through a late brachiating stage.

REFERENCES

Dart, R. A. 1937. Racial origins, pp. 1–31. In Schapera, I. *The Bantu-speaking tribes of South Africa.* London: Routledge.

———. 1939. Population fluctuation over 7000 years in Egypt. *Trans. Roy. Soc. S. Afr.* 27:95–145.

Dreyer, T. F. 1935. Skulls of racial hybrids in South Africa. *Zeits. f. Rassenk.* 2:207–208.

Galloway, A. 1937. A contribution to the physical anthropology of the Ovambo. *S. Afr. J. Sci.* 34:351–364.

Gates, R. R., and G. E. Darby. 1934. Blood groups and physiognomy of British Columbia coastal Indians. *J. Roy. Anthrop. Inst.* 64:23–44. Pls. 5.

Gear, J. H. 1929. Cranial form in the native races of South Africa. *S. Afr. J. Sci.* 26:684–697. Pl. 1. Fig. 1.

Hug, E. 1940. Die Schädel der frühmittelälterlichen Gräber aus dem solothurnischen Aaregebeit, etc. *Zeits. f. Morph. u. Anthrop.* 38:359–528. Pls. 36.

Kappers, C. U. A. 1927. Indices for the anthropology of the brain applied to Chinese, dolicho- and brachycephalic Dutch, foetuses, and neonati. *Proc. Roy. Acad. Amsterdam.* 30:81–94. Figs. 8.

———. 1928. The influence of the cephalization coefficient and body size upon the form of the forebrain in mammals. *Proc. Roy. Acad. Amsterdam.* 31:65–80. Figs. 10.

Morant, G. M. 1925. A study of Egyptian craniology from pre-historic to Roman times. *Biometrika.* 17:1–52. Figs. 8.

———. 1935. A study of predynastic Egyptian skulls from Badari based on measurements taken by Miss B. N. Stoessiger and Professor D. E. Derry. *Biometrika.* 27:293–309.

———. 1947. A study of the first dynasty series of Egyptian skulls from Sakkara and of an eleventh dynasty series from Thebes. *Biometrika.* 34:18–27.

Parsons, F. G. 1924. The brachycephalic skull. *J. Roy. Anthrop. Inst.* 54:166–182. Figs. 10.

Pearson, K., and L. H. C. Tippett. 1924. On stability of the cephalic indices within the race. *Biometrika.* 16:118–138.

Pickering, S. P. 1930. Correlation of brain and head measurements and relation of brain shape and size to shape and size of the head. *Am. J. Phys. Anthrop.* 15:1–52. Figs. 2.

Sergi, G., 1899. Specie e varietà umane. *Riv. di Sci. Biol.* 1. pp. 20. Pl. 1.

———. 1914. *The Mediterranean Race.* London. pp. 320.

Weidenreich, F. 1945. The brachycephalization of recent mankind. *Southwest J. Anthrop.* 1:1–54. Figs. 11.

Woollard, H. H. 1938. The antiquity of recent man. *Sci. Progress.* 33:17–28.

6

Local Evolution of Modern Racial Types: From Pithecanthropus to the Australian Aborigines

In the last chapter we were dealing with the evolution of the Hominidae from the prehominid stage, as represented mainly by Pithecanthropus, Sinanthropus, and Africanthropus, to the paleo-hominid level as indicated by the various Neanderthaloid species in Europe, Asia Minor, and Africa. We now take a step forward, in an endeavor to trace further lines of descent from the Neanderthaloid level to modern man in various parts of the world. Fortunately, there is evidence that some of these lines take their rise back at the prehominid level. This is particularly clear in the case of the Australian aborigines, who may be credited with having the longest line of known ancestral types in their pedigree. Keith (1931) has particularly recognized the principle of local evolution in the development of modern races of man, and Weidenreich (1943) clearly accepts the same point of view. He agrees, for instance (1939), that the conception of one center of human evolution is incorrect.

Modern (neanthropic) man belongs to the genus Homo. His skull represents a toning down of the frontal and occipital regions as seen in Neanderthal man, a general smoothing of the skull surfaces and thinning of the bone. In these respects evolution has moved forward, apparently in a main path which has taken parallel lines in different parts of the world. For it is impossible to imagine that man in all these regions formed a single breeding population, and there is much evidence to the contrary. This line of development has not been accompanied by increase of brain size since Neanderthal, and probably such further increase could not take place without some reconstruction of the female pelvis. These parallel developments have been accompanied by much divergence in different types of modern man.

What lies behind this progressive development of the human head from Pliocene to recent times? To call it orthogenesis does not furnish an explanation. Man's brain appears to have continued to increase in size because of the advantages of additional intelligence which the increase conveyed, until the limitations of the birth-canal prevented further development in this direction. From another point of view, the evolution of the human brain is an example of nomogenesis (Berg, 1926), for Neanderthal man had a brain at least as large as modern man, yet he could make relatively little use of its potentialities in the conditions under which he lived. In modern man the brain has undergone some increase in structural complexity and efficiency, but there is no reason to suppose that Homo has yet fully exploited the possibilities of his brain. It appears rather that many potentialities for further mental development are present in man's brain as it exists. We thus have an example of nomogenesis, or the development of an organ with prospective potentialities which only began to be fully realized much later.

Selection within each race could no doubt also improve the mental standards within each class of society, for there is no question of the inheritance of mental abilities and disabilities (see Gates, 1946). But no modern society can claim to have surpassed or even equaled the concentration of intellect and artistic ability found in the small population of ancient Greece. Whatever human progress can be claimed in the last three thousand years has not been due to any improvement in intellect or even in moral qualities, but is simply the result of the accumulated knowledge and experience of mankind. This has been made possible by written records, which prevent the loss of knowledge once gained, on which fresh discoveries can be based. In the past, however, much knowledge has been lost through human tumult, as in the burning of the great library of Alexandria, not to mention the black-out of classical knowledge for a thousand years which followed the downfall of the Roman Empire. An atomic war might leave civilized man in an equally parlous condition. It remains to be seen whether the Caucasian race will, by self-extermination, make way for the development of some colored race in the

Northern Hemisphere, where the climatic conditions are most favorable for the flowering of mankind.

From paleoanthropic to historical times many different human types evolved in many parts of the world. Their development can only be considered in summary form here. Coon (1939), in his *Races of Europe*, has systematized and coördinated the vast amount of anthropological as well as historical and linguistic knowledge on which our understanding of these racial relationships is based. I have considered elsewhere (Gates, 1946) the inheritance of many anthropological characters from a genetical point of view. These include eye-color and hair-color, stature, nose shape, hair characters, head shape (especially dolichocephaly and brachycephaly), skin color and the sacral spot, as well as a number of other features of less significance. The analysis of most skeletal characters in genetical terms remains for the future.

We may consider next some of the lines of evolution which emerge from the studies of fossil skulls and skeletons in different parts of the world. As pointed out in Chapter 4, many of the descendants of Pithecanthropus spread over the region of Sundaland in the southeastern extensions of Asia during the Pliocene and Pleistocene, and one branch of them entered Australia by water when it was less isolated from New Guinea than now. They took with them the dingo,[1] the only other placental mammal in Australia when the white man arrived, and they spread over the Australian continent. Plate IX(c, b) shows the skull of an Australian aboriginal woman and the head of a man. Note the deep glabellar notch in the latter and the low, sloping forehead in both. The discovery of fossil remains indicates a clear line of descent, Pithecanthropus ⟶ Palaeoanthropus (*Javanthropus*) soloensis ⟶ *Homo wadjakensis* ⟶ *Talgai skull* ⟶ Modern Australian aborigines (*Homo australicus*). The ancestry of no modern race of man is better authenticated than this. The first two steps in the series were considered in Chapter 4.

Wadjak man is represented by two skulls found in Java, Wadjak I by Mr. Van Rietschoten in 1889 and Wadjak II by Dubois (1922) in 1890. They are clearly similar and resemble Australians rather than

---

[1] This dog has been found in a fossil state in the Pleistocene of Australia, along with Thylacinus and Diprotodon. It is possible, as Elliot Smith suggests, that the latter, a giant marsupial, was exterminated by the dogs of the early Australians.

Papuans. Dubois also found here a skeleton (III) of much later age and covered with red ocher. All the bones were broken, probably by rock falls. The skulls I and II were dolichocephalic, like all early men, while III was brachycephalic. Wadjak I, a woman, had a cephalic index of 72.5, nasal index 60, cranial capacity 1550 cc.; Wadjak II 1650 cc. and a jaw nearly as massive as the Heidelberg jaw. These heads are much larger than the modern Australian aborigines, but they show the same characteristic subglabellar notch, less marked, and superciliary ridges, more marked. The occiput jutted backwards more than in the modern Australian and the frontal was slightly less developed. The teeth were like those of *Homo australicus* and not taurodont as in Neanderthal man. Keith (1931) regards the type as not merely proto-Australian but showing relations to the type of Rhodesian man. This is not unlikely, since the Keilor skull (described below) from southern Australia is of the same type. Wadjak man would then have moved south into Australia and also north and westwards to survive as Rhodesian man, unless the survival in Africa represents a parallel development. The Wadjak type is no nearer to Neanderthal man than is the Australian, showing again an independent line of evolution from that in Europe. The palate was huge, very broad, even larger than in Rhodesian man, the upper jaw surprisingly like that of Pithecanthropus. These finds demonstrate two important points, (1) that Pithecanthropus evolved to the level of the genus Homo in Java, (2) that Wadjak man was a proto-Australian, representing the optimum development in that line of evolution. He is no more primitive than the modern Australian, who declined in size after reaching Australia, possibly because of the less favorable conditions.[2] Here is a clear case of negative evolution in the Australian continent, from *Homo wadjakensis* to *Homo australicus*. It has been suggested that this took place because conditions for man were less favorable in Australia, but it seems more likely that these changes represent the smoothing and refinement of the skull which appears to have taken place in other types of man isolated in different parts of the world during the interval between the paleoanthropic and the neanthropic stages.

The more recent finding (Fenner, 1941) of an Australoid skull of some geological antiquity in Pleistocene beds at Aitape on the

---

[2] Birdsell (unpublished) finds that the southeastern tribes under more favorable conditions, in Gippsland and adjacent areas, are of larger size.

northeastern coast of New Guinea is another step on the way to Australia. This skull nearly corresponds with the modern Australian. It is also stated that an Australoid type occasionally occurs among the modern Melanesian population of New Guinea. In a recent study of thirty-one skulls from Ambrym island in the New Hebrides, Hambly (1946) finds that the skulls not only of Ambrym but also of New Britain were predominantly Australoid. The skulls show marked sex differences, the males being more rugged with high vault, massive brow ridges, depressed nasal root, a tendency to formation of an occipital torus, and large mastoids. Photographs of living natives of Ambrym show Negroid as well as Australoid traits, but there are three marked differences from Australoids and seven from Negroids.

That Wadjak man was the type which reached Australia has been shown recently by the discovery of the Keilor skull in southeastern Australia, ten miles northwest of Melbourne. The geological conditions of its finding and the age of man in Australia have been discussed in detail by Mahony (1943), who concludes that the age of the river terrace in which it was found is Riss-Würm Interglacial. Afterwards portions of a second skull were found six feet away but at the same level. The skull is large, cranial capacity 1593 cc., cranial length 197 mm., index 72.6. Wunderly (1943) describes the skull and concludes that it combines Australoid and Tasmanoid characters in equal proportions. Adam (1943) describes the palate and dental arch, which he finds to be wide like the Tasmanian. Weidenreich (1945) shows the extreme similarity of the Wadjak and Keilor skulls. Both have a horse-shoe shaped maxilla, contracted in the premolar region. Both have small incisors, large premolars, and molars. From measurements, Weidenreich finds that the "likeness could not be greater if the skulls belonged to identical twins." Yet Wadjak man has a supraglabellar torus, which is absent in the Keilor skull. The Wadjak forehead is also less vaulted and more sloping, the jaw slightly more prognathous and the teeth larger. It is of much significance that the transition from the Wadjak type to the near modern Australian occurred in isolation on the Australian continent. In the Keilor skull the superciliary ridges are rather light and the glabellar notch shallow.

The age of the Keilor skull is in dispute. It was found in 1940 in a sand pit eighteen feet below the surface and eighteen inches above the floor of the pit, in the highest of three river terraces, sixty feet above low-water mark. This terrace is correlated (Zeuner, 1944) with the Last Interglacial forty- to fifty-foot level in Tasmania and the Main Monastirian (18.5 meter) level in Europe. The latter belongs to the first half of the Last (Riss-Würm) Interglacial, some 500,000 years ago. Mahony (1943) says, "The skull was unearthed beneath undisturbed strata at 18 ft. below the surface of the terrace," and that skull and terrace are evidently contemporary. He further states (p. 23) that "geological evidence of the age of the Keilor skulls and bones seems irrefutable." If this age is correct it means that man in Australia has practically stood still in an evolutionary sense for half a million years. Kanjera man, of similar age in Kenya, was also of "modern" type (see Chapter 7).

At Keilor, one limb bone and other fragments were present, the skull was undisturbed and an artifact quartzite flake was close by in the sand, but Weidenreich disputes the contemporaneity of the skull and the river terrace sand deposit in which it was found. He believes that "we cannot exclude with absolute certainty that it did not get into the sand long after the terrace was formed." If the skull was contemporary with the terrace, then it is probably contemporary with Neanderthal man in Europe. Considering all the other evidence of different rates of evolution in nearly related forms this does not appear to be a serious difficulty. It merely means that man in Australia had nearly reached the modern level when Neanderthal man in Europe was still at a lower level of evolution. If it is admitted, as it surely must be, that the Australian and Bushman races of modern man are more primitive in some respects than other living races, then there is surely no reason why Neanderthal man in Europe should not be contemporary with early Homo in Australia. This is only recognizing the general principle that the speed of evolution varies at different times and places, whatever the causes may be. Moreover, even in Eurasia there is evidence that Paleoanthropus and Homo were in part contemporary, as well as the evidence that the Steinheim skull, which was earlier than Neanderthal in age, was already later in type. There is also the evidence (if more is needed) of the Piltdown, Swanscombe, and Galley Hill skulls, which will be referred to later. In any case, it seems clear that modern man (*Homo australicus*) has evolved in isolation in Australia quite independently of the modern species produced in Europe and elsewhere.

There is one important point in which the superciliary ridges of the Australian aboriginal differ from those of Neanderthal. The ophryonic groove in the Australian skull runs obliquely over the eye, separating an outer temporal from an inner medial portion. According to Sollas (1908) this is not present in the Neanderthal skull. It represents a further stage in the disintegration of the frontal torus so conspicuous in Sinanthropus, Pithecanthropus, and the Rhodesian skull. In this respect the Australian is definitely more advanced than Neanderthal. Mahony (1943) cites all the finds of Australian remains which have yet been made.

The Talgai skull, found in 1884 in Queensland in a fossilized state, encrusted with limestone colored with iron salts (Plate IX, a) is another in the Australian series. It was described by Smith (1918) as a youth with the third molars unerupted. The skull was nearly complete except the mandible, though much broken, and it had the characteristic (Australian) infraglabellar notch, which is not present in the Wadjak and Keilor skulls. The palate is oblong, the teeth not taurodont, so there is no resemblance to Neanderthal. The Pleistocene age of this skull was regarded as established. Hellman (1934) restored the palate and found it of a generalized human type which occurs infrequently in the modern Australian aborigines. The Cohuna skull is another which, according to Keith, corresponds exactly with the Talgai.

Everyone appears to be convinced that this series, whatever may be the age of certain members of it, links Javan Pithecanthropus directly, via Sundaland, with the modern Australian aborigines. This Australoid type also spread northeastward, where it is represented by the Sakai of Malaya, the Veddas in Ceylon, and certain jungle tribes in southern India, such as the Paniyans, and also probably by the Toala of Celebes and other groups in Borneo, Java, and Sumatra. Among these is a little-known tribe, the Che Wong, in the province of Pahang, Malay peninsula, about 150 miles north of Singapore. They are briefly described by Ogilvie (1940). Their skin is light yellow-brown, hair fine, black and wavy, their language differing greatly from that of neighboring tribes. The photographs (Plate x) show heavy brow ridges of Australoid type and broad noses depressed at base, thus indicating a surviving Australoid remnant with some Malayan or Mongoloid admixture.

Birdsell (1947) has recently recognized three waves of race in Australia: (1) the Tasmanians, of Negrito origin, (2) the Murrayian race, allied to the Ainu, (3) the tall, dark, wavy-haired linear Carpentarians, related to the aborigines of India. The Murrayians arrived in glacial times and are represented by the Wadjak, Aitape and Keilor skulls. The Carpentarians in northeastern Australia arrived in early post-Pleistocene times.

Bean (1910) described a man of Australoid type from near Taytay, Luzon (Philippines). He had protruding brow ridges, prominent cheeks, a wide, massive nose depressed at base, and full lips. His C.I. was 73.7, N.I. 102.2, stature 156.8 cm. The mandible was massive with a square ramus and a receding chin. From the photograph, the hair looks more Tasmanoid than Australoid. Bean found this Australoid type in all parts of the Philippines. In a study of Filipino students (Bean, 1909a) he found 30 of Australoid type among 377, but all except one had straight hair. He concluded (1909b) that the Australoid was one of the aboriginal types in the Philippines, the other being presumably Malayan.

While in India in 1938, I took a number of photographs of native tribes, when arranging to have them blood-grouped. These arrangements were unfortunately terminated by the war. A few of these photographs are reproduced here. Plate XI (a) shows a group of men from the Kanikar jungle tribe, which lives in a malarious area near the coast of Travancore, a native state in southern India, in the foothills at an altitude below 2000 feet. The Deccan or southern tableland of India appears to have been the original home of these Australoid tribes. A century ago they made up a third of the population of India and they now number probably over twenty million, or one-twentieth of the population.[3] Krishna Iyer (1936) gives the measurements shown in Table 11.

Iyer records in the Kanikars rare cases of albinism, with white skin, yellow hair, and dark blue iris. He finds a Negrito strain in South

---

[3] These numbers are only approximate. The Indian census, in 1929, gave 9,774,611 as animist in religion (Santals, Bhils, Gonds, etc.). Many jungle tribes are being transformed into Hindu castes and some are being Christianized.

TABLE 11
MEASUREMENTS OF AUSTRALOID TRIBES IN INDIA

	Number	Stature	Cephalic Index	Nasal Index
Kanikars	189	152.9 cm.	74.2	89.6
Kurubas	50	163.6	77.3	73.5

India, in Kadars and Pulayas of Cochin as well as some Uralis and Kanikars with frizzy hair. The Negrito element is believed to have been early displaced by proto-Australoids. Some of these tribes formerly used the boomerang, which was in use in early Egypt but has only survived in the Australians. An early account of aboriginal tribes in the Nilgiri Hills of southern India is that of King (1870). It includes the Todas, Kurumbas, Khotas, Erulas, Niadis, and Vadacas.

Note the conspicuous superciliary ridges of the second from the left (in Plate xi, a), and the projecting malars and zygomatic arches of three others, also the very wavy hair in all except the man on the left, who may have some Malay or Mongoloid ancestry. Their trunks appear hairless, unlike the Australians. Plate xi (b) shows a group of Kanikar women and a child with curly hair. Plate xii (a) represents several families of Pulayas in front of their thatched dwelling in the suburbs of Trivandrum, the capital of Travancore. The wavy hair and slightly negroid features (broad nose and somewhat thick lips) are characteristic. The Betta Kurubas (or Kurumbas) are in the elephant jungles of Mysore at Bandipur. They are valuable trackers in keeping guard over the herds of wild elephants. A drum and three or four crude wooden flute-like instruments constituted their band. As their photographs had to be taken at noonday when the sun was directly overhead, they were not good enough for reproduction. Plate xii (b), the Uralis, are a tribe isolated in the jungle at the end of a lake in the Nilgiri hills in southern India. The hair is very wavy and the features somewhat Australoid. The chief is on the left. The Uralis build huts in the trees, where they sleep to escape from the wild elephants who sometimes raid their primitive gardens at night. It will be seen that these three jungle tribes—the Kanikars, Uralis, and Kurubas—all show certain Australoid characters in different degrees. They also show some evidence of Negrito ancestry, in the short stature and the somewhat kinky hair.

Local Evolution of Modern Racial Types 153

In a preliminary classification of Indian types, Eickstedt (1933) recognizes two Veddoid (Australoid) types: (1) a Gondid autochthonous jungle people with dark brown skin, wavy hair, and a totemic matriarchal culture; (2) a Malid jungle type with black-brown skin and curly hair, which probably contains a Negrito element.

Mackay in his (1938) excavations at Mohenjo-Daro (Mackay, 1938) in northwestern India, dated the city about 2800–2500 B.C. This civilization was contemporaneous with early Egyptian and Babylonian civilizations, but appears to have grown up to some extent independently. Piggott (1946), however, shows that the closely related Harappa city culture in the Indus valley during the third and fourth millennia B.C. derived its pottery styles from Iranian and Sumerian sources. At Chanhu-Daro there are five occupation levels, the three earliest culture stages being those of an Harappa town. The Amri culture and probably the Quetta ware were earlier than Harappa. While the origin of Harappa is not known, it was connected with Elam and Sumer, probably by sea routes. Al Ubaid, in Iraq, preceded the earliest Sumerian dynasties and is antedated by agricultural villages having no metal and dating from about 5000 B.C. Al Ubaid belongs to the period of the flood at Ur, about 4000 B.C.

Guha and Basu (in Mackay, p. 613) studied fifteen skeletons, nine of them probably killed by raiders, from excavations at Mohenjo-Daro. Two races are indicated, with some mixture between them. Race A had a big brain, long head (C.I. 70.5), high cranium, and two individuals at least had prominent brow ridges, a deep subglabellar notch and enormous post-auricular occipital development. Keith demurs to calling these proto-Australoids because of their affinity to Sumerians and Caucasians. They were broad-nosed but not Veddoid, since Vedda skulls are smaller and less massive. The frontal area was low, the occiput large for attachment of neck muscles. Race B, with a high narrow nose, is Mediterranean (Brown), the same as now inhabiting the Indus valley. One skull was of type A but with a high narrow nose, indicating racial mixture of the two types.

That more primitive genes are carried in the germplasm of some Australian aborigines is shown by the description of a skull by Burkitt and Hunter (1922). The skeleton of this aboriginal woman, Jenny, was exhumed in 1880. There is an enormous development of the superciliary ridges and glabella, more massive than in any other modern skull, resembling the "Neanderthal" (Javanthropus) condition and having no ophryonic groove over the orbits. The forehead was exceedingly low and sloping and the occipital torus was massive, having a pithecoid conformation with lateral tubercles. This Neanderthaloid cranium was combined with a modern facial skeleton, orthognathous with a leptorrhine nose (N.I. 47.2). The skull length was 203 mm., C.I. 65, cranial capacity 1211 cc. The bones of the vault were very thick and there were resemblances to Pithecanthropus.

A figurine from the prehistoric city of Mohenjo-Daro shows an Australoid face. The skulls from here were both dolichocephalic and brachycephalic (Marshall, 1931), one being proto-Australoid. The islands of New Britain and New Caledonia also have some of the Australian type. In the Malay peninsula, skull fragments were found in a shell-mound and described as Australoid by Huxley many years ago. In 1889, Dubois and van Riedschoten found skeletons of a man and woman in a mountain slope near Wadjak, Java. They were tall and dolichocephalic, characterized as proto-Australian.

It is significant in this connection (see Gates, 1946) that the Australian aborigines have about 40 per cent of the A blood group, 60 per cent O, and very little or no B, except in the northeast nearest Papua (the Papuans being high in B). The Paniyans (another jungle tribe in southern India) were found to have 62 per cent A and only 8 per cent B, whereas surrounding Hindus show 30 per cent or more of B. This is a case where the blood groups strongly support the evidence of physical anthropology, confirming an ancestral relationship between the Australian aborigines and these surviving pre-Dravidians of India.

Six skulls were excavated in earthenware urns sometime before 1900 in the Tinnevelly district of southern India, and were deposited in the Madras Museum. They are regarded as proto-Dravidian, somewhere between 400 and 4000 years old. Zuckerman (1930) found one to be Australoid, but not of the most primitive character; the other described was Mediterranean. Mixture of Australoids with northern invaders had already occurred. Further excavations in southern India and further determinations of the blood groups in the jungle tribes should throw more light on the history of the south Indian and Ceylon peoples.

In parts of southern Arabia and along the shores of the Indian

Ocean all the way from the Straits of Babel Mandeb to the mouth of the Indus, Coon (1939, p. 425) finds evidence of a Veddoid element with Australian affinities, more or less modified by crossing with the native populations. Whether this represents a very ancient or a later extension of the Veddoid type is not at present clear. In their account of the Veddas of Ceylon, the Seligmans (1911) found them less primitive than the Australians. The base of the nose was less depressed, the nose less broad, the chin more pointed, but the hair long and wavy. The cephalic index ranged from 64.9 to 75.9, with the mean at 70.5. The women have much less ugly faces than the Australian women. The coastal Veddas have a large mixture of Tamil blood.

The question of the relation between the Australian and Tasmanian aborigines need not be entered into in detail here. The latter had at least one Negrito character in the woolly hair. They were of medium height, the skin almost black, eyes small, deep-set, nose short and broad, mouth and teeth large, hair black and woolly, the beard peppercorn around the border. The skull was dolicho- to mesocephalic, pentagonal (Sergi), mean cephalic index 74.9, somewhat broader than the Australian (ranging from 69.1 to 79.9), cranial capacity 1060–1430 cc., mostly about 1250 cc. Wood Jones and Campbell (1924) described six Tasmanian skulls with cranial capacity 1100–1320 cc. Wood Jones (1935) concludes that their average cranial capacity was 1353 cc., some 50 cc. larger than the Australians. Among 60 skulls, 12 were chamaecephalic, 42 orthocephalic, and 6 hypsicephalic (Turner, 1908). The glabella and superciliary ridges were marked in some and slight in others. The nose was short and concave, platyrrhine, the N.I. ranging from 49.1 to 69. They were megadont like the Australians and some were prognathous. Photographs (see Wood Jones, 1935) show that they had conspicuously large ears, generally with a large lobe. They frequently had a retreating chin, the lips were full but not thick and the mouth large like the Australians. Wood Jones shows that the authenticated skulls indicate a pure race. He is probably correct in concluding that the Tasmanians are descendants of a few waifs cast on the shores—an early Melanesian group modified through long isolation. He assumed that they came from New Caledonia but it is now highly probable that they came from Australia. They were shorter and stouter than the Australians. Like some other

Melanesians, they had the habit of wearing a human jaw attached to a necklace around their neck. In an account of their stone culture, Pulleine (1929) calls their simple stone implements archaeoliths, meaning tools, especially scrapers, made from formed flakes with marginal chipping but not of conventional form. They were generally chipped on one side only, away from the bulb of percussion.

Elliot Smith (1911) described a Tasmanian brain. The original population in Tasmania is estimated at 7000 (Crowther, 1934). Their stone implements resembled the Mousterian (Roth, 1890) and their simple culture has been classed by some as Mesolithic. There was no hafting, but wooden spears pointed and hardened in the fire were thrown, and the waddy was a club thrown like a boomerang. They made baskets of reeds and grasses, using a method exactly like that of the Swiss Lake Dwellers, with types of twisting like some of the stitches in fancy work. Primitive canoes were made from three rolls of Eucalyptus bark, and pitchers from kelp. Their habitations were simple wind-breaks, and some wore a kangaroo skin over the back.

Berry and Robertson (1910) made a study of forty-two newly found Australian calvaria. In a comparison with other skulls they concluded that they were much more advanced in an evolutionary sense than Neanderthal man, they were also higher than the Kalmucks but lower than the Veddas or Cro-Magnon man. In a further analysis they concluded that Tasmanian and Australian came from a common Pliocene or Pleistocene stock and that the latter were a cross between Tasmanian and either Polynesian (Sergi) or Dravidian (Mathew), neither being directly related to Neanderthal man. The Tasmanians they regarded as a pure race formerly widely distributed in the Pacific, while the Australians, as hybrids, were more variable. The conclusion that the Australians were a cross between Tasmanians and a Dravidian (Australoid) type has now been abundantly confirmed. Wood Jones (1929) drew norma for the Tasmanian skull, based on measurements of fifty. He found the cranium of a somewhat less low type than had been supposed.

In a study of the natives of Arnhem Land, Howells (1937) emphasizes the remarkable homogeneity and the archaic character of the Australian race. From its isolated origin, tracing back directly to Pithecanthropus, we are fully justified in recognizing it as a separate species, *Homo australicus*. It is in effect a survivor from an earlier evolutionary level. Birdsell (1940) has recently found in the Cairns region at the base of the Cape York peninsula in northeastern Australia a bloc of tribes which deviate widely from the Australian norms and show a marked affiliation with the Tasmanians. They inhabit an elevated tableland covered with tropical jungle. The full account of this work will be awaited with interest. It supports the view that the Tasmanians passed through eastern Australia before reaching Tasmania. If the Keilor skull is also partly Tasmanoid, this would be supporting evidence of the same movement, and of the accompanying mixture which would probably take place between the two types. In a later communication, Birdsell (1941) finds an element, most resembling the Veddas of Ceylon, in natives of Arnhem Land, and, in the Murray River basin, a group showing Ainu resemblances. He tentatively concludes that Australia had three invasions: earliest, a branch of the Oceanic Negritos, then a primitive Ainu-like people, and finally a dark-skinned, linear Veddoid stock, the present population showing different proportions of these elements in different parts of Australia. It may be pointed out that the Ainu and the Australians agree in having a high frequency of the N blood type, but the Ainu are also high in B.

Tindale (1940), in a field survey of all the Australian tribes, finds a high correlation between tribal limits and ecological and geographical boundaries. Such features as mountain ranges, watersheds, rivers, straits, peninsulas, plant associations, and micro-climatic zones often furnish stable boundary lines. In the deserts, waterholes are frequently tribal centers, waterless tracts delimiting many tribal boundaries. In the Cairns district of Queensland about a dozen small tribes of Tasmanoid peoples are known; they are described by Tindale and Birdsell (1941). About 1500 inhabit an area of dense jungle. They are of relatively short stature and have crisp, curly hair. Some of their burial customs (including the use of red ocher) and material culture (basket ware) are closely similar to those of the Tasmanians. They apparently are the product of hybridization with the southern type of Australians.

The occasional occurrence of tow or ginger hair in Australian aborigines is well known. Hrdlička (1926) records several cases of light tawny hair in children of pure aborigines in different parts of Australia. It is probably a local mutation. There is evidence that in some at least the hair begins to darken while they are still in childhood.

In one small child with ginger hair the roots of the hair were already brownish-black.

Since the Tasmanians are now extinct, their blood groups cannot be taken, but there are a considerable number of Tasmanian-white mixed-bloods on islands off the southeast coast of Australia and adjacent to Tasmania.

TABLE 12
BLOOD GROUPS AND TYPES OF AUSTRALIAN ABORIGINES

	No.	O	A	B	AB	No.	M	MN	N
Australians	805	53.2	44.7 *	2.1	0	730	3.0	29.6	67.4
Tasmanian x White	31	38.7	61.3	0	0	31	9.7	51.6	38.7
Trihybrids	49	59.2	38.8	2	0	49	16.3	47.0	36.7

* All A_1.

As shown in Table 12, from Birdsell and Boyd (1940), thirty-one Tasmanian-white hybrids, who are descended from six aboriginal Tasmanian women, had no B and were even higher than the Australians in A. The Tasmanians, if they differed at all from the Australians in blood groups, may have had an even higher percentage of A. The trihybrids were mixtures of Tasmanian, Australian, and white. The very high frequency of the N blood type in Australians is shown in the Table.

On Kangaroo Island, off the coast of south Australia, primitive stone implements are found, which Tindale (1937) suggests are similar to those first brought to Australia from Malaya by men of Tasmanian race. There were no dingoes on this island. It is to be hoped that the Tasmanoid mixtures on this island will also be blood-grouped.

Graydon and Simmons (1945) have recently obtained the following blood groups from Papuans:

Number	O	A	B	AB	M	MN	N
200	45.5%	28.5	19	7	7	24	69

It is significant that, as with the Australian aborigines, the A were all of subgroup A_1, all were Rh positive and the frequency of N was extremely high. Other recent findings show that the subgroup A_2 is practically absent from eastern Asia and the Amerinds, and these peoples are also practically all Rh positive. By contrast, in Caucasians some 22 per cent of the A's are A_2 and in Negroes the proportion is still higher. The Rh subgroups already show important racial differences. Thus Rh_0 reaches 40 per cent in Negroes but only 2 per cent in whites; Rh_I is only 20 per cent in Negroes, over 50 per cent in whites, 60 per cent in Chinese, still higher in Indonesians and Filipinos.

As regards the brain of Australian aborigines, Woollard (1929) made a study of the anatomy of four brains preserved in 10 per cent formalin, three of which may have been of mixed descent. He found them distinguished from Europeans in three main features: (1) the brain and cerebral hemispheres are smaller; (2) the paracalcarine fissure or sulcus lunatus is retained; (3) the insulae of Reil tend to be exposed. The skulls were ultradolichocephalic. In a later study of the growth of the Australian brain, Woollard (1931) found that in the newborn infant the occipital area is most developed; the frontal area lags behind and the temporal area is still more retarded, compared with Caucasian infants. The cerebral cortex, examined histologically, shows the cell stratification less well established than in whites, except in the visual area. Of course the aborigines are well known for the keenness of their vision. The cortex was also thinner than in whites everywhere except in the visual area, and the number of cells in the cortex was fewer.

The brain of a Bushwoman was studied in a remarkably clear paper by Marshall (1864), and seventy years later Shellshear (1934) published the original photographs of this brain with a more detailed characterization based on modern knowledge, in comparison with the Australian and the Chinese brains. He found the sulcal pattern (Plate XIII) of the Bushwoman brain very primitive and simple, the main features being: (1) in the occipital region a wide lateral extension of the area striata with fully developed sulcus lunatus and interstriate sulcus of primitive form, the sulcus occipitalis anterior clearly defined and failure of development of the cuneal and precuneal areas on one side; (2) in the parietal region the branching of the posterior part of the parallel sulcus is comparable with many gorilla and chimpanzee brains; (3) the temporal lobe is under-developed, narrow and lacks fullness, failing to extend forward to cover the posterior part of the orbital surface of the frontal lobe, the island of Reil being exposed on both sides; (4) the frontal region is primitive in form and in sulcal pattern. There is a narrowing of the frontal lobes, with prominent antero-external angles very similar to the condition in Pithecanthropus.

In a further detailed study of the Australian brain, Shellshear (1937) finds the striate cortex vertically folded, frequently in association with a broken condition of the sulcus lunatus. For the Australian aborigines as a whole he finds that "the new cortical areas are not developed to the extent that they are in higher races." The most fully developed of Australian hemispheres would look ill-formed and under-developed compared with a fully developed Chinese brain, but he is uncertain whether one would see a definite difference between the most highly developed Australian and the least developed normal Chinese. He is inclined to think that racial differences in the brain will probably manifest themselves as tendencies, not showing an absolute gap between one race and another. Shellshear mentions that of twenty-five hemispheres in Sydney not all had the sulcus lunatus, but eight in Brisbane all had it. The narrow temporal lobe not extending forward was found to be the same in the Australian aborigines, the Bushwoman, and the Rhodesian skull.

We may add here, in anticipation of the next chapter, further studies of the Bushman brain by Wells (1937). He studied the endocranial casts of seven male and four female skulls. The volume of the brain was smaller than in the Rhodesian skull; it was characteristically long and narrow with short and narrow frontal lobes. The temporal lobe was tapering and isolated from the frontal lobe, the insulae were exposed. The Bushman brain was regarded as intermediate between that of Rhodesian man and a European. The larger Boskop brain is more primitive than the Bush, the latter being juvenile in morphology, like a Bantu infant. The small size results from defective expansion. Wells concluded that "the Bushman must be considered definitely inferior in cerebral development to the European."

In this careful study the endocranial volume was found to vary from 1250 to 1350 cc. for males and from 1000 to 1150 cc. for females, the means being 1281 cc. for males and 1062 cc. for females. All were microcephalic. The estimated brain weights (using Kappers' formula) were as shown in Table 13.

Thus the brain weight of the Bushmen and the Australian aborigines is the same; the Basuto brains were scarcely larger, but in other Bantu they were definitely and markedly larger. It is not without significance

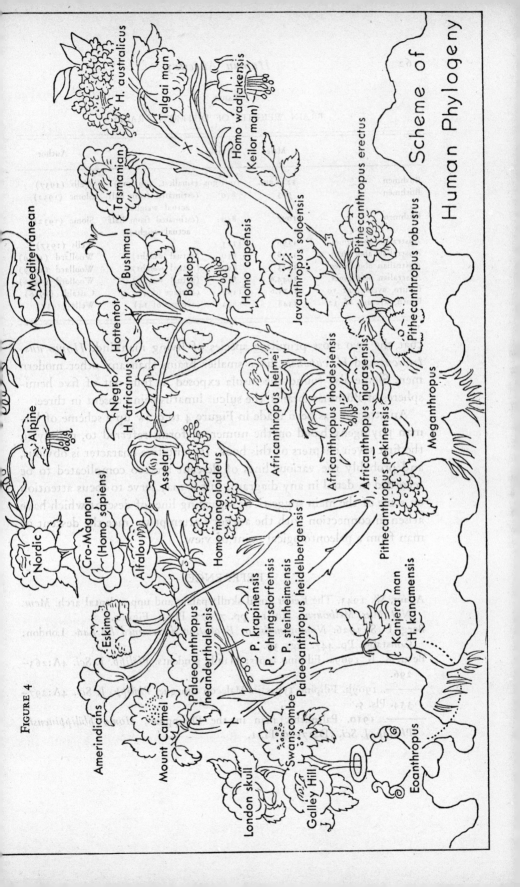

FIGURE 4

TABLE 13
BRAIN WEIGHTS OF DIFFERENT RACES

	Male	Female		Author
Bushmen	1212 gm.	1005 gm.	(smallest, 949 gm.)	Wells (1937)
Bushmen	965	850	(estimated from half actual weight)	Slome (1932)
Bushmen	1097	850	(estimated from half actual weight)	Slome (1932)
Australian aborigines	1196	1123		Wells (1937)
Australian aborigines	1138	850	(actual weight)	Woollard (1929)
Australian aborigines	1123	850	(actual weight)	Woollard (1929)
Australian aborigines	1044.6	850	(actual weight)	Woollard (1929)
Basuto, average of 10	1215.6	1165	(average of 4)	Castaldi (1936)
Bantu, average of 32	1348	1249	(average of 14)	Wells (1937)

that the two most primitive species of living mankind, *Homo australicus* and *H. capensis*, have smaller brains than any other modern men. Slome also found the insula exposed in four out of five hemispheres of Bushman brains, the sulcus lunatus being absent in three.

An attempt has been made in Figure 4 to draw up a scheme of human phylogeny based on the numerous forms referred to, mainly in the first seven chapters of this book. Its tentative character is obvious, and probably the various lines of descent are too complicated to be recorded in detail in any diagram; but it may serve to focus attention on some of the main suggestions regarding lines of descent which have arisen in connection with the author's attempt to trace the descent of man from a paleontological point of view.

REFERENCES

Adam, W. 1943. The Keilor fossil skull: palate and upper dental arch. *Mem. Nat. Mus. Melbourne*. No. 13. pp. 71–77. Pls. 2. Figs. 4.

Berg, L. S. 1926. *Nomogenesis, or Evolution determined by Law*. London: Constable. Pp. 447. Figs. 33.

Bean, R. B. 1909a. Filipino types: Manila students. *Philipp. J. Sci.* 4A:263–296.

———. 1909b. Filipino types in Malecon morgue. *Philipp. J. Sci.* 4A:297–334. Pls. 5.

———. 1910. Paleolithic man in the Philippines, *Homo philippinensis*. *Philipp. J. Sci.* 5D:27–29. Pl. 1.

Berry, R. J. A., and A. W. Robertson. 1910. The place in nature of the Tasmanian aboriginal as deduced from a study of his calvarium. Part. I. *Proc. Roy. Soc. Edinb.* 31:41–69. Fig. 1. Part II. *Proc. Roy. Soc. Edinb.* 34:143–189. Figs. 10.

Birdsell, J. B. 1940. The Tasmanoid tribes of northeastern Australia. *Am. J. Phys. Anthrop.* 27:supp. p. 16.

———. 1941. A preliminary report on the trihybrid origin of the Australian aborigines. *Am. J. Phys. Anthrop.* 28:No. 3, Supp. 1. p. 6.

———. 1947. New data on racial stratification in Australasia. *Am. J. Phys. Anthrop.* 5:232.

Birdsell, J. B., and W. C. Boyd. 1940. Blood groups in the Australian aborigines. *Am. J. Phys. Anthrop.* 27:69–90. Figs. 2.

Burkitt, A. St. N., and J. I. Hunter. 1922. The description of a Neanderthaloid Australian skull, with remarks on the production of the facial characteristics of Australian skulls in general. *J. Anat.* 57:31–54. Pl. 1. Figs. 5.

Castaldi, L. 1936. Studi su encefali di Basuto. *Scritti Biologici.* 11:339–344.

Coon, C. S. 1939. *The Races of Europe.* New York: Macmillan. pp. 739. Pls. 46. Figs. 38. Maps 16.

Crowther, W. E. L. H. 1934. The passing of the Tasmanian race. *Med. J. Austral.* 1:147–160. Figs. 4.

Dubois, E. 1922. The proto-Australian fossil man of Wadjak, Java. *Proc. K. Akad. Wetens. Amsterdam.* 23:1013–1051. Pls. 2.

Eickstedt, E. F. 1933. Die Rassengeschichte von Indien mit besonderen Berücksichtigung von Mysore. *Zeits. f. Morph. u. Anthrop.* 32:77–124. Pls. 8.

Elliot Smith, G. 1911. Le cerveau d'un Tasmanien. *Bull. et Mem. Soc. Anthrop. Paris.* 2:442–450. Figs. 11.

Fenner, F. J. 1941. Fossil human skull fragments of probable Pleistocene age from Aitape, New Guinea. *Rec. S. Austral. Mus.* 6:335–356. Pls. 2. Figs. 9.

Gates, R. R. 1946. *Human Genetics.* New York: Macmillan. Vols. 2. pp. 1518. Figs. 326.

Graydon, J. J., and R. T. Simmons. 1945. Blood groups in the Territory of Papua. *Med. J. Austral.* 32:77–80.

Hambly, W. D. 1946. Craniometry of Ambrym Island. *Fieldiana: Anthropology.* 37: No. 1. pp. 150. Pls. 30. Figs. 7.

Hellman, M. 1934. The form of the Talgai palate. *Am. J. Phys. Anthrop.* 19:1–15. Figs. 8.

Howells, W. W. 1937. Anthropometry of the natives of Arnhem Land and the Australian race problem. *Peabody Mus. Papers.* 16:1–97. Pl. 1.

Hrdlička, A. 1926. Light hair in Australian aborigines. *Am. J. Phys. Anthrop.* 9:137–139.

Iyer, L. A. Krishna. 1936. Anthropometry of the primitive tribes of Travancore. *Proc. Indian Acad. Sci.* 4:494–513. Pl. 1.

Keith, Sir A. 1931. *The Antiquity of Man.* Philadelphia: Lippincott.

King, W. R. 1870. The aboriginal tribes of the Nilgiri Hills. *J. of Anthrop.* 1:18–51. Pls. 3.
Mackay, E. J. H., 1938. *Further excavations at Mohenjo-Daro.* Delhi. Vols. 2. pp. 718. Pls. 146.
Mahoney, D. J. 1943. The Keilor fossil skull: geological evidence of antiquity. *Mem. Nat. Mus. Melbourne.* No. 13. pp. 79–81.
———. 1943. The problem of the antiquity of man in Australia. *Mem. Nat. Mus. Melbourne.* No. 13. pp. 7–56. Pls. 3. Figs. 7.
Marshall, John. 1864. On the brain of a Bushwoman; and on the brains of two idiots of European descent. *Phil. Trans. Roy. Soc.* 154:501–558.
Marshall, Sir John. 1931. *Mohenjo-Daro and the Indus Civilization.* London. Vols. 3. pp. 716. Pls. 164.
Ogilvie, G. S. 1940. The "Che Wong": a little known primitive people. *Malayan Nature J.* 1:23–25. Figs. 8.
Piggott, S. 1946. The chronology of prehistoric north-west India. *Ancient India.* No. 1. pp. 8–26. Figs. 3.
Pulleine, R. H. 1929. The Tasmanians and their stone culture. *Rept. Austral. Assn. Adv. Sci.* 1928. pp. 294–314. Pls. 10.
Roth, H. Ling. 1890. *The Aborigines of Tasmania.* London. pp. 224.
Seligman, C. G. and B. Z. 1911. *The Veddas.* Cambridge University Press. pp. 463. Pls. 71.
Shellshear, J. L. 1934. The primitive features of the cerebrum, with special reference to the brain of the Bushwoman described by Marshall. *Phil. Trans. Roy. Soc.* 223B:1–26. Pls. 3.
———. 1937. The brain of the aboriginal Australian. A study in cerebral morphology. *Phil. Trans. Roy. Soc.* 227B:293–409. Pls. 23.
Slome, I. 1932. The Bushman brain. *J. Anat.* 67:47–58. Pls. 5.
Smith, S. A. 1918. The fossil human skull found at Talgai, Queensland. *Phil. Trans. Roy. Soc.* 208B:351–387. Pls. 7. Figs. 27.
Tindale, N. B. 1937. Relationship of the extinct Kangaroo Island culture with cultures from Australia, Tasmania and Malaya. *Rec. S. Austral. Mus.* 6:39–60. Figs. 16.
———. 1940. Distribution of Australian aboriginal tribes: a field study. *Trans. Roy. Soc. S. Austral.* 64:140–231.
Tindale, N. B., and J. B. Birdsell. 1941. Tasmanoid tribes in North Queensland. *Rec. of S. Austral. Mus.* 7:1–9. Pls. 4.
Turner, Sir William. 1908. The craniology, racial affinities, and descent of the aborigines of Tasmania. *Trans. Roy. Soc. Edinb.* 46:365–403. Pls. 3.
Weidenreich, F. 1945. The Keilor skull: a Wadjak type from Southeast Australia. *Am. J. Phys. Anthrop.* N. S. 3:21–32. Figs. 3.
Wells, L. H. 1937. The status of the Bushman as revealed by a study of endocranial casts. *S. Afr. J. Sci.* 34:365–398. Figs. 6.
Wood Jones, F. 1929. The Tasmanian skull. *J. Anat.* 63:224–232. Figs. 4.
———. 1935. Tasmania's vanished race. Australian Broadcasting Commission. Pp. 32. Figs. 12.

Wood Jones, F., and T. D. Campbell. 1924. Six hitherto undescribed skulls of Tasmanian natives. *Rec. S. Austral. Mus.* 2:459–469. Pls. 3.

Woollard, H. H. 1929. The Australian aboriginal brain. *J. Anat.* 63:207–223. Figs. 9.

———. 1931. The growth of the brain of the Australian aboriginal. *J. Anat.* 65:224–241. Pls. 3. Figs. 4.

Wunderly, J. 1943. The Keilor fossil skull: anatomical description. *Mem. Nat. Mus. Melbourne.* No. 13. Pp. 57–69. Pls. 6.

Zeuner, F. E. 1944. *Homo sapiens* in Australia contemporary with *Homo neanderthalensis* in Europe. *Nature.* 153:622–623.

Zuckerman, S. 1930. The Adichanallur skulls. *Bull. Madras Govt. Mus.* N. S. 2:1–24. Pls. 3.

7

Evolution of Man in South and East Africa

In passing from Australia to Africa we enter a new area, where human evolution has taken a different course—naturally more complicated because of its less complete isolation, but with even more numerous fossil documents to support it. Because of the greater opportunity for crossing, we must expect to find the various types, even in South Africa, flowing into each other to some extent. An amazing wealth of fossil skeletons and skulls from many parts of the African continent, but especially from southern and eastern Africa, throws much light on the evolutionary course of events and shows the complexities of the racial relationships. The skulls of South African races available for study must now number several hundred. Only a few of them can be referred to here. A very active school of anatomists, anthropologists, paleontologists, and archaeologists under the leadership of Broom and Dart has been busy in describing and explaining these remarkable finds.

In South Africa the evolutionary series reads *Africanthropus njarensis* → *A. rhodesiensis* → Florisbad ⟨ Boskop / Bushman ⟩ (*Homo capensis*) → Hottentot, but as we shall see, some of these types overlap and have interbred to a considerable extent. Whether Africanthropus is more primitive than the Rhodesian skull can well be doubted, although it is probably much older chronologically. Dart (1940) believes that Rhodesian man might have been derived from the Australopithecinae, but we have seen that they may be a parallel development.

The next step in our evolutionary series is the Florisbad skull discovered in the Orange Free State twenty-five miles north of Bloemfontein in 1932 by Dreyer (1935), nearly one thousand miles south of Broken Hill. It consists of most of the calvaria with parts of the face and palate, the bones being very thick. I am able to reproduce here, in Figures 5 and 6, original drawings of this skull kindly sent by Dr. R. Broom, and similar views of the Cape Flats skull (see below) as well as the modern Korana and Bush for comparison. The Florisbad skull was found in the eye of a former spring, covered with sand to a depth of twenty feet; this sand containing Mousterian-Levalloisian implements and the bones of many extinct mammals, including *Equus helmei* and *Bubalus antiquus*. It is regarded as at least of mid-Pleistocene age, the oldest human skull yet discovered in South Africa. Dreyer proposed for it the name *Homo helmei*. The skull shows marked resemblance to Rhodesian man (Plate VII, b). It is very large (*ca.* 200 mm. in length), the supra-orbital width being great (136 mm., as compared with 125 mm. in Neanderthal man, 139 mm. in the Rhodesian skull, and 138 mm. in the gorilla), minimal frontal width 120 mm., but differs from the former in that the frontal lobes (in the endocranial cast) are far less pointed and more flattened at the frontoventral end. Dreyer (1938) shows that the frontal lobes (in the endocranial cast) are very different from those of Rhodesian man, the latter differing but slightly from Europeans in this point. The face is extremely prognathous and the cranium is rather flat dorsally. There seems no reason why the Rhodesian-Florisbad series should not have given rise to the African Negro. In the shape of the orbits and the presence of deep depressions under them the Florisbad skull resembles some modern types of man. Perhaps *Africanthropus helmei* is the best name for it. Drennan (1937, 1938) calls it *Homo florisbadensis* and regards it as a variant of Neanderthal man, nearest La Chapelle-aux-Saints. He recognizes, however, the great lateral extension of the supraorbital torus in the intact frontal region. It is surely soundest to compare the Florisbad skull with Rhodesian man on the same continent. Anthropology appears to have suffered at times from too wide comparisons, between forms that are so far separated by geographic distance or barriers that they are unlikely to be genetically related. Keith (1938) regards the Florisbad skull as transitional between the Rhodesian and the Boskop type, the vertex resembling the Boskop skull while the malars and the facial part of the maxilla are like those of the Bushman. Galloway (1937) also recognizes Florisbad man as the pro-

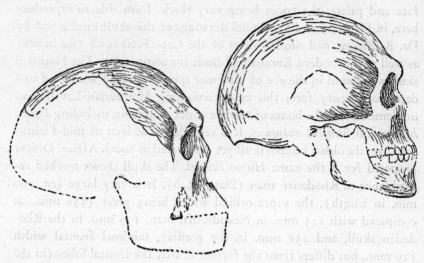

FIGURE 5. Side view of a) Florisbad and b) Cape Flats skull.

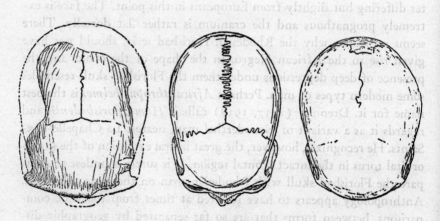

FIGURE 6. Vertical views of a) Florisbad skull, b) Korana male, c) Bush female skull.

genitor of Boskop. He finds resemblances to Wadjak man in Java, but there is no need to suppose that these are based on a common ancestry. Galloway finds that the skull agrees with Neanderthal metrically but differs in non-metrical characters, the endocranial cast being very similar to the Rhodesian endocast, but the supra-orbital torus being already broken into three parts, as in the modern Australian aborigines (Galloway, 1937b). Dreyer (1936) concludes that the Florisbad skull is a more primitive member of the same race as the European Steinheim skull. He finds, in harmony with the viewpoint of the present author, that by the Middle Pleistocene human stocks were already evolving independently in Europe, Africa, and Java.

The Boskop skull, found in the Transvaal in 1913 (Dart, 1923), represents another surprise in the anthropology of South Africa, for all the evidence from numerous other skulls found since shows that the big-brained Boskop race was widespread in southern and eastern and even northern Africa and played an important part in the evolution of the Bushmen and Hottentots and quite possibly of other races. In fact, we may regard the Hottentots as the modern descendants of Boskop man. The Bushmen, though closely related, are coeval with the Boskop race in early post-glacial times and they or their derivatives may have entered southern Europe. Dwarfism still persists in the southern Italians, but is probably not of Bushman origin.

A farmer unearthed the Boskop skull fragments near Potchefstroom. The original consisted of a right parieto-occipital and three other pieces which articulate to form a large part of the left side of the skull. The bones are very thick. Further excavations (Houghton, 1917) a year later yielded part of a mandible and fragments of limb bones. The skull was of enormous size, far larger than in modern races, the length being about 210 mm. and the width estimated at 150 mm. The cranial capacity has been variously estimated, 1832 cc. by Houghton, 1960 cc. by Broom, and 1750 cc. by Dart, although the skull is probably female. The height and width of the skull is greater than in Cro-Magnon man, who was a fine specimen of humanity. The forehead was practically vertical, the supra-orbital ridges moderate, the occiput projecting strongly with a torus apparently broken into parts. No artefacts were found with the skull, but the age is probably Pleistocene.

Broom (1918) named the skull *Homo capensis*. This specific name

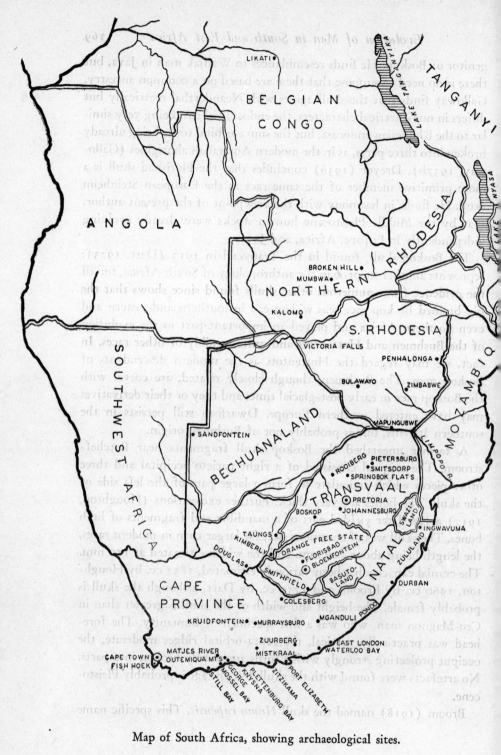

Map of South Africa, showing archaeological sites.

may well apply not only to Boskop man but to his modern descendants, the Hottentots, as well as the Korana and Bushmen. In other words, it applies to all the people of South Africa with yellowish skin and peppercorn hair. From a study of the endocranial cast, Elliot Smith emphasized the great size of the brain in contrast to the modern natives of South Africa, and the curious blend of primitive and modern features. The general form resembled that of Neanderthal man, yet in essentials it was like Cro-Magnon, but it has a depression in the interparietal region. Elliot Smith and Broom both recognized that this macrocephalic African type may be ancestral to both Cro-Magnon and Neanderthal man, since it has features of both. In that case, the later and more specialized phases of Neanderthal man would represent a type which became extinct. In the same way, the modern specialized Bushmen type is now nearing extinction.

The Boskop type of skull is well represented by the skull (Plate XIV, b) of a modern Hottentot from the dissecting room in Capetown (Galloway, 1937). It had a cephalic index of 68.7, a length of 211 mm., and a cranial capacity of 2000 cc. Galloway regards it as an almost complete "throwback" to the Boskop type. From a genetical point of view this enormous skull is the result of inheritance, but further analysis is necessary to determine how that inheritance has come about. It should be possible to determine, for instance, whether it is dominant or recessive to the Bushman, and whether it behaves as a unit in crosses. These large skulls are interpreted as an expression of infantilism. The endocasts show that the brain is deficient in the regions generally considered necessary for intelligence.

The next fossil discovery of the Boskop type was by Mr. Fitz-Simons at Zitzikama in one of the rock-shelters on the coast near Port Elizabeth. Three skulls were described in a preliminary account (Lang, 1924). Two of them (Za1 and Za2 in Table 14) were of the light

TABLE 14
CRANIAL CAPACITY OF BOSKOP SKULLS

	Cranial Capacity	C.I.
Za1	1270 cc.	77.8
Za2	1280 cc.	78.5
Za3	1550 cc.	75.8

Bush type while the third (Za3) was larger and primitive, with heavy supra-orbitals and prominent parietals. They were all mesocranial. Za 1 showed evidence of hybridization between the Bush (San) race and one which had heavy supra-orbitals and prominent parietals. Later, Laing and Gear (1929) reported on eight skulls and a complete skeleton, probably of a chief, in a sarcophagus of stones. Usually the body was placed on its side with knees flexed and two or three large flat stone slabs were laid over it. The remains were fairly recent, belonging to the kitchen midden or Strandlooper period and within fifteen feet of the surface. The teeth, even in that skull which was of a child ten years of age, were much ground down.

From the deeper level at Zitzikama, Laing and Gear (1929) found that the early inhabitants of the midden were of Boskop type. The larger group of skulls confirmed that the later dwellers were of mixed race. Two were Boskopoid, four intermediate in some characters and Boskop or Bush in others. Gear (1925) made a study of further skeletal materials collected by Mr. FitzSimons from the midden floors of several rock-shelters near Port Elizabeth. These included a very broad shoulder blade, a very massive hipbone, a vertebral column, with a lumbar curve only, an anthropoidal type of sacrum, and a femur with excessive platymery and pilastry. Thus could be reconstructed a race of man which was neither Bushman, Negro, nor European. Gear (1926a) also showed that six skulls from the deeper level at Zitzikama were of Boskop type. He described (1926b) a Boskop skeleton from Kalomo, Northern Rhodesia, which was found at a depth of six feet, associated with pottery fragments, two hundred yards from an old native kraal.

The Kalomo skeleton included thick skull fragments, an incomplete mandible, portions of both femora and both tibiae, an incomplete sacrum, parts of both humeri, a portion of an ulna, and fragments of pelvis, scapula, ribs, and vertebrae. Although not prehistoric, the skeleton shows a mixture of Boskop and Negro elements. The mandible is massive and the occipital, frontal, and parietal fragments conform to the Boskopoid type. The mandible articulated perfectly with an upper jaw arcade from Zitzikama (see below), but departed from the Boskop type in the large (Negroid) mastoid process and the digastric groove of the temporal bone. The huge sacrum was not pure Boskop, the tibiae showed no platycnemy and the femora were pilastric, the tibio-femoral length index being Negroid. The mandible had a broad ramus, a thick body, and primitive symphysis, the glenoid cavity also being primitive. This skeleton shows the persistence of Boskop characters in the native races of South Africa.

The people in the forested region near Port Elizabeth lived in a paradise for primitive hunters. There was an abundance of fruits, berries, edible roots, and honey as well as wild animals, but their animal subsistence was largely shellfish. Mr. FitzSimons (1926), director of the museum at Port Elizabeth, in an expedition to a large cave a few miles inland in the Outinequa range, found fifty-one skeletons by excavating the cave floor. The burials were around the cave margins while it continued to be occupied. With the skeletons were placed crude stone weapons, ostrich-shell drinking vessels, stone mortars, and pottery. They evidently made both Paleolithic and Neolithic implements. Notwithstanding their fine physique and big brains, they were overcome by the dwarf Bushmen, who had bows and poisoned arrows. Skeletons of both types are found side by side. Probably the men were killed and the women taken as wives, because later generations were evidently hybrids at least in part. But we do not know how the later mixtures would segregate genetically. Evidently the term Strandloopers should be given a geographic and not a racial significance. In Plate xv are compared a skeleton of the tall cliff dwellers of Zitzikama and a typical inland Cape Bushman. There is a difference of five inches in height.

In an important paper, Dreyer, Meiring, and Hoffman (1938) have introduced genetical conceptions into the interpretation of South African skull shapes. They contend that the Boskop skull is not an ancient ancestral type but the product of hybridization. They apply their analysis to a variable series of skulls recently obtained from Kakamas, which are mostly Hottentot-like, and to the Matjes River series (see below) which are mostly Bush-like. The increased cranial size and decreased facial size, which constitute the main features of pedomorphism, they represent by a gene, D. Another gene, S, determines a long cranium. Thus the Hottentots are SSdd, the Bushmen, ssDD. Hybridization and recombinations give SSDD, SsDd, SSDd, and SsDD, all larger heads than either parent race but of diverse sizes. The segregates ssDD, ssDd, and ssdd represent the small (more Bush-like) crania. On this basis the Matjes River crania approach the ssdd condition, the Kakamas series of graded sizes SSDD. The straight (Bush) or sloping (Hottentot) forehead is another pair of genes. They find that frontal breadth (F) and parietal breadth (P) are independent Mendelian factors, the ancestral Bush being FFPP (short oval), the Hottentots ffpp (long oval). The hybrid will then be FfPp, later generations producing such segregates as FFPP, ffpp, FFpp, Ffpp, ffPP. In this nomenclature the original Boskop skull is SSDDffPP, the broad parietal region making the skull pentagonoid. The large Boskopoid skulls (length 200 mm.) on this basis may be SSdd or SsDd, the small ones ssdd. The Boskop skull, and several other large ones like it which have since been found, would thus represent a recombination hybrid type from Bush x Hottentot rather than a race *sui generis*. Some of the derived hybrid types could, of course, give rise to sub-races if they multiplied in relative isolation. The later Boskop and Boskopoid types would represent some of these hybrid combinations which have become relatively homozygous in some of their skull characters through inbreeding, rather than an independent race ancestral to the Hottentots and Bush. The Alfalou type in northern Africa may be such a derived type.

In a more recent paper, Cassel, Harris, and Harris (1943) make a detailed comparison of three sets of crania, showing that many characters are independently inherited, but without assigning to them genetical formulae. One series of twelve adult skulls were from the southwest Kalahari, a second series (twenty-two crania) from southwest Africa, these representing two recognized groups of Bushmen. The third (seventeen crania) from Knysna, Cape Province, represent a Strandlooper population. The tracing of the various characters cannot be detailed here, but a few examples may be cited. In the Kalahari group, seven skulls were dolicho-, one hyperdolicho-, two mesati-, and two brachycranial. All twelve were microcephalic, but six were chamae-, five ortho-, and one hypsicephalic. Seven showed trigonism and seven had the ophryonic groove, ten were platyrrhine, one leptorhine. The group as a whole showed a mixture of Bush and Boskop characters, the latter predominating, while certain skulls showed some Negro features. In the other Bushman group of skulls, fourteen were dolicho-, eight mesaticephalic; six were chamae-, eleven ortho-, one hypsicephalic; eighteen were acrocephalic; eight meso- and fourteen microcephalic. By contrast, in the Knysna group seven were dolicho-, six mesati-, three brachy-, and one hyper-brachycephalic, the C.I. ranging from 70 to 87. Two were chamae-, fourteen ortho-, and one hypsicephalic, the auricular height index 56–64. As regards the nose, seven were platy-, four meso-, and one leptorrhine, N.I. 62–48.

The origin of the brachycephaly in these skulls seems to present the same problem as in other parts of the world. All the Knysna brachycrania have foetal parietal bossing with maximum cranial breadth high in the parietal region, whereas in non-African brachycephaly the bossing is rare and the maximum cranial breadth tends to be lower down. It seems clear that Mongoloid ancestry has played no part in the evolution of this early South African brachycephaly. Galloway (1936), in a study of several skulls from the Natal coast showing Bush-Boskop mixture before the Bantu came in, has found that a relatively broad skull could result from crosses between the small Bush head-length and the large Boskop head-breadth.

As regards height of skull, the Knysna Strandloopers tend to be orthocranial, the southwest African Bush hypsicranial, and the Kalahari Bush chamaecranial. Prognathism is found only in the southwest African group; larger cranial capacity in the Knysna skulls. All these group differences can thus be referred to differences in the representation of genes, depending on the proportions of Boskop, Bush, and Negro ancestry, the Knysna group having apparently no Negro genes. Studies of this kind show that there is a good prospect for the ultimate analysis of all cranial characters in terms of genic factors.

Kohler (1943) has recently made a study of nineteen adult skulls from kitchen middens in the vicinity of Port Elizabeth. The earliest were the prehistoric Strandloopers—Bushmen with some Boskop (more robust) features, but less Boskopoid than the Zitzikama skeletons. Later came Negro infiltration. The Hottentot tribes in this area were the Damaqua and the Damasonqua, Dama meaning black. They were mixed, with Negro characters. These skulls thus furnish a demonstration of population changes from prehistoric to historical times, the last skulls being nearly or fully Negro. At the end of the sequence, some individuals were Eurafrican, probably following the European settlement at Port Elizabeth.

There was now abundant evidence that the Boskop type had extended at least from the Transvaal to the Cape and that they had mixed with the Bushmen.

A diversion is necessary at this point, to refer to the remarkable discoveries in East Africa, especially by Leakey (1931, 1935). In this great area there were pluvial (wet) periods when the climate was cooler, corresponding to the glacial periods of Europe. The permanent glaciers in East Africa are now at an altitude of 14,000 feet, but in pluvial periods they come down to 10,000 feet on Mount Kenya and Kilimanjaro. Low level areas which are now desiccated were, in the pluvial periods, fertile and sustaining a large animal and human population.

Leakey (1931) recognizes seven phases corresponding roughly with those in Europe. (1) A long (Kamasian) pluvial period when the ice was down to 10,000 feet on the mountains. In this long Lower and Middle Pleistocene period rough (Oldowan) choppers represented the only culture. They evolved into crude (Chellean) hand-axes in the latter part of this period. (2) A dry period with great earth movements, during which the eastern branch of the Great Rift Valley was formed and the Victoria Nyanza, which formerly stood at a much higher level, became the source of the Nile. This marked the end of the Middle Pleistocene. The high beaches of Lake Victoria frequently contain implements belonging to the Kamasian pluvial period. (3) The second (Gamblian) pluvial, with a wetter climate, equated with the Riss and Würm in Europe. (4) A short arid period. (5) The post-pluvial Makalian wet phase with increased rainfall. Two upper Aurignacian cultures belong to this period. (6) A short arid period. (7) The second (Nakuran) post-pluvial wet phase, about 850 B.C.

In Leakey's chronology, remains of Elephas, Bos, and Equus are regarded as defining the Pleistocene deposits. This makes some "Pliocene" deposits late Pleistocene, for example, the East Anglian sub-crag contains Elephas and so becomes, by definition, Lower Pleistocene, having the earliest Chellean culture. From the Chellean to the end of the Kamasian pluvial is regarded as Middle Pleistocene; while Upper Pleistocene includes the Gamblian pluvial and the Makalian wet phase, the animals being mostly species now living. Leakey recognizes five stages in the evolution of Chellean culture, and six stages in the succeeding Acheulean phase. At the close of the Middle Pleistocene (Kamasian pluvial) there were four contemporary cultures—the sixth Acheulean, early Mousterian, basal Aurignacian, and Nanyukian. There is some evidence that, as in Europe, the Acheulean sequence of stages was accompanied by a Levalloisian sequence.

a. The undeveloped Talgai skull found in Queensland

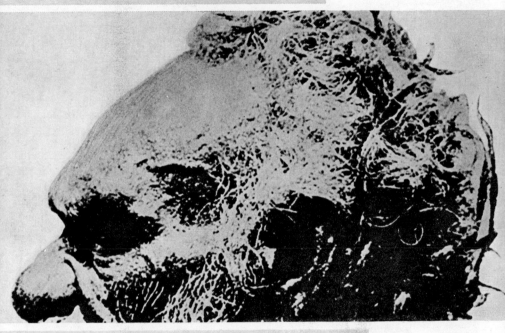

b. Head of an Australian man

c. Skull of an Australian aboriginal woman

A group of Che Wong, a primitive tribe in Perak, Malaya. Note the broad nose, depressed at the base, and the brow ridges and sunken eyes

a. (upper): A group of men of the Kanikar jungle tribe in Travancore, southern India. Note the depressed base of the nose and heavy brow ridges in several, also the wavy hair and rather short stature. The man on the left with the straight hair may have Mongoloid mixture

b. (lower): A group of Kanikar women and a child. Note the somewhat Australoid features and the curly hair of the little girl

a. (upper): A group of Pulayas at Trivandrum on the Malabar coast of India. Note the slightly wavy hair and the slightly Negroid features

b. (lower): A group of Uralis, a jungle tribe in the Nilgiri Hills of southern India. The hair is very wavy. The chief is on the left

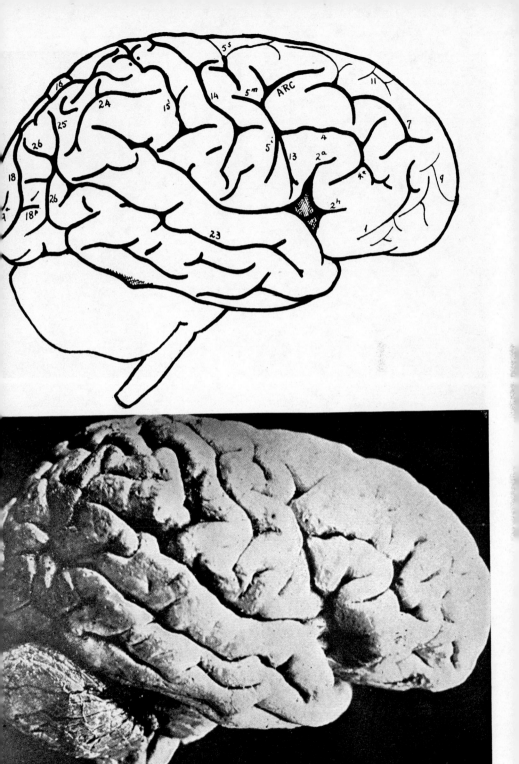

a. (upper): Line drawing showing the sulcal pattern in (b)
b. (lower): Marshall's photograph of the brain of a Bushwoman, showing the simple and primitive sulcal pattern

XIII

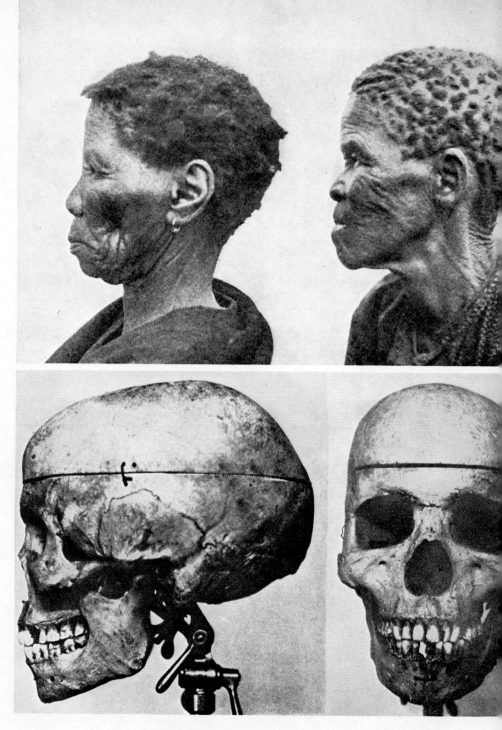

a. (upper): Two old Korana women at Kimberley. Note the peppercorn hair, especially on the right
b. (lower): Boskop type of skull in a modern Hottentot

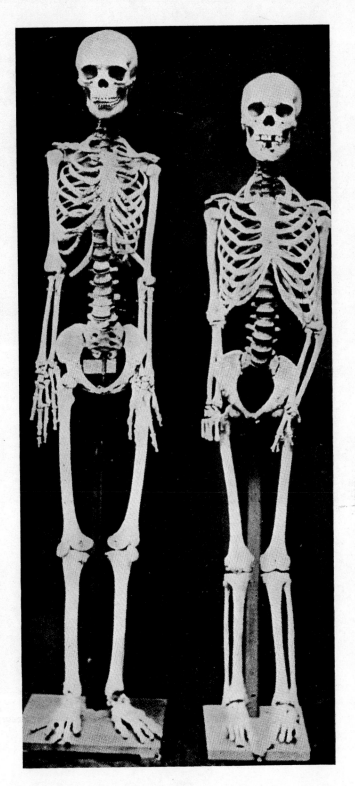

Skeletons of (a) a cliff dweller from Zitzikama, (b) an inland Cape Bushman

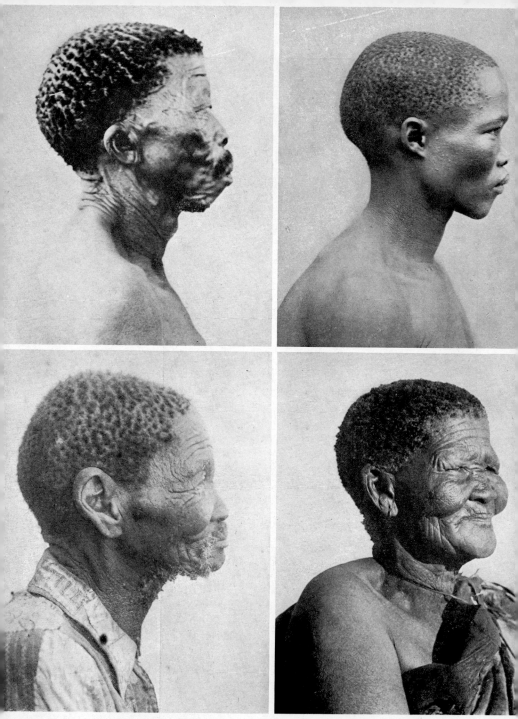

a. (upper left): Typical old Bushman, "Dial," living at Postmasburg. Note the low parietal, the small ear, and the marked prognathism

b. (upper right): Young Bushman from Langeberg. Note the very small ear without a lobe, the low parietal region, and the distinct prognathism

c. (lower left): Typical Korana, Dirk Lucas, at Kimberley. Note the very large ear, the moderately high parietal, well-developed chin, and slight prognathism

d. (lower right): Old Korana woman living at Pnial. Note the large ear, the broad nose, and the very prominent chin

After the earth movements marking the end of the Middle Pleistocene, only two cultures—the Kenya Mousterian and Kenya Aurignacian—survive. The Mousterian gives rise to the Kenya Stillbay culture, which, in the succeeding dry period (4) gives place to the Magosian. In the Makalian wet phase (5) the Upper Aurignacian gives rise to the Elmenteitan and Wilton A cultures, the Wilton B being derived from the Magosian. In the Nakuran wet phase (7) two Neolithic culture groups appear, the Gumban derived from the Wilton, and the Njoroan, possibly from the Sudan. A fairly complete series of cultures in Kenya, from Chellean to the Neolithic, has thus been worked out (see Table 16, p. 191).

In a study of past climates and early stone cultures, Smuts (1938) concludes that the high veldt of the Transvaal must have had four pluvial periods separated by dryer inter-pluvials. The first pluvial probably corresponds with the Gunz-Mindel interglacial in Europe and the last pluvial was followed by two small humid phases in Holocene time. Stone implements are found in all the pluvial periods, but the inter-pluvials are barren. Man evidently migrated elsewhere during the desert phases. The artifacts of the first pluvial are believed to be over a million years old and are regarded as perhaps Pliocene in age.

Before considering the numerous skeletons unearthed in Kenya, some of which were clearly of Boskop type, we will continue the story of developments in South Africa.

Here mixtures of races have occurred in all degrees, so it will be simpler if we now move forward to historical times and then endeavor to fill in the interval with the various mixtures of types which have occurred involving the Boskop race, the Bushmen, and Hottentots. There is still naturally some difference of opinion regarding the exact course of events and regarding the origins of these three races.

In South Africa there have been intrusions from the north from time to time, but I believe Sir Arthur Keith is right that the later evolution of the Bushman type in the broad sense was in relative isolation mainly in this area, although a number of local types developed, and as we shall see, the race probably originated farther north. The climate seems to have shown relatively little change in the last million years except in rainfall. The Kalahari desert may have served as a barrier right through the Pleistocene, except on the east coast, even when (in pluvial periods corresponding to the periods of glaciation in Europe) the Sahara was a well-watered country and a favorable home for man.

When the Dutch landed at Cape Town in 1652 they found two native peoples, the Bushmen and Hottentots, who are obviously related; but the nature of that relationship has been much debated and is still somewhat obscure. They agree in having the peppercorn type of hair, which is not found outside of Africa so far as I am aware, a flat face and broad nose. They also have a yellowish-brown skin color, paler than that of Bantus or Negroes, which has often been ascribed to Mongolian infusion but for which a better explanation is available; namely, that this is a persisting foetal character, arising as a part of the phenomenon of pedomorphism. The Hottentots were of moderate height and they were pastoralists, having herds of cattle and sheep and a more complex organization. They used copper and knew how to smelt iron. The dwarf Bushmen were primitive nomadic hunters who lived more in the mountains. The Bushmen had an essentially Paleolithic, the Hottentots a post-Neolithic culture. These two races were constantly at war with each other, but there was also much intermixture. They spoke different, but related languages. Schapera (1926) discussed this relationship at length. The Hottentot tribes were found all over what is now Cape Colony and in the southern half of southwest Africa. Pure Bushmen were found, but also every degree of mixture. Stow (1905) believed that the Bushmen were aboriginal in South Africa from very remote times and that the Hottentots and Bantu had come down from the north (central Africa) in comparatively recent times. He recognized three main tribes of Hottentots: the Namaquas, the Cochoqua or Grigriqua, and the Kora (Korana).[1]

The Hottentots were believed to have been driven southwards (along the west coast) in the fourteenth and fifteenth centuries, by the Bechuana, passing along the south coast and finally northwards as far as the Orange and Vaal rivers. The Koranas in the Cape region fought a war of extermination with the Bushmen. Broom (1923) points out that Bushmen brought up on the farms are taller than the Kalahari Bush, indicating that the latter are stunted by the hard conditions. He recognizes, from skeletons, that the early Bushmen at the Cape (the so-called Strandloopers) were also well developed, taller and with broader heads, but the skull shape is characteristic and unlike that

---

[1] Broom, however, regards the Korana as an independent line of descent, the Hottentots being derived from crosses between Korana and Bush, with all degrees of mixture. He has probably over-emphasized the separateness of the Korana from the Hottentots.

of any other race. The parietal region is depressed and the occiput relatively short.

The Koranas spoke a dialect of Hottentot. They apparently came to the Vaal and Orange river valleys from the west province of Cape Colony over a century ago, and many of their graves, in the form of cairns, are found in the Douglas district. Broom considers that their darker skin and coarser features, with more hair on the head and face, are derived from crosses with Bantu before their southern migration.

In an important early study of South African craniology and skeletons, Shrubsall (1907) recognized three yellow-skinned groups: the Bushmen, Hottentots, and Strandloopers. The last were from caves and middens along the southeast coast, which had been inhabited by Bushmen when the Dutch arrived. Shrubsall regarded these Strandloopers as the purest and most primitive Bushmen—dwarfs pushed down to the coast by pressure of the Hottentots. The latter he regarded as a cross between the Bushmen and the east African Bantus, which had taken place long ago because the Hottentot skulls were now uniform, intermediate between Strandloopers and Bushmen with many characters from central Africa. Shrubsall could not distinguish Hottentot skulls from Bushmen. He found not only dolicho- but mesati- and brachycephalic skulls in each race. It is now generally recognized that the Strandloopers were simply Boskop people who had taken up a coastal existence, living mainly on sea food, and crossing with the dwarf Bushmen. Broom (1941) finds that Shrubsall's confusion arose through assuming that all old skulls were of Bush type.

Broom (1923, 1941) recognizes another yellow-skinned race, the Korana, who have been regarded by others as a tribe of Hottentots. At Douglas, near Kimberley, he found living Bushmen (Plate XIV, a; Plate XVI, a, b) and Koranas (Plate XVI, c, d; Plate XVII, a), and also made a large collection of skulls and skeletons, some of mixed type, from early graves. The Korana skull had a C.I. of 70.8, while one skull of the northern Bushmen, not quite so pure as the Strandloopers, was 182 mm. long, 137 mm. wide, the C.I. of Bushmen from the Kalahari and southwest Africa ranging from 72–77.6, face flat, brows prominent, parietal eminences, the parietal region rather low and flat. The Korana skulls by contrast were large and oval, dolichocephalic, with no parietal eminences (see Fig. 6, p. 168).

Broom also finds among these Korana a type which he calls Australoid, with well-marked supra-orbital ridges (Plate xvii, b). They present a striking variation in type, but as this character is directly on the road between the frontal ledge of Rhodesian man and the modern type in which the brow ridges are lost, this is obviously parallel to the Australian condition but not in any way genetically connected with it. It is simply an intermediate link between early and late African types. The use of the term Australoid in any but a purely descriptive sense is therefore definitely misleading. The term pseudo-Australoid is preferable. This is no doubt what Sir Arthur Keith meant when he wrote (1933, p. 151), "On the evidence at present available it seems most probable that the prehistoric type of South Africa has come into being south of the Zambesi, and that in the course of time local races have arisen." Broom may be right in his view that the big-eared Korana are a separate race, but the heavy-browed type of Korana probably represent a simple genetic difference (perhaps a single gene) from the ordinary Korana who have no brow ridges. If there are no intermediates then a single gene basis would be indicated. It would appear to represent about the same degree of skull difference as that between dolichocephalic and brachycephalic skulls, and has probably been derived from crosses with descendants of Rhodesian man.

The work of Pijper (see Gates, 1946, p. 713) on blood groups shows that the Bushmen and Hottentots must be separate races, the former having about 56 per cent of O and only 7 per cent of B, while the latter have only 35 per cent of O but nearly 30 per cent of B. If both are descended from Boskop man, as generally believed, they must have remained separate to produce such a large divergence in blood group frequency. From this point of view the Bushmen are more primitive, and the high frequency of B in the Hottentots allies them with other African peoples. The blood groups thus confirm the view of Broom that the Bush and Hottentots are very different races. We will return to this subject again.

Various cases of "Australoid" skulls in other races have been described. They may be of similar character, but there is no certainty that such is the case. In the American Indians the Punin skull in Ecuador and the Nebraska skull (Bell and Hrdlička, 1935) are examples. They are sporadic and may represent mutations possibly of atavistic character, or more likely an inherited condition in the strain. In the Nebraska skull, which was found in the top layer of a burial mound, with pottery, the heavy supra-orbitals are stated to be between those of Neanderthal and modern man, the medial third of each being more developed than the distal third. In Pecos Pueblo, Hooton (1930) recognized a "pseudo-Australoid" type of skull, which probably has no relation to the modern Australians but is at best a parallel variation. This term is preferable to Australoid, as the latter seems to imply a genetic relationship which does not exist.

Equally significant is the skull of the aboriginal Australian woman already cited. In describing this skull, Burkitt and Hunter (1922) point out the enormous development of the superciliary ridges, harking back to Pithecanthropus, the projecting glabella, the heavy occipital torus, and the low, sloping forehead, yet the facial skeleton was more modern and the cranial capacity 1211 cc. The cephalic index was 65, a very extreme dolichocephaly. Although modern genetics gives little support to the conception of atavisms, it does not exclude the possibility that they may occur. In fact, in genetic experiments with guinea pigs an atavistic mutation produced extra toes which had been lost in the species for perhaps a million years, and from this mutation a race was derived. Many supposed atavistic mutations, however, have no such meaning.

More recently Wells (1942) has described the Bayville skeleton. It appears to be a member of the Boskop race, large-headed and hyperdolichocephalic but with heavy supra-orbitals of the Australoid type. Broom regards this character as derived from the Koranas by hybridization. It is not surprising to find it in any human species descended from Rhodesian man.

Keen (1943) reports on two Korana skeletons from burials near Upington, north of the Orange River. Each grave was marked by a heap of stones and the bodies were flexed, with the head pointing to the east. He regards the Korana as mixed, not only with Bush but with Bantu, as shown by the low C.I. and the tall stature of the males. He finds historical evidence that the Korana married with their Bantu and Bush neighbors. They are a vanishing tribe, probably not numbering now more than a thousand. Measurements of a few living Korana showed that they are smaller than Hottentots and more like the Bush in size. They are historically Hottentots but anatomically Bush. This is explained by the fact that they have lived close by the Bechuana at least since 1778. The Bechuana took Korana women as wives and the resulting children became Bantus. The Korana in turn took Bush wives and their children were "Koranas." This explains the physique of the present Koranas. The two earliest skeletons were less like the Bush and very close to the Hottentot type as now established.

The Boskop type of skeletons continued to be found in various parts of South Africa. At Zuurberg, about fifty miles north of Port Elizabeth, a collective grave was found in open ground enclosed by a circle of stones. It contained skeletons and chipped implements (Wells, 1929). Here was a sequence of racial types and cultures, the upper burials being found at a depth of three and a half to four feet, the lower at seven feet. The upper level included Boskopoid Bushmen and a pseudo-Australoid Bush skeleton with a skull resembling the Rhodesian type and leg bones like Neanderthal man. The culture was Smithfield. Wells believes that Bushman history in South Africa includes an earlier arid period when the people crowded to the coastal belt and a later humid time when the game moved inland and the Bushmen followed them. It is suggested that the arid period lasted from about 6000–2000 B.C. to the present time. Wells also recognized among the skeletons three no doubt interbreeding types which he regarded as distinct: Boskopoid Bushmen, "Australoid" Bushmen, and the Springbok Flats type.

In 1929, Broom published an account of the Springbok Flats skeleton, found in the Transvaal eighty miles north of Pretoria. It was in tufa three feet from the surface; the skull and bones were badly broken; the man had probably been killed and trampled by a wounded buffalo of extinct species (*Bubalus Bainii*), the bones of which were also found here, together with the bones of an antelope. The skeleton was impregnated with lime and was that of a powerful man resembling both the Cro-Magnon and Boskop types. The skull was 195 mm. long and about 144 mm. in width, having small supra-orbital ridges and small teeth, the parietal region low. This skull agrees with the Koranas in many respects, and is of mid-Paleolithic cultural affiliation. The Springbok Flats are littered with Mousterian implements. A special study has been made of the extraordinary mandible (Schepers, 1941), which is very large with a wide ramus, longer than the Mauer jaw, but, like it, having small teeth. The chin is projecting and it is suggested that the peculiar shape of the mandible is due to a marked reduction of the dental arcade and teeth. There is no feature of the South African Negro, but the face is long like the east African Hamites. Keith (1931) suggested that the Springbok Flats type is ancestral to the northeast African Bantu, but the latter have large teeth while Springbok man is microdont. Keith's suggestion that crosses between the Springbok Flats type and the South African Boskopoid type might produce the Cape Flats man (see below) meets with the difficulty that the former are both microdont while the latter is macrodont. Crosses between the Cape Flats and the Springbok type would probably produce strange mandibular forms like some of the Bush-Hottentot mandibles. The Springbok Flats mandible has many features in common with the Bushman, Korana, and Hottentot types in South Africa, so his type may have contributed to the ancestry of these nearly related races. Schepers thinks the reduction in the facial elements may be pedomorphic in nature. The subject of pedomorphism will be discussed later. The relations of Springbok Flats man to Cro-Magnon, Heidelberg, and Neanderthal remain obscure because of the large gaps involved, but the discovery of this extraordinary man in the Transvaal as well as Boskop man helps to link the African with the European types.

Leakey (1935) finds that the Springbok Flats man resembles the Elmenteita skulls from Kenya (see below) and the associated *Bos bubalus bainii* is closely related to the fossil buffaloes found in Kenya with Upper Pleistocene (Mousterian and Aurignacian) cultures. Oldoway man is of similar type.

A skeleton recently described by Cooke, Malan, and Wells (1935) from the Lebomo mountains in the Ingwavuma district of Zululand includes limb-bones as well as an infant and an adult cranium. Many skeletons were found later. The skull is related to the Springbok-Fish Hoek and the Florisbad types, with a capacity of 1450 cc. The limb-bones show both platymeria and pilastery. The accompanying artifacts are Levalloisian of South African Middle Stone Age, with animal bones of the same age.

The Cape Flats skull (Fig. 5, b) was discovered by Drennan (1929ab) in a sand quarry on the low isthmus near Capetown and is probably of considerable antiquity, as it comes from an old land surface under four feet of sand. It is macrodont, pseudo-Australoid, with an estimated capacity of 1230 cc. and length of about 191 mm., C.I. 69.1, and the face decidedly Negroid. Broom considers it a skull of his Australoid Korana type, showing that the Koranas are not a modern introduction to South Africa and that the race has long carried this pseudo-Australoid variant. He also points out resemblances to the Rhodesian skull. In addition to the main parts of the skull, Drennan found a femur and fragments of another thick-walled skull, as well as limb-bones of three individuals, some of Bushman type. The stone tools were partly of kitchen-midden (Bushman) type and partly earlier Stillbay, probably belonging to the pseudo-Australoids.

Drennan (1929b) regards the Cape Flats skull as almost Neanderthaloid owing to the supra-orbital ridges, almost a torus, but with large frontal sinuses unlike Neanderthal man and the Australians. The forehead is sloping and there is a post-orbital depression. The face has some characters of early Europeans and Negroes. The teeth are cynodont, larger than the average in the Bantu. The complete femur, which probably belongs with the skull, shows platymeria and pilastery. In an account of the accompanying artifacts, Goodwin (1929) points out that the sand-pit was excavated to a depth of ten feet or more, but the skull was unearthed by workmen and its original exact position is unknown. This Neanderthaloid skull was probably associated with a Stillbay industry, and microlithic cores belonging to the Wilton culture were also found.

It is impossible to refer here to all the ancient skeletons which have been unearthed in South Africa, but a few more must be mentioned. The Mistkraal group (Allen, 1926), found on a river about forty miles north of Zitzikama, is another of the Australoid Korana type. The skull is pentagonal ovoid, with a narrowing of the frontal region and prominent supra-orbital ridges. It is therefore a mixture of Bush and Korana features. The skeleton in Skildergat Cave, of "Stillbay" archaeological age, belongs to the large-headed Bushman type. The Fish Hoek skeleton from the coast near Capetown is probably some thousands of years old; Keith suggests 15,000 years. It was found in 1927 with implements of Stillbay culture. The head was large, skull 200 mm. long with a capacity of 1600 cc., the teeth and mandible small, the chin prominent and the occiput projecting. It is quite different from the Florisbad skull and belongs to the Boskop type, but is very narrow with long face like the modern Nilotes. Keith pointed out similarities between Fish Hoek man and Cro-Magnon man in France. They both had a big brain, but the latter was tall whereas the former was almost a dwarf. Broom (1943) regards the Fish Hoek skull as Korana with a little Bush mixture, and Drennan (1937a) emphasizes its pedomorphic character. Keen (1942), in describing two skeletons from the Fish Hoek cave, found them to be recent but prior to the European occupation. Both were females, one adolescent. They were taller than Bushmen or Hottentots, but he regards them as Strandloopers, that is to say, beach rangers, belonging to the Hottentot race; in other words, Boskop survivors having the pseudo-Australoid frontal ridges.

The Plattenburg Bay skull found by Drennan in a cave is Bush with a very large brain, according to Broom. Another skull, discovered by the Rev. Mr. Sharples at the Plattenburg Bay rock-shelter (Drennan, 1931), is that of a big-headed pre-Bushman, but infantile (pedomorphic) to an exaggerated degree.

A collection of eleven skulls from Kruidfontein, about 150 miles north of Mossel Bay, were all but one from graves, the individuals being in a sitting or flexed position (Klopper, 1943). Two skulls were of Kalahari Bush type, five belonged to Drennan's Oakhurst type, dolichocephalic with cephalic index 69.5–73.3, and four were broad-headed (C.I. 79.4–83). They are partly comparable with Shrubsall's Strandloopers but are nearly all apparently derivatives from hybridization between Bush and Boskop.

The Oakhurst rock-shelter at George, on a farm four miles from the Indian Ocean near Mossel Bay, has been fully studied by Drennan (1938). Nine skeletons (5 ♂, 4 ♀) were found buried six feet deep with much late pottery, the age being later than the Matjes River cave (see below). The Oakhurst people were very close to the modern Hottentots, taller and with bigger heads than the Bushmen. Drennan concluded that the Hottentots are descended from the Boskop type,[2] not as was formerly supposed, from modern crosses of Bushmen with Bantu. If the latter had been their origin they would have darker skin color and would show genetic segregation in various features. There is the same extreme sex difference in these Oakhurst skulls that Shrubsall (1907) found in Hottentot skulls. The children of these cave-dwellers were represented by fifteen skeletons, six of which were brachy-, three mesati-, and one dolichocephalic. At six years of age the skull was already of adult type. Study of the pottery by

---

[2] Coon suggests that the fine-type Hottentots were derived from crosses between Boskop and Hamites.

J. F. Schofield showed that it was the industry of a Hottentot clan of no great antiquity but not later than the last quarter of the eighteenth century.

One of the most important discoveries made in South Africa was at the mouth of the Matjes River near Mossel Bay and not far from Zuurberg. Keith (1933) gives a splendid detailed account of some of the skulls from this cave, with measurements of seven skulls and nine mandibles. One skull was found thirty feet deep with a Mossel Bay culture, others were found at a ten or twenty foot level. In addition to the Boskop type, another skull in the same stratum shows links with the Strandloopers and Hottentots, and in upper (Wiltonian) deposits was a Boskopoid child's skull of Zitzikama type. The racial types may be regarded as intermingling, much as the Nordic, Alpine, and Mediterranean races intermingle in modern Europe. Dreyer (1934) thinks the Wilton people a separate stock because of their different culture, but there appears to have been in South Africa a remarkable mingling of cultures as well as stocks. The occupation in this cave was evidently continuous over a long period, there being no break in culture. Keith explains the changes as a gradual evolutionary transformation of one stock through a long period, and not a sudden replacement of stocks except through local exchanges. The Strandloopers would then be the lineal descendants of the Boskop here, the Bushmen and Hottentots arising elsewhere from the same stock. This Matjes River type is peculiar in showing trigonocephaly (Fig. 7) due to the trigonic form of the frontal bones, a condition seen to a less degree in the modern Bushmen and Hottentots. These people were of moderate height, not dwarfs; the cephalic index ranged from 66.6 to 73.4, so all were dolichocephalic, some extremely so. True Bushmen were found in the surface layer (Drennan, 1934). The chin was well marked as a low, rounded triangular elevation. The molar teeth showed taurodontism (see Gates, 1946, p. 369), in which the tooth body becomes enlarged at the expense of the roots. Taurodontism also occurs in Neanderthal man, but Keith shows that it manifests itself differently in these two races.

Keith regards the Matjes River people as a branch of the prehistoric type which developed in South Africa and is found in no other continent. He included in this line of evolution the Boskop, Bushmen (living and prehistoric), Strandloopers, Matjes River people, and Hottentots, all having certain distinctive features in common. He believed that they probably originated south of the Zambesi and that a number of local types differentiated, the Springbok Flats skeleton, the Boskopoid type, and the "Australoid" Bushmen being merely variants of one stock. The more northerly representatives of this race, such as the Kalomo skeleton of Boskopoid character, and the Lake Nyasa Bushwoman skull found by Dr. Stannus, Keith regarded as

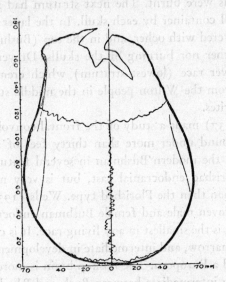

FIGURE 7. Outline of the Matjes River skull, showing the trigonocephaly.

strays from the south. South Africa would then be the home of this pedomorphic stock, with a number of local races. The type had large brains, characteristic eyebrow ridges, cheek bones, and lower jaw, with a tendency to retain juvenile characters which are usually transient. These pedomorphic characters include (1) flexure of the skull base with consequent flattening of the suture between the temporal and parietal bones, (2) the contour of the sagittal arc of the parietal, (3) the forward position of the upper frontal, (4) the tendency to dwarfism, (5) the tendency to retain the less pigmented skin of the foetus, and (6) the tendency to hairlessness. There are records of modern Bushmen with an enormous brain, presumably derived from some Boskop ancestor.

The Matjes River cave has been fully excavated and Dreyer (1933) gives a full account of the burials, the implements, and the animals, all of which were of living species except one. Unfortunately some of the skulls were not sent to Sir Arthur Keith. These skulls and the forms of burials might have altered his opinion about the racial relations involved. The lowest burials had the skulls embedded in powdered ocher, which evidently had a preservative effect. In the next layer the skulls were burnt. The next stratum had no ocher but an ostrich eggshell container by each skull. In the layer above the skulls were again covered with ocher, and in the top (Bushman) layer there was neither ocher nor burning of the skulls. Dreyer concludes that the Matjes River race (lowest stratum) which cremated the skulls, was distinct from the Wilton people in the middle stratum, who had no cremation rites.

Meiring (1937) made a study of the frontal convolutions in one of these skulls, found under more than thirty feet of ashes and shells. It differs from the modern Bushman in several features and approximates the Florisbad endocranial cast, but is very much nearer the modern Bushman than the Florisbad type. Wells (1937) made a similar study of eleven male and female Bushman endocranial casts. The Bushman brain is the smallest in any living race. It is characteristically elongated and narrow, and intermediate in development between Rhodesian man and a European. The Boskop brain is more primitive than the Bush, being intermediate between Bush and Rhodesian. The Bush brain is juvenile in morphology, like that of a Bantu infant, with various regions incompletely expanded.

A preliminary correlation of archaeological types in Europe and South Africa is given by Lowe (1932) and reproduced in simplified form in Table 15. The foundations of South African archaeology were laid in two extensive accounts of excavations under the aegis of the South African Museum in Cape Town, by Péringuey (1911) and Goodwin and Lowe (1929). Goodwin (1925) recognized four stone industries in South Africa, one earlier and three (Eastern, Smithfield, and microlith) later. The Smithfield and microlith industries he considers the same as the north African Capsian, and he believes that the bearers of the Capsian and Aurignacian cultures introduced the bow into north Africa and Europe. The Capsian culture of north Africa is the same as the Aurignacian of France. According to Good-

TABLE 15

COMPARISON OF EUROPEAN AND SOUTH AFRICAN ARCHAEOLOGICAL AGES

	South Africa	Europe
Late Stone Age	Neolithic Kitchen middens Wilton Smithfield	Neolithic Mesolithic Azilian-Tardenoisian Capsio-Aurignacian
Middle Stone Age	Stillbay Pietersburg Howieson's Poort Glen Gray	Solutrean Mousterian + Solutrean Mousterian + Mousterian
Old Stone Age	Fauresmith Victoria West Industry Stellenbosch	La Micoque Combe Capelle Levallois Acheulean-Levalloisian Chellean-Acheulean

win (1929) in north Africa it lasted through the Solutrean and Magdalenian in Europe, and forms with them the Azilian and Tardenoisian of France. Goodwin concluded that the South African microlith and Smithfield cultures of the Bushmen represent a form of Capsian culture which broke away from north Africa. That the Bushmen paintings associated with the Smithfield culture strongly resemble the art types in caves of northern Spain and southern France is well known. Goodwin suggests that a common racial element is necessary to account for the resemblance. The modern Kalahari and Cape Bushmen are widely divergent and according to several writers are mixed with Hottentots and Bantu. But a few pure Bushmen survive in the Kalahari desert as nomads who cannot be approached.

Maingard (1932) shows that the Hottentots did not know the bow and arrow when the Dutch and English landed at Cape Town. They threw stones (like the Tasmanians) and used fire-hardened staves as weapons. The bow they borrowed from the Bushmen after the arrival of Europeans. The Hottentots did not then use iron, at least not at the Cape, but the Bantu on the east coast had the assagais. The Hottentots learned the use of iron either from the Bantus or from the Europeans. The Bantu in their original home in northeast Africa did not use the bow, but two of their tribes, the Bechuana and

Herero, adopted it later after advancing southwards into large new areas.

Many writers believe that the correlation of European and South African cultures will be determined in north Africa. Old Paleolithic (Stellenbosch) implements have been found in the most recent Vaal gravels in the Transvaal, with mammoths and other animals. This Acheulean industry persisted for a long period, until relatively recent times.

Smuts (1932) has published a very lucid and masterly exposition of the successions of cultures in South Africa, their relation to climatic changes during the Pleistocene, and their correlation with east Africa and Europe, as given in Table 16. In a discussion of the thermodynamics of glaciation, Lewis (1946) agrees that it is preceded and accompanied by a cool, dry climate, but concludes that any pluvial period must come at the beginning.

The glacial periods have been extended in years, according to more recent knowledge (see Table 3, p. 104). The provisional story in South Africa is that Pleistocene times began with a long pluvial period of heavy rainfall, a favorable climate with full rivers even in the region of the Kalahari. Under these conditions man appeared and developed the Chellean type of industry over a long period. In the following long dry period the Kalahari developed a sand deposit over a hundred feet deep and man was apparently driven from this part of South Africa to more favorable marginal areas. The second great pluvial period followed during which man developed a series of new cultures of the Middle Stone Age. In the short Achen dry period which followed, man's activities again ceased in South Africa. The succeeding Bühl period of rainy conditions saw the development of the Smithfield and Wilton cultures of hunters and rock painters, since which time the climate has been dry and perhaps increasingly so.[3]

---

[3] While it appears to be established that pluvial periods in Africa correspond with glacial advances in Europe and while the glacial episodes in North America appear to correspond in general with those in Europe, yet it does not follow that ice ages in one region of the earth are always contemporary with pluvials in another. The development of an ice age depends upon whether precipitation is mainly in the form of snow or rain. Knoche (1941) concludes that an Ice Age may be caused by either a rise or fall of say 5° in temperature. Cooling can produce glaciation in temperate regions, and he believes that a warmer period is generally identical with a pluvial age in the same area. As we have seen, the glacial periods of Europe are generally equated with a pluvial period further south. But at high altitudes in the Andes, for example, a higher temperature through increased solar radiation could increase the glaciation. In certain cases an increase in total precipitation may be involved.

TABLE 16

CORRELATION OF EUROPEAN CLIMATES AND CLIMATES AND CULTURES IN SOUTH AFRICA

European Pleistocene	S. African climates	Stone Age	S. African cultures and man
A.D. to present	Slowly decreasing rainfall (Same in E. Africa)		Bantu advance and racial mixtures
Sub-Atlantic phase 1000 B.C. to A.D.	Second post-pluvial period (Nakuran in E. Africa)	Late Stone Age	Early Bantus Wilton Smithfield
Sub-boreal phase 6500–1000 B.C.	Dry period (Same in E. Africa)	Late Stone Age	Wilton Smithfield
Bühl minor ice advances 9500–6500 B.C.	First post-pluvial period (Makalian in E. Africa)	Late Stone Age	Wilton Smithfield
Achen oscillation 13,500–9500 B.C.	Very dry period (Intensely arid in E. Africa)		
Riss-Würm glaciation 72,000–230,000 B.C.	Second great pluvial period (Gamblian in E. Africa)	Middle Stone Age	Stillbay (Fishhoek man) Howieson's Poort Pietersburg (Springbok and Boskop man) Glen Gray Early Mousterian (Rhodesian man) Fauresmith
Second interglacial 230,000–435,000 B.C.	Long desert (Kalahari) period (Arid volcanic period in E. Africa)		
Mindel-Gunz glaciations 435,000–500,000 B.C.	First great pluvial period (Kamasian in East Africa)	Lower Stone Age	Upper Stellenbosch (Acheulian) Middle and Lower Stellenbosch (Chellean) Lower Stellenbosch (pre-Chellean)

The most remarkable difference from European conditions is that in South Africa men of modern or light-browed type were making Mousterian tools while the same kinds of tools were being made by Neanderthal man in Europe. The fact that men of different genera, one probably not even derived from the other, could thus make and use implements of similar crude construction, shows that extensive cultural contacts must have taken place somewhere, perhaps in Asia Minor. It seems most reasonable to conclude that these types of tools were invented by men of the Homo type and later transferred to the Neanderthalers by culture diffusion. On the other hand, it is possible that these crude Mousterian tools were independently developed by different races.

At Mumbwa, near Broken Hill in Northern Rhodesia, Dart and Del Grande (1930) found a Boskop human type with Bush mixture and a Mousterian culture which remained uniform over a long period. They suggest that the more primitive Boskopoid and Australoid types were responsible for conservation of the very crude Old Stone Age cultures. The Bushmen either secured or developed the Mousterian (Fauresmith) culture in South Africa. Later levels at Mumbwa were Neolithic and even iron-smelting furnaces of simple type were found. Habitation ceased probably about the time of Christ when the Bantu hoe and axe culture was introduced. The Wilton culture in the south belongs to the coastal Bushmen. In the last 6000 years the Mousterian, Late Stone Age, Neolithic, Copper, Bronze, and Iron Ages in South Africa were thus all telescoped more or less into one.

There are many other records of mixed racial types in South Africa. Drennan (1932) reported on a skeleton broken in fragments, found in trenching in the Matoppo Hills south of Bulawayo, Southern Rhodesia. Only the mandible could be restored. It belonged to a child about eleven years old. The skeleton was about four feet below the surface with an occupation layer (burnt earth and charcoal) three feet above it. The mandible was of Bush type with some Bantu mixture. The Bantu, Bush, and Boskop all differ in their teeth. In the Bushmen, Neanderthal, and most apes the second molars erupt before the premolars, while the reverse is the case in Bantu and Europeans. In primitive man, as in simians, the second molar is generally larger than the first.

We have already seen that in the kitchen middens of the South African coast the Strandloopers of Bushman type were preceded by the Boskopoid type, found in deeper excavations. The latter had not only larger skulls but larger teeth and jaws. Drennan (1945) describes, from Kalk Bay near Cape Town, a skull excavated from under a sterile kitchen midden at a depth of four feet; this skull is typically Bush but has the large Boskop dentition. At a depth of six feet in an adjoining cave was a typical macrodont Boskop skull, the mandible as large as in the Kalomo skeleton and next in size to that of Springbok Flats. This Bush skull with large Boskop teeth is probably a result of racial crossing. The same large dentition is found in Hottentots and Bantu. The molars were shown by X-rays to be taurodont in both the Bush and Boskop.

The Canton Kop skull is a variety of pseudo-Australoid Korana 205 mm. in length and fairly high, and the Valsfontein skull is similar. The Ramah skull is regarded as a Korana with a little Bush mixture and a little Australoid.

The region around the junction of the Vaal and Orange rivers has been a melting pot area for the Griqua and Korana Hottentots. The Bathlaping are the southernmost of the Bechuana (Maingard, 1933). They are physically mixed but culturally mainly Bantu, but they borrowed the bow and arrow and reed pipes from the Korana. These central and east coast Bantu were both in contact with the Hottentots, each borrowing some elements of physique, language, or culture.

In a symposium on the skeletal remains from the Cape coast, Bernstein (1935) described the remains of at least ten individuals in a cave at the mouth of the Keurbooms River in Plattenburg Bay, now two hundred feet above sea level. There were twelve feet of midden deposits on the rock floor, the human skeletons of all ages were partly on the surface and partly in the upper foot of the midden. The bones were considerably mineralized, not recent, the dressed quartzite flakes, two pebble grindstones, and ocher indicating a Late Stone Age culture homologous with the C layer at Matjes River. Two skulls were like the Boskop Matjes River race, less pedomorphic than the Zitzikama people. The Boskop type thus persisted side by side with the slender Bushmen. Another skull, from Waterloo Bay, was of recent age, associated with pottery of two different types, of which one was Bushman, the other, of higher quality, Hottentot with Bantu influence; but the skull was pure Boskop, like Zitzikama. As the Bantu did not reach here before the seventeenth century, this pure Boskop type must have survived until that time. Two skulls from Mqanduli, Transkei, demonstrate intimate mingling between the aboriginal Bush-Boskop population and the immigrant Bantu. The third skeleton is predominantly

Bantu. Similarly on the Natal coast Galloway (1936) shows that the Late Stone Age population was a Bush-Boskopoid mixture. Later, when the Bantu brought down iron-working, the Bush-Boskop population shows hybridization with them. Wells (1934) found similar conditions in eleven skeletons from the coast at East London which were mostly recent prehistoric. They were Bushmen with Boskopoid features in more than half, and one European. Of three other skulls from a Bantu cemetery of about 1860–1870, one was typical Bush and two had mingled Bush-Bantu characters.

Mapungubwe is built with retaining walls of loose, uncemented stones after the fashion of Zimbabwe. It is on a sacred hill by the Limpopo on the borders of Southern Rhodesia, the Transvaal, and Bechuanaland. Zimbabwe is in the same area, east of Bulawayo, and Miss Caton-Thompson (1931) placed its age at A.D. 700–1500. She places its zenith in the tenth to eleventh century and believes that the Bavenda, who have since migrated to the Transvaal, were probably connected with its construction. Gold beads, tacks, and wire were found near the grave area of Mapungubwe. The pottery shows that two peoples lived here simultaneously and the age is the same as Zimbabwe. Twenty-four fragile skeletons were found, one of them with a gold plate on the skeleton, others with quantities of gold bangles. Galloway (in Fouché, 1937) found the skeletons to be all Negro, Bush, and Boskop types, with no evidence of any other element, but there was very little Negro influence in the skulls. Mapungubwe thus represents a very early stage of infiltration of Negro (Basuto) features into the homogeneous Bush-Boskop population, which was akin to the post-Boskop inhabitants of the coastal caves and survived as such to a very late date. Phaup and Drennan (1939) describe other remains of prehistoric gold miners in Southern Rhodesia.

A study of the hyoid bone by Schepers (1937) shows that it has characteristic structural features in different races. In the Bushmen two types of hyoid were found. The Zitzikama hyoid differs from most of the Bush. The Mapungubwe hyoid is related to type 1 of the Bush and to that of the South African Negro. Certain other features agree with the Boskopoid character of the mandible. Europeans and South Africans also show characteristic features of this bone, the racial significance of which has received little attention.

At Colesburg, Cape Province, fifty-three skeletons of Bushmen of relatively recent date were exhumed. Their osteology was studied by Slome (1929). The bones are all of small size, the cranial capacity ranged from 995 cc. to 1390 cc. and the cranial index from 69.9 to 80.5, half the skulls having a C.I. over 75, one of thirty being definitely brachycephalic. The dentition of the same skulls was studied by Drennan (1929). Forty-five were regarded as pure Bushmen, nine having some Hottentot or Bantu blood.

In more recent years evidence of the Bushmen has been found farther north. The Nebarara skull (Galloway, 1933) was unearthed from a tomb raised slightly above the ground level, in the Masai steppe country of northern Tanganyika. The tomb was sixty-three paces across and twenty feet deep. Thirty tons of limestone were removed and the body was found in a crouching position facing east. The skull was small, light, and delicate, resembling the modern Bush skull. It was orthopentagonoid, the frontal and parietals were strongly bossed (a foetal condition), the occipital infantile, the cranial capacity 1350 cc., but the zygomatic arch and the temporal bones are Boskopoid. Galloway concludes that the Bush, Strandlooper, Hottentot, and Korana are not separate races but belong to one main type. This view, as we have seen, cannot be maintained, although these races are all nearly related.

Five skeletons from the northeast Transvaal were described by Berry (1935). One was Boskopoid, one Bush, two Bantu (one with some Bush mixture), and one Bushmanoid. MacLennan (1935) reports two collections from the Pietersburg district of northern Transvaal. Skull fragments, a mandible, and some long bones were found four feet deep on a farm thirty-four miles southeast of Pietersburg. They were essentially of Bush type, the second premolar and second molar erupting simultaneously, as in the Bushmen, the molars taurodont. The second collection, thirty miles northwest of Pietersburg, consisted of two skulls and some fragments, unfossilized and recent. One skull showed an intimate mixture of Bush-Boskop and Bantu characters. Fifty miles southwest of Pietersburg a human skeleton was found (DeSaxe, 1935) buried sitting (the Bantu custom) facing west and surrounded by Bantu potsherds and charred animal bones. This skeleton combines Bantu with very strong Boskopoid features and was not recent but was unfossilized. One hundred miles southeast of Pietersburg, skeletal material (Dearlove, 1935) of relatively pure

Bantu was found. This antedates European settlement and may represent forebears of the present Basuto of the Transvaal (Plate XVIII, a, d).

Cooke, Malan, and Wells (1945) have recently described human remains found in Zululand. They consist of a partial adult cranium and the skeleton of an infant. The associated fauna included two extinct species, and the rich lithic industry was of Middle Stone Age (Pietersburg culture). There were indications of primitive stone-tipped arrows at this early period.

At Rooiberg in the central Transvaal, on Rookpoort farm in a belt of dolomitic limestone in a large cavern, Mr. A. J. Burger found a human skull with a femur and a clay pot. In an account of this skull and of a subsequent visit to the cavern, Dart (1924) gives many further details. The floor had a thick layer of bat guano and was littered with clay pots. From the entrance the floor descended thirty to forty feet until progress was hindered by a stalagmite eight feet in diameter extending from floor to roof, and another three to four feet across. A tunnel about twenty-five feet long was the shaft of an abandoned native mine. The huge stalagmites must have been formed since the mining operations ceased. The large cavern beyond the stalagmites had several recesses, in three of which there was evidence of mining operations by men of small size, one recess containing a large wicker basket. Besides the skull, further search revealed several vertebrae and other human bones as well as many animal bones, melon seeds, and calabash husks. The skeleton was recent and there was evidence that it belonged to a medicine man who had perished by falling into his own fire, perhaps while under the influence of a drug. The skull had a very massive occiput and well marked superciliary ridges, as in Boskop man, but otherwise it resembled the Bushmen. It was like the Bantu in having huge frontal air sinuses and a narrow mandibular fossa. It was therefore a hybrid product of Bush and Bantu with some ancient Boskop features.

The Bushmen race includes all in South Africa who are of small stature, with peppercorn hair, pentagonoid cranial form, yellowish or pale brown skin color, and steatopygia. They were formerly supposed to be confined to South Africa, but the discoveries quoted above, as well as others mentioned by Galloway (1933), show that they were formerly indigenous much further north, even in northern Tanganyika. These discoveries include evidence of former Bushmen in the Penhalonga district on the borders of Southern Rhodesia, and Portugese East Africa and the skull found by Dr. Stannus in an old beach of Lake Nyasa. West of "Lake" Eyasi, which is really a vlei, in other words, a lake in the wet season and a mass of reeds in the dry, Miss Bleek (1931) found a hunting tribe, the Hadzapi, north of Mkalama, some of which have moved southeastwards and mixed with the Isanzu Bantu. The Hadzapi have no physical resemblance to the Bushmen, except in steatopygy, which is most marked in the men. But they live the life of Bushmen, hunting, gathering food, building bush shelters, using bows and arrows like some of the Angola Bushmen, and having no villages or chiefs. Their language resembles that of the Kalahari Bushmen and the Hottentots. Rock paintings are also found here.

Wilson (1932) also found in the Great Rift valley remnants of a primitive race living in holes in the ground, feeding on game and honey beer, short, dark, and big-buttocked. They would appear to be closely related to the Hadzapi. Wells and Gear (1931a) describe from the Riet River valley, in the southwestern part of the Orange Free State, two entire skeletons, not mineralized. One of them shows an even blending of Bush and Bantu characters, the other is predominantly Bantu, while both had some Boskopoid affinities. Mr. Fitz-Simons made an extensive collection from a cave far up in the Outeniqua mountains. The floor midden deposit was eight to ten feet deep and belonged to the Late Stone Age, with red ocher and ostrich eggshells, showing no cultural change. The cave was twelve miles from the sea but the people lived almost wholly on shellfish. Twelve skeletons and four mandibles were selected for study by Wells and Gear (1931b). The skulls were divided into two groups, based on mineralization. The earlier group show Bush and Boskopoid features and some characters regarded as Mongoloid. The later group show the same features plus "Australoid" (Korana) and Bantu characters. Yet this site is two hundred miles outside the limits of Bantu occupation. The Mongoloid element is represented by broad skulls with broad, flat faces, the skeleton being non-Negroid. Brachycephaly by itself was found in the Matjes River group. The authors refer to the "Chinese Hottentots," a racial group on the southeast African coast in historic times, and to historical evidence (Paver, 1925) of Chinese contact with east Africa in the period between A.D. 1000 and 1300. Dart also directed attention to Bushman paintings of figures in Chinese dress.

Wilson (1932) has mapped the distribution of terraces and irrigation in great areas of east Africa, including Tanganyika, Abyssinia, Uganda, Kenya, and Northern Rhodesia. The terraces are parallel, following the contour lines of the hills, and terracing is still practiced by certain tribes in Tanganyika. The evidence indicates that the whole highlands area was terraced at one time, and Wilson assumes that the Rift Valley region was under the domination of a single race at some time before 1500 B.C., the terracing at Mufindi being abandoned at least nine hundred years ago. He suggests that an ancient civilization arising in the north spread through the Rift Valley and into the Great Lakes region, perhaps reaching Zimbabwe. They are believed to have developed into a nation with trade routes, ports (Rhapta), and communications with the Red Sea and the Persian Gulf. The evidence below suggests that these terraces are considerably later, and of Chinese origin.

In a study of early Chinese connections with east Africa, Schwarz (1938) has built up considerable evidence of such contacts. Fleets of Chinese junks sailed from Cail, at the tip of the Indian peninsula, where there were formerly quantities of Chinese pottery fragments and cinnabar used as red pigment. At this time Chinese ships are believed to have sailed along the Arabian coast and down the east African coast, calling at all ports as far as Sofala. The "Book of Marvels of India" says that in A.D. 945 the Chinese set out to conquer the island of Kambalu, which he identifies as Comoro. Their fleet was a thousand sail; the larger ships were said to carry one thousand men, six hundred of them sailors, the rest soldiers. The Phoenicians in the time of Ptolemy Necho are said to have landed on the south coast of Africa, sowed millet or barley, reaped the crop, and then sailed around the Cape, through the Straits of Gibraltar and back to Memphis.

Near the Portuguese border of Southern Rhodesia and also on the Usambara Plateau southwest of Mombasa "the whole country for dozens of square miles is covered with terraced gardens" and the remains of irrigation channels. Schwarz attributes these to Chinese temporary rice cultivation between A.D. 900 and 1200. He states that Sung pottery in enormous quantities has been found on the coast, and that Ming pottery has been dug up at Zimbabwe (Southern Rhodesia). In Rhodesia there are said to be legends of the departure of the Chinese about A.D. 1250, when they were driven from the Indian Ocean by the Arabs. According to Baxter (1944), Babylon influenced east Africa as early as 3000 B.C. and the sail was introduced about 1200 B.C. The Chaldean King Nabonidus, 606 B.C., opened sea routes between Babylon, China, India, and east Africa and Hindus then migrated to Africa. Greek coins, dated after the defeat of Persia by the Greeks, 336 B.C., have been found on the coast of Tanganyika between Tanga and Lindi. Baxter states that the Chinese reached Mafia Island, off this coast, in A.D. 742 and that Chinese Ting ware (1068–1086) has been found at Pangani on Pemba Island, while the Chinese and Arabs traded on this coast in the twelfth and thirteenth centuries. East Africa thus had connections with civilization from very early historical times, but Egypt alone produced a civilization. Toynbee (1934), in his *Study of History*, lists twenty-one human societies or civilizations in all history. These include two in America, the Mayan and Andean, but not one in Africa except the Egyptian, which was of Mediterranean origin, yet Central Africa was in touch with Egypt from very early times.

The Bahurutsi (Barotse), whose history begins about A.D. 1250, Schwarz suggests were half-caste children of Chinese men and native women in east Africa. He quotes Bretschneider, "The first black African slave was brought to the Chinese court in 976 where he occasioned the greatest surprise." The Bahurutsi multiplied into thirty or forty tribes with kings, dating back to about 1250. The Bechuana adjoin the Kalahari desert and their oldest section is the Bahurutsi. Schwarz suggests that when the Chinese connection ceased the Bantu drove the half-castes, under the name Batwa, into the hills where they formed clans, took the baboon as a totem and became the Ba-Chwena and related tribes, such as the Baralong with their capital at Mafeking. His attempt to account for the origin of the Hottentots as a Chinese-African cross appears to be negatived by innumerable facts of earlier Hottentot-Boskop ancestors; but if there is a Mongoloid element in the Bechuana it could be accounted for by such recent crossing between Bantus and Chinese (see Plate XVII, c.).

Dart (1939) gives further evidence of Chinese trade on the east African coast between A.D. 1030 and 1644. This trade resulted in the spreading of Negroes through the East Indies and into Melanesia. The

Bantu language, Swahili, became the *lingua franca* of the Indian Ocean.

All the previous records show that the Bushman and the Boskop race were formerly widespread north of the Zambesi, at least as far as northern Tanganyika. Everywhere they are so intermingled that their separate histories are very difficult to disentangle. They represent two closely related types, but while the Bush race has survived as such, the Boskop persisted as the Hottentot. Galloway (1933) believes that the gigantic systems of terracing and irrigation found in eastern Africa were the work of earlier Bushmen, who are shown by their burial modes to have been more prosperous; but that these Bushmen were ever capable of such productions may well be doubted. Their pedomorphic degeneration is a unique phenomenon in human history, apparently shared by Bush and Boskop in different ways. Drennan's theory of pedomorphism in South Africa has undoubtedly thrown light on the nature of these races and has been widely applied to the interpretation of South African racial conditions. Nevertheless, much remains obscure. Dreyer (1932) has discussed the many views of the Bush-Boskop-Hottentot-Strandlooper relations. He describes Bush skeletons only four feet high and refers to a number of other skulls without reaching a solution of the problem. He regards the Hottentots as Hamites descended from a Mediterranean type of Cro-Magnon man, the Boskop Matjes River group as an ancient Cro-Magnon race using Mossel Bay stone implements. But the unique peppercorn hair of the Hottentots clearly unites them with the Bushmen and there seems to be only confusion from ranking the South African Boskop race as a part of the European Cro-Magnon, although the north African Alfalou may be a connecting link and they all certainly appear to be nearly related.

There is also the problem of the relation between the dwarf Bushmen of South Africa and the central African Pygmies of the Ituri forest, which can only be touched on. An essential difference of the tropical forest Pygmies of the Congo is that they are in my view achondroplastic dwarfs, with short stout limbs but large joints. There is no indication of this character in the Bushmen. It is dominant in some modern Caucasian families and recessive in others (see Gates, 1946, for a discussion of dwarfs in various races). There is also a great difference in the blood groups, the Pygmies having 25–29 per cent of B while the Bushmen have only 6-8 per cent, showing their more primitive character. The Hottentots, however, also have 29 per cent of B. This is one of the differences from the Bushmen which requires an explanation. The Pygmies of the Ituri forest live in a condition of vassalage to the Negroes and undergo considerable crossing with them. The west African Negroes range in B from 33.3 per cent through the 20's to as low as 12 per cent in Liberia where there must be considerable white ancestry. The Bantu in southeast Africa range from 17 per cent B on the coast to 21.6 per cent in the Zulus and 24 per cent in the Bechuana, showing a northward gradient of increasing B. In all these records the most striking feature is the low frequency of B in the Bushmen—a condition which undoubtedly reflects their very primitive character. We may well suppose that, like the Australian aborigines, they originally had only the A and O blood groups.

Various writers have supposed that the Bantu are Negroes who have some Hamite ancestry. The evidence of the blood groups is strongly against this view, for the Negroes are high in the B blood group. The Bantu appear to be rather Hamites with a slight admixture of Negro blood. The Bantu differ from the Arabs in being higher in O and lower in A. Roughly they are as shown in Table 17. The higher

TABLE 17

BLOOD GROUPS IN BANTU AND ARABS

	O	A	B	AB
Bantu	52%	27	19	2
Arabs	40	35	20	5

frequency of O in the Bantu may indicate greater isolation. Northeast Africa would appear to have been a center of evolution of the Bantu-speaking people. They are partly of Mediterranean blood but their type is not easily accounted for.

To his extensive blood grouping of forty-six tribes (some 20,000 individuals) in southern and eastern Africa, Elsdon-Dew (1939) has added all the results from other parts of Africa and subjected the whole to analysis by means of Cartesian coördinates. Much in the analysis depends upon how mutations have occurred, whether (1) in certain centers only; (2) whether, for instance, B has arisen through mutation from A or from O or from both; or (3) whether certain mutation rates are general, occurring throughout the continent. Inbreeding will also affect the proportion of genes in a population. Other unpredictable distortions can arise if a small group migrates and afterwards multiplies, and of course the blood group genes have frequently been transferred by intercrossing.

Elsdon-Dew gives reasons for supposing that Egypt has been a separate mutation center from India for the B blood group. Since the percentage of B is lower in Egypt it would have begun later unless the mutation rate here has been lower. The Pygmies also appear to have been a center of the B mutation. Elsdon-Dew assumes that the blacks throughout Africa were originally devoid of both A and B, but, as pointed out elsewhere (Gates, 1946), there is reason to believe that early man already had some A from his anthropoid ancestry, although it could be lost in local migrating groups, as in the American Indians. It is clear that the blood groups can only be used in combination with physical characters in interpreting racial histories. Elsdon-Dew's suggestions solely on the blood group evidence, that the Bushmen are less "ancient" than the Bantu, and that the Hottentots arose from the Egyptians, fly in the face of the anthropological evidence and show the danger of drawing even tentative conclusions from a limited field of facts.

The accompanying photographs taken by the author in 1929, show a few of the South African native types. Plate XVIII (b) is of two workers in the Premier mine near Pretoria. The one on the right has marked superciliary ridges. Both are wearing the characteristic blankets, having just emerged from the mine and donned them. Members of nearly all the tribes in South Africa come to the mines to work. Plate XVIII (c), a young Zulu woman, was taken in the market at Durban, Natal. Note the coiffure, which was to announce her betrothal; also the ear plugs. Plate XIX (a) is a typical kraal in Southern Rhodesia. They characteristically dot the landscape in Zululand.

Plate XIX (b) shows a group of Bavenda women at Pietersburg, northern Transvaal. The features show almost a family resemblance, with prognathous jaw, small face, and depressed nose. Plate XX (a) is of a group of young Sesutos in Pietersburg. They had come in from the country to engage in litigation. Plate XX (b) represents some

"Kaffir" women and children in the Kruger Game Reserve, northern Transvaal. The ringlets of hair in two women are probably of hybrid origin. They are neither Negro, Bushman, nor Bantu. Plate XXI (a) shows some Negro children at Letaba in the Kruger National Reserve. Note the "skull-cap" distribution of the hair and the peppercorn hair in two of them. Plate XXI (b) is of two Matabele women at Bulawayo, Southern Rhodesia. Plate XXI (c) is of two women near Mafeking, Bechuanaland, showing evidence of Hottentot ancestry. Plate XVII (c) is of Barotsi Negroes at Victoria Falls, Northern Rhodesia. They may have some Mongolian ancestry.

An extreme case of microcephaly in a Basuto woman is described by Dru-Drury (1920). She died in a mental hospital at the age of thirty-two years. She had a prognathous, ape-like face, was the size of a child of twelve years, and weighed only sixty pounds. Her skull was very long and narrow, length 123 mm., breadth 86 mm., cephalic index 69.9, cranial capacity 340 cc., equal to that of an infant two weeks old. She had great intelligence for the size of brain, with the vault of the skull scarcely rising above the eye-level.

The skeleton described by Drennan (1942) from Likasi in the Belgian Congo, partly impregnated with copper salts, represents a male about twenty-five years old. The skull is complete except for the mandible. The features are Negroid, the teeth large and it might represent the present Congo population except that it is microcephalic with a low cranium, having a capacity of 1250 cc., length 181 mm., C.I. 72.4. It is very similar to the Asselar skull except that the head is small, not large, which is one of the marked differences also between Bushmen and Hottentots.

The Kohl-Larsen expedition to Lake Nyasa found a series of skeletons and skulls, some of which were described by Bauermeister (1940). Near the site of Taungs skull were found five skulls and parts of skeletons. The median incisors had been removed in youth. These skulls were negroid with some Cro-Magnon characters. The nose was broad but there were superciliary ridges, the orbits were square (not round), the mandible primitive, deep, with a nearly erect ramus. One skull was typically negroid, pentagonal in vertical view with strong alveolar prognathy. Crucial evidence is still lacking regarding the place and time of development of the Negro race, but clearly it must have taken place in Africa.

DWARF RACES

The problem of the relation, if any, between the Bushman and the Pygmies remains an open one. They may have had an independent origin, but Dart (1939) regards them as a part of the Bush (Pygmy) race. We have already pointed out that they may be achondroplastic dwarfs. Schebesta also supported the view that the Bambuti Pygmies are Bushmen with a little Boskop ancestry. It is evident that the B blood group filtered into South Africa from the north, races like the Bushmen, which already had at least 30 per cent of A, gradually absorbing 6–8 per cent of B. This confirms the evidence from the Australian aborigines that man originally had A and has since acquired B, probably through repeated mutations. The relationships of the Pygmies have been discussed elsewhere (Gates, 1946, p. 320).

A pygmy race of chimpanzee which was first recorded by E. Schwarz in 1929 as *Pan satyrus paniscus* has since been fully described and ranked as a different species, *Pan paniscus,* by Coolidge (1933). It is smaller throughout, with more slender bones than the chimpanzee, and shows many youthful characters. Coolidge suggests that it may represent an ancestral type, but it seems more probable that it is a mutational derivative from chimpanzees of normal size. Its characters apparently agree with those of human ateleiotic or miniature dwarfs. In this respect it agrees with the Akka Pygmies of central Africa, who are negroid dwarfs of normal proportions, not achondroplastic like the Ituri forest Pygmies.

There is another consideration regarding the dwarfs, which can be mentioned here. We have already referred to the general evolutionary tendency, abundantly exemplified in the mammals, to increase of size in each group as it evolves. If such a tendency applied to human phylogeny we should expect the earliest men to be dwarfs, and that some living dwarf races were survivors from an earlier age. However, the Pygmies, as achondroplastic dwarfs, appear rather to be a mutational derivative from a taller race; and the Bushmen appear to be a pedomorphic development from a taller ancestry presumably without infantile characters. Many races of modern man as well as Cro-Magnon are indeed taller than Sinanthropus, Pithecanthropus, or Neanderthal. This may be regarded as an advance in the usual evolutionary direction, though it amounts to little and is an increase in stature rather than bulk. Weidenreich (1945), in discussing the subject, in relation to Meganthropus and Gigantopithecus, believes that man's whole phylogeny has involved decrease in size, from Gigantopithecus, the size of a giant gorilla, to puny modern man. This is contrary to all the probabilities. If Gigantopithecus stands at the head of a declining series, his ancestry would have to represent a corresponding increase in size. He is probably an aberrant gigantic anthropoid contemporary with Pithecanthropus, just as the dwarf Pygmies are contemporary with taller man. Osman Hill (1939) has described a gigantic orangutan from Sumatra which, whether it proves to be a separate species from the Borneo orangutan or not, is at any rate a contemporary giant of that species.[4]

Without entering here into the problem of the numerous pygmy races of mankind (see Gates, 1946), it seems probable that they represent mutational dwarf derivatives of different races and not (except perhaps the Bushmen) survivors from an early ancestral stage of man. This conclusion would be still more certain if one of man's ancestral lines began with a giant form, such as Meganthropus or Gigantopithecus. There are many observations still to be made on pygmy races. Lord Moyne (1936) described the Aiome Pygmies near Mount Hagen in New Guinea. Two men were 4 ft. 6 in. and 4 ft. 9 in. in height respectively and three women measured from 4 ft. 2½ in. to 4 ft. 5 in. They were entirely different from the adjacent Ramu River natives—nearly a foot shorter, with a light-brown hairless skin. From the photographs, the hair appears wavy in some, woolly in others. The lower lip was pendulous and the chin receding, the nose broad, low at the root. They were more lively and intelligent than most river tribes, and very cheerful, with a sense of humor. Their bodies were filthy and their large mops of hair were enclosed in a bark covering. Lord Moyne [5]

---

[4] Selenka (1896) made an extensive study of the orangutans in Borneo. He recognized *Pithecus sumatranus* in Sumatra, with two subspecies, as distinct from *P. satyrus* in Borneo. Six geographic races of the latter were recognized, separated by the rivers which radiate from the mountains. The orangutans could not cross these boundaries but the lighter and more agile gibbons could swing across on the tree branches. The geographic subspecies were distinguished chiefly on the basis of skull size, length of mandible, and prevailing color of hair. Four varieties were microcephalic (male skull 400–450 cc. capacity) and two macrocephalic (capacity 440–540 cc.). The presence or absence of cheek-pads was another feature, and the degree to which the last molar is developed.

[5] As High Commissioner for Palestine, Lord Moyne was afterwards murdered, under the false, but too frequent, assumption that a grievance, supposed or real, is sufficient justification for any crime.

made observations of several other pygmy tribes, including the Aeta in the Philippines, the Semang in Malaya, and the Ongé in the Andamans, who are somewhat steatopygous. The hair of the Ongé is short and woolly, appearing to be peppercorn in some of his photographs.

In his wide-ranging studies of the Bush and Boskop races, Dart (1939) has collected together evidence that the dolichocephalic, low-vaulted, pentagonoid microcephalic Bushmen extended at one time from the Cape of South Africa to the Red Sea. He points out that microcephalic pygmoid skeletons have also been found in the Neolithic of Switzerland. Sergi studied forty-seven skulls, mainly from ancient skeletons in southern Italy, indicating that the short stature of the population of southern Italy is probably from this source. The crania are, however, of Mediterranean, not Bush type. He also examined 145 microcephalic skulls from Russian Kurgans of Early Iron Age and found the microcephalic dwarfish race in Sicily and Sardinia. Sergi thus found convincing evidence of the penetration into southern and eastern Europe of a race which is naturally identified with the Bushmen. The early steatopygous statuettes found in southern and eastern Europe support this view. It appears, however, that this dwarf race were not Bushmen, though possibly derived from Bushmen. A branch of the Boskop race entered Europe later from the west. Verneau (1924) states that as a medical man he has encountered cases of steatopygia in Paris. It is known (Gates, 1946, p. 776) that steatopygia can arise in Caucasians as a pathological condition, perhaps arising from endocrine derangement.

The records of these dwarfs in Europe appear to have been somewhat neglected by anthropologists. It is therefore desirable to refer to them in greater detail. The evidence is found mainly in several Swiss publications. In 1892 a large early-Neolithic burial place was discovered near Schaffhausen. There were remains of twenty-seven forest-dwelling Neolithic people, earlier than the period of the Swiss Lake Dwellings; fourteen of these were adult male skeletons and thirteen were children under seven years of age. Five of the adults were pygmies, less than 160 cm. tall. Neolithic Pygmies have been found in five places in Switzerland: Dachsenbüel and Kesslerloch near Schaffhausen, Chamblandes bei Pully (1880), in a Lake Dwelling at Moosseedorf bei Burgdorf (1901), and at Ergolzwyl near Basel (1902).

In the cave of Dachsenbüel (Nüesch, 1903) a grave was found with two skeletons, oriented east and west with their heads to the east. The grave was an oblong rectangle lined with stones. The two skeletons in this primitive stone cist were, one, a man of normal size, the other a pygmy woman. A necklace made of Serpula tubes and a drilled tooth of a wild boar were ornaments. In scattered positions around the grave were six other skeletons (three adults, one male pygmy, and two children of from one to three years), some of the bones showing signs of fire. The whole group thus included four of the large race, two pygmies and two young children. The male pygmy was about forty years old and 146 cm. high, the female in the grave only 130 cm., the male 160 cm.

This intimate association of the tall and dwarf races strongly indicates that they intermarried. Indeed, the relationship would appear to have been similar to that between the Bush and Boskop race in Africa. From Kollmann's (1903) comparison of the skulls, femur, humerus, tibia, atlas, and patella in the tall and dwarf types it is not clear, however, that anything more than a difference in stature and size is involved. A Sicilian pygmy skull had a capacity of 1031 cc. compared with 1460 cc. in an ordinary European. Kollmann also cites a pygmy skull from Abydos, upper Egypt, probably 4,000 years old.

The Kesslerloch cave, also described by Nüesch (1904), was of late Paleolithic (Magdalenian) age, belonging to the end of the mammoth and the beginning of the reindeer period. From remnants of a pygmy race found here in 1874, the height is estimated at only 120 cm. One wonders if some of these Swiss dwarfs could possibly be cretins. The associated animals in the Kesslerloch cave included thirty-three species of mammals and ten birds. *Elephas primigenius, Rhinoceros tichorhinus,* and *Equus caballus* were characteristic. Art reached its best, including a finely etched reindeer by a pool. The other abundant artifacts included bone needles, harpoons, arrow points, and stone implements.

Lapouge in 1883 described pygmies from Neolithic caves in the Cevennes in southern France, under the name *Homo contractus*. In 1896 he described another "pygmy race" from Soubès, Hérault. They were still smaller and lacked a chin. He regarded these people as aboriginal races in Europe, probably derived from the steatopygous Bushmen. Drawings and sculptures such as the Venus of Brassempouie, from a cave in the Dordogne, confirm the presence of steatopygia. In the Cave aux Fées (Seine-et-Oise) Manouvier described five Neolithic dwarfs (only 142 cm. tall) with 56 tall skeletons. Verneau recorded two female dwarf skeletons 148 and 152 cm. tall in the Grotte Mureaux. In 1892, three pygmy Neolithic skeletons at Châlons-sur-Marne were estimated to be 144–154 cm. high. Similar skeletons from Silesia were 142.9–152.3 cm. and four from Alsace 120–152 cm. From these records it is clear that in Neolithic France and Switzerland there existed a widespread "pygmy race," but it is probable that they were nothing more than very short Mediterraneans like the modern people of southern Italy. Early writers assume that all other dwarf races in various parts of the world were from the same original source. It seems more probable that many dwarf races have arisen independently from different tall ancestors. Wirth (1898), for instance, describes a tribe of pygmies only 130 cm. high, on Okinawa Island in the Lu Chu archipelago south of Japan. He says that in the southern islands of the chain, nearly half the population were dwarfs. On Okinawa, men whose height was about 1.30 m. and women 1.25 m. were not uncommon. These pygmies had a very dark skin, very much wrinkled, hair different from the Japanese, and "wild" eyes. There was also a tendency to steatopygia. These dwarfs he regarded as the autochthonous race. He concludes that they were conquered by a taller race, excavations in northern Okinawa and on Oshima in the eighteenth century having uncovered skeletons 1.90 m. tall. There is also archaeological evidence (Gates, 1946, p. 721) of Ainu in the Lu Chu islands during the Neolithic, and Wirth found storehouses of Ainu type still existing. He says the Ainu declared that the sun once came up on the opposite side, meaning that they formerly lived south of the equator. This is confirmatory evidence that they originated in the south and not in the west. Another type in Okinawa Wirth compared with Hindus. Some aboriginal tribes in neighboring Formosa have straight hair and features resembling the Amerindians (Gates and Darby, 1934). Chinese and Japanese were also represented in Okinawa, and all these elements have doubtless been intercrossing during the last half century, so that the original dwarf element, which may have been merely a branch of the Ainu, is now largely obliterated. An intelligent service man who was in Okinawa during the war says the people were markedly small but scarcely dwarfs. A recent article (Steiner, 1947) gives a general account of these people. Newman and Eng (1947), in a detailed anthropometric study of the inhabitants, show that they are shorter than the Japanese.

Another negroid skeleton was described by Boule and Vallois (1932) from Asselar in the Sahara [6] about 400 kilometers north of Timbuktu. It is of Mesolithic or early Neolithic age, apparently post-glacial, and nearly complete, the accompanying industry being more or less north African Aurignacian, in other words, Capsian. The country at this time was well watered and must have supported a rich savannah vegetation and fauna. This man was tall, over 170 cm., with long limbs and slender hands. The skull was dolichocephalic, cranial index 71. The upper incisors had been removed, a custom still surviving in certain races. The authors regard it as intermediate between the Grimaldi (Mentone) and the Hottentot (that is to say, Boskop) race. It is Negroid but not Negro, hypsicephalic with a large nose and some prognathism. It is like Cro-Magnon man in the craniofacial dysharmony, low orbits, very long lower legs and arms. They suggest that Cro-Magnon man on the one hand and Grimaldi and Asselar on the other came from a common ancestry. Also that the Hottentots and Bushmen were derived from this source, the former less specialized, the latter becoming more degenerate and infantile. Boskop, as the ancestral Hottentot, would be between the primitive Negroid and the modern Bushman. This would account for the obviously close relation of all three. They believe that the Negroid-Boskopoid stock of Asselar mostly disappeared, traces remaining in pre-dynastic Egypt, the Negroes proper developing later by a sort of orthogenesis which produced their special racial characters. There is reason to believe that the broad nose and thick lips evolved in central Africa as an adaptation to the moist tropical conditions.

The finds at Alfalou-Bou-Rhummel, near the coast in Algeria, are on a larger scale and still nearer to Cro-Magnon man. Six adult skeletons were found together. From the account of Arambourg, Boule, Vallois, and Verneau (1934), the Alfalou type differs from Cro-Magnon and

---

[6] That man formerly occupied the Sahara is shown by the abundant Paleolithic and Neolithic stone artifacts of numerous different types found in many parts of this vast desert region. Acheulian axes as well as Mousterian, Aurignacian, and Tardenoisian implements have been found (Boule, 1923). All over the Sahara occur pictorial representations of animals, some of them extinct, as the great buffalo, *Bubalus antiquus*. Some are Neolithic, others Pleistocene. They include the elephant, giraffe, and rhinoceros. In style they by no means equal the Aurignacian cave drawings but resemble some of the later decadent rock-paintings of Spain.

the Canary Island skulls in having a broader nose, a sloping forehead, and heavy brow ridges, but all are related and from a common source. Some, however, show brachycephaly. With the human remains were found *Rhinoceros Mercki, Bos primigenius,* and *Bubalus antiquus.* The upper incisors of the human skulls were always knocked out. Alfalou man in north Africa apparently represents the real source of the Cro-Magnon race in Europe. The peppercorn hair of southern Africa was probably exchanged for wavy hair and a lighter skin, though when this substitution, through negative mutations, took place we do not know.

In this chapter we have mainly considered the line of human evolution leading from Africanthropus or at least the Florisbad type to Boskop and the modern Hottentot and Bushman (*Homo capensis*). Everywhere there has been some mixture of related and geographically adjacent types, the Bantu invasion southwards extending over some two thousand years. Several local types, such as the Zitzikama and Matjes River groups, developed through isolation in the south coast region. The Bushman and Boskop type seem everywhere to have hybridized and yet to have remained distinct types, the smaller Bushmen with their bows and arrows being the aggressors. The pedomorphy of the Boskop expressed itself in their enormous skulls with some infantile characters, that of the Bushmen in their small size, while both possessed the "foetal" characters, yellowish skin, hairlessness, and, I think we may add, peppercorn hair.[7] This last condition is seen in some Negro children and is quite possibly a character of the Negro foetus. The Bush and Boskop races also agree in having low (chamaecephalic) skulls. The three races, Bush, Hottentot, and Negro, also agree in having small ears without a lobe, although the Negro head is typically orthocephalic while the others are chamaecephalic.

The similarity of the peppercorn and woolly hair-forms shows a close ancestral relation between Bushman and Negro. If the pedomorphic theory is true we would expect the Bushmen to have been derived from Negroid ancestors somewhere in Africa; not from any modern form of Negro but from proto-Negroids who had not yet developed

---

[7] Quatrefages and Hamy (1882), in their great work, *Crania Ethnica*, figure a Greek vase belonging to the third century B.C. with a side-view of a head showing peppercorn hair, although the features apparently are neither Negroid nor Bushmanoid. At the tomb of the Kings, Biban-el-Molouk, Egypt, are figured the four types, Libyan (fair), Semite, Negro, and Egyptian. In the same work (p. 165) an Etruscan vase is figured showing clearly a Negro head in relief, with thick lips and broad nose.

the prognathism, thick lips, and other specialized features of the modern Negro race. Some writers point out that the Kanjera negroid type (to be described later) may have evolved direct from Rhodesian man, but his greater age indicates that he belongs to the orangoid stock without heavy-browed ancestors. In any case, evolution on the African continent south of the Sahara appears to have been an African matter, little influenced from outside except by the late arrival of the east African Bantu, whose home is believed to have been on Lake Chad, but who appear to be to a large extent of extra-African origin. They only reached South Africa within historical times. The Bantu movement from the Great Lakes region is believed to have begun about the ninth century. In the tenth century they were south of the Zambesi. There were three main waves, a westward movement into Bechuanaland, the central Barotse near the Zambesi, and the east coast invaders, the Tembus, Pondos, and Xosas, and behind them the Swazis and Zulus of northern Natal. As a part of this movement the Kaffirs invaded Pondoland, killed the Hottentot men and kept the women. This accounts for Hottentot characters in the present Pondo (Broom, 1943). Some old Kaffirs still living had Hottentot grandmothers.

This southern movement of the Bantu-speaking peoples is considered in some detail by Fantham (1936) but need not be entered into here. It led to nine Kaffir wars with the whites in the century between the years 1779 and 1877. The earth shook with the impis of Chaka, but the Zulus were subdued in 1879. The western stream ended with the Hereros and the Ovambos in southwest Africa, but the eastern coast stream was the largest, in four linguistic groups. Many collisions and some intermingling of tribes took place as this human deluge swept southward. A century ago ninety-five different tribes were known. Many are now detribalized and the natives of Cape Province have little idea of their origin. Practically all tribes in South Africa now show intermixture. Fantham states that in crosses between Bushmen and Bantus the Bush characters are dominant.

Another racial movement, probably beginning earlier and still going on, has received much less attention than this southern progress of the Bantu. The Seligmans (1932) have considered it at length. It involves the populations of northeast Africa and their movements westward in contact with the Negroes. The Seligmans regard the Hamites as originating either in southern Arabia or further east, or in the Horn of

Africa itself. When the Sahara was a fertile, watered country in the pluvial periods of the Pleistocene, the same must have been true of the great Arabian desert, which may well have been the seat of some human evolution but which is archaeologically unexplored. This may have been the home and source of the Hamites, who belong to the Mediterranean (Brown) race. They are regarded as purest in predynastic Egypt and in such modern tribes as the Beni-Amer in the Red Sea province of the Sudan, the Somalis, Gallas, and Beja. They are about 64–66 in. high, with a cephalic index of 74–75, the skin color a yellowish or coppery red-brown, the hair wavy or frizzy (not woolly), the beard scanty and the nose straight and narrow. In all these features except skin color and dolichocephaly they contrast strongly with the Negro.

Whether these Hamites [8] originated in Arabia or, more probably, in the Horn of Africa, they pushed westward into the territory of the Negroes in successive waves, beginning perhaps at the end of the last pluvial. Thus were produced many groups of more or less Negroid Hamites, who were pastoral Caucasians and asserted their domination over the agricultural (hoe culture) Negroes. An advancing wave would intermarry with the Negroes. The resulting mixture, being superior to the pure Negro, would be pushed farther west by the next wave. This process of infiltration, with the Hamite element remaining the aristocrats, continued over a long period, the people becoming more pastoral. Such tribes as the Masai and the Baganda are the result of this mixture. The Bahima are tall, cattle-owning aristocrats with a narrow nose and a long face but woolly hair like the Negroes. The Bahera, a Negro people living in the same country, are shorter with broader faces. They supply the Bahima with grain and thus the two tribes live in a form of symbiosis.

In the Sudan region of the Upper Nile there are thus two great groups of Hamiticized Negroes, (1) the Nilotes, and (2) the Nilo-Hamites mainly in Kenya (including the Bari, Lotuko, Lokoiya,

---

[8] To me the origin and status of the Hamites remains very uncertain. They appear to be autochthonous in the Horn of Africa, but they may be hybrid in origin. T. H. Huxley and others have recognized an Ethiopian race, which could include the Hamites and many of the Abyssinians. E. Fischer (1929) believes that the Somali type of hair, frequently in long ringlets, is the result of a cross between the wavy (*schlicht*) hair of the Mediterranean race and the closely spiral (kinky) hair of the Negro. He finds the Somalis intermediate also in width of nose, thickness of lips, breadth of jaws, prognathy, and skull form, and accounts for the larger size of some as a possible case of hybrid vigor.

Masai). Both groups are dolichocephalic and very tall, these characters being most marked in the Nilotes. The Seligmans "leave little doubt that stature greater than either parent stock is one of the results of the mixing of Negro and Hamite." If this is the reason for the tall, stork-like Nilotes and their relatives it is the best example of hybrid vigor (heterosis) in mankind. In fact there is no case of heterosis in man which is genetically authenticated, but in the Nilotes this appears to be the most likely explanation. The subject is not without difficulties, however; for instance, no one has recorded hybrid vigor in crosses between Negroes and whites. The authors suggest that heterosis only occurs in the Negro-Hamite cross when their bloods are mixed in certain proportions. In all experimental results with plants and animals, however, the maximum hybrid vigor occurs in the first hybrid generation and it rapidly diminishes and disappears as heterozygosity becomes less. Thus, on the basis of present knowledge, hybrid vigor is only a superficial explanation of the tallness arising from Negro-Hamite crosses. It seems clear that some form of selection must have occurred in the descendants of these crosses, as the present stature is so uniformly very tall. This cannot depend on the swampy conditions in which the Nilotes live, for the Bari are equally tall and the Batusi of the Ruanda uplands are tallest of all (average 1.80 meters or nearly 71 in.).

The Nile-Congo watershed marks the limit of Hamiticized Negroes, but even among the Azande in the southwestern Sudan the aristocracy appear to have some Hamite blood. The king of one tribe had a tolerably thick beard, an orthognathous profile, a Caucasian nose, and a bright brown skin, while in many of the Zande the skin has a reddish coppery color.

Mousterian implements have been found in the Sudan, but the Neolithic was much later than in Europe. Grinding stones occur, as well as microliths. There is evidence of Egyptian influence, in the form of scarabs and plaques, from 700 B.C. Although contact with Egyptian civilization dates from this early period, there was no permanent development as a result. Even if there was physical heterosis, it was accompanied by mental stagnation through the ages. In the northern Sudan (Nubia), which was in direct contact with southern Egypt, the Meroitic Kingdom developed from about 700 B.C. to A.D. 350. Extensive ruins on the eastern bank of the Nile near Khartoum include palaces, temples, and the pyramids of the kings. The Meroitic language was a debased form of Egyptian hieroglyphics but it lasted only about four centuries. The rest of the Sudan never acquired even a temporary civilization.

Iron was probably introduced from the Meroitic Kingdom about two thousand years ago. The Nubians are much the same now as they were 2000 B.C. It must be pointed out that the vast swamps of sudd on the White Nile rendered any direct communication with Egypt by this route impossible; but Egyptian influence passed southwards by a more easterly route up the Blue Nile into Abyssinia and thence to the Great Lakes. The modern Nilotes distort the horns of their cattle into the same bizarre shapes that the Egyptians used in the Pyramid Age of Saqqara 2700 B.C. The Congo burial customs also show Egyptian influence. It is said that no Negroes are represented in Egypt until after the Age of the Pyramids.

Rates of diffusion of culture elements vary enormously, depending on many conditions. The diffusion rate can seldom be determined with any accuracy except in modern times. As an ancient example, the domestic fowl, derived from the jungle fowl of India and Ceylon, is said to have reached Egypt about 700 B.C. and the Congo about 400 B.C.

Three of the chief tribes of Nilotes are the Shilluk, Dinka, and Nuer, all of which have been much studied. They are not only very tall and dolichocephalic, with a very dark skin, but their hair is woolly and their nose usually broad. Some of their measurements are compared in Table 18.

TABLE 18

MEASUREMENTS OF THREE TRIBES OF NILOTES

	C. I.	N. I.	Stature
Shilluk (21)	71.3	93.3	177 cm.
Dinka (85)	72.7	91.6	178 cm.
Nuer (40)	73.5	100.	179 cm.

Some of the Shilluk, especially in the aristocracy, have a long, shapely face, thin lips, a fine nose, and à well-modeled forehead. Temperamentally the Nilotes are apart from all other people in the Sudan. They are proud, aloof, intensely religious introverts. The cradleland of the Nilotes was east of the Great Lakes. Here, in a welter of partially Hamiticized Negroes owning cattle, was differentiated a group with black skin and woolly hair, retaining the Hamite outlook regarding cattle, but with some elements of Negro culture, such as the absence of circumcision. Thence emerged two great waves of migration. The first wave moved northwards, producing the modern Dinka and Nuer. The second wave, moving northwards, gave rise to the Shilluk-speaking tribes (Shilluk, Luo, Anuak). They made contact with the mesaticephals from west of the Nile-Congo divide. The new mixed people so formed have some elements of Negro culture, such as rain-stones and a notched stick over graves. The Shilluk and Dinka rain-makers are "divine kings" (Fraser) killed ceremonially for the people's benefit. The Nilo-Hamites, related to the Masai and Nandi of Kenya, have more Hamitic blood and language. The Bari reached the Nile by migration and adopted rain-stones from the western Negroes, but were not affected physically. The Acholi were affected both culturally and physically. The Bari in Mongalla Province are frequently pseudo-Mongoloid. They have high cheek bones, a broad flat face, eyes with a narrow palpebral fissure and often with the epicanthic fold. Seligman suggests that the Mongoloidism is a mutation, and Weninger indicates that the epicanthus may have come from the Bushmen. Further study of this pseudo-Mongolism is desirable.

All these peoples apparently have uniformly woolly hair and black skin. As I shall show in a future book, genetic segregation is just as characteristic of human racial crosses as of any Mendelian differences in animals and plants. This is true notwithstanding all the propaganda to the contrary. Indeed, it was clearly shown over fifteen years ago to be of general occurrence in mankind (Gates, 1929, chapter XVI).

Since these tribes of mixed ancestry now breed true, or at any rate do not show marked segregation in regard to the above-named characters, a homozygous condition of these characters must have been attained. The net result is the transfer of certain characters, and not of others, to the resulting race of mixed origin. The same process of transfer can be seen in some of the hybrid skulls from South Africa already considered. Whether any process of selection is involved in these phenomena remains for the future to determine. Extensive blood-grouping of the Sudan peoples would throw considerable light on their history and their relation to the Negroes and the eastern Bantu.

Returning to South Africa: Broom characterizes the Korana as of moderate size, with a darker skin than the Hottentot, a rather broad nose, sloping forehead, and a very long, narrow skull. Some Koranas, both living and from burials, had broad brows with marked supra-orbital ridges (Plate xv). We have already seen that this pseudo-Australoid character is an intermediate condition probably descended from crosses with the Rhodesian type of man. There is no reason to suppose that it implies any genetic relationship whatever to the Australian aborigines, but it is a very definite parallel in the African continent. The Koranas are found mainly in the Orange Free State and East Cape Colony (Gonaquas). They are a vanishing race, having mixed with Bantu and Bush as far back as records go. Broom confines the term Hottentot to the Bush-Korana mixture, found now mainly in West Cape Colony and Namaqualand. He suggests that since the Bushmen are very low in B blood group and the Hottentots very high, the Korana must have had only O and B. It is to be hoped that the Korana will soon be blood-grouped. If they prove to be as he suggests, it would be a strong confirmation of his hypothesis.

Dart (1938) recognizes the Boskop race as essentially ancestral to the Hottentots. In Broom's view the Hottentots are hybrids of Korana and Bush, while the Korana carry factors for heavy eyebrow ridges which make them descendants, not of the Boskop but of the Rhodesian type of man. Boskop man must in either case already have had the peppercorn type of hair, from which the woolly Negro hair may be regarded as a further development offering better protection from a tropical sun. Dart (1939) goes further and identifies early Boskop with Cro-Magnon man. This seems justified on the basis of skull shape and stature, but it appears that there must have been certain differences between the European and the African branches of the Boskop race. The peppercorn hair must have been replaced by wavy European hair and yellowish-brown by white skin color. It has been suggested elsewhere (Gates, 1946, p. 1358) that peppercorn hair is genically hypostatic to woolly hair. These genetic changes probably took place in North Africa, perhaps in the Sahara, before Cro-Magnon man ventured across the straits into Spain.

The continent of Africa has seen the evolution of one segment of Hominidae from the in some ways prehominid Rhodesian man and Africanthropus to the Boskop type (*Homo capensis*) which spread south to the Cape and northwards to the Mediterranean. This evolution appears to have produced the African Negro in the heart of the continent through specialization and adaptation to tropical conditions, the similarity between the peppercorn hair of the Bushmen and the kinky hair of the Negroes being a close connecting link. The evident relation of Boskop man to the European Cro-Magnons, makes Boskop appear as ancestral to them but unrelated to Neanderthal. As already mentioned, the dark skin and peppercorn hair were probably shed in northern Africa before this species of Homo entered Europe. What the hair and skin characters of Neanderthal man were like we have no means of knowing.

Dart (1937ab) has considered further the remarkable genetic relations between the Bush and Boskop races, as seen in the remnants of living Bushmen. At the time of the Empire Exhibition in Johannesburg in 1937, seventy-seven Bushmen of the Auni-Khomani group, regarded as the purest, were brought from the Kalahari Desert to Johannesburg and carefully studied somatologically. They divided themselves into eleven hut groups, or families, all but two of which were found to be closely interrelated by marriage and highly interbred. Marriages were permissible between cousins and also between uncle and niece as well as half-brother and half-sister, provided they did not have the same mother; but actually no marriages between half-sibs were found. The families had little or no knowledge of interrelationships even in the generation preceding their own, but the coefficient of inbreeding must have been very high, every individual being related several times over to the rest of his generation. It is believed that not over two hundred Bushmen are left in the Union of South Africa, though some having the same names remain in Angola, a thousand miles northward.

From the evidence already considered in this chapter, it is clear that the ancient Boskop race was almost continent-wide in Africa. The Bush and Boskop elements or races within this line of descent have apparently intercrossed from an early period; yet they remain as more or less distinct elements, and Dart (1937b) believes they can be genetically disentangled even in the present Kalahari Bushmen, because they hybridize like two genetically distinct types.

The Boskop people could not have produced the Neolithic culture of the historical Hottentots. The latter must therefore have acquired it by contact with Bantus or Hamites in northeast Africa. Schultze (1928) made a study of seventy-four Hottentots and fifteen Bushmen and suggested the name Koisan for the two yellow-skinned groups. His observations extended over twenty years. In Dart's observations and measurements, which are a continuation of this work, he contrasts the Bush and Boskop elements in all their features. Without entering into detail, a few of the results may be considered. In cephalic form, Dart finds the Bush short-pentagonoid (170–180 mm. long), the Boskop long-pentagonoid (190–200 mm. long), while the Negro skull is typically ovoid. Of the 77 Kalahari natives, he found 48 (62.3 per cent) pentagonoides acutus, 27 (35 per cent) pentagonoides obtusus, 2 ovoid, and 2 "Australoid." The average Boskop head measured 193.2 × 147.9 × 147.1, the average Bush head 178.5 × 137.2 × 138.6 mm. In the whole group 17 (22.1 per cent) of the heads were Boskop, 18 (23.4 per cent) Bush, the rest (54.5 per cent) being Bush-Boskop or Boskop-Bush, in other words, showing recombinations of some of the factors for head shape. In a similar analysis of facial features he finds that while both racial elements have a short, broad face the Boskop nose is larger in all three dimensions, also broader and more elevated than the Bush. In facial type, 55.8 per cent were dominantly Bush, 44.2 per cent Boskopoid. A similar analysis of the facial elements in the Bantu is shown in Table 19.

TABLE 19

ANALYSIS OF FACIAL ELEMENTS IN THE BANTU

	Authors	Negro face	Bush	Brown	Mongoloid
98 Bantu ♀	Orford & Wells	48.9%	32.7	15.3	1%
977 Bantu ♂	Wells & DeSaxe	51.3	25	22.3	1.5%

The facial features were found to be frequently of other racial types. Thus, in 23.4 per cent the features were Mediterranean, in 15.6 per cent Armenoid, in 24.7 per cent Mongoloid—a remarkably high frequency. In a further analysis the Mediterranean and Mongoloid features were more often associated with a Bush face, the Armenoid with a Boskop face, suggesting genetic linkage. Then by comparing cranial type with facial type the relations in Table 20 were found.

TABLE 20.

ASSOCIATION OF CRANIAL AND FACIAL TYPE IN BUSHMEN

Number	Cranial type	Boskop face	Bushman face
17	Boskop	13	4
18	Bush	4	14
18	Boskop-Bush	10	8
24	Bush-Boskop	7	17
38 Males		24	14
39 Females		10	29

Thus Dart found the Boskop cranium more frequently associated with Boskop face and vice versa. More interesting still, the males generally had a Boskop face, the females a Bush face. Also, of 13 with Boskop cranium and face, all were males but one infant; and of 14 with Bush cranium and face, all were female except one infant. The linkage of cranial and facial type with sex is thus of much interest.

The short narrow Bush head in crosses with the long wide Boskop head gives short and wide (mesaticephalic) heads. Dart attributes the origin of brachycephaly in Africa to crosses with Orientals, but this is doubtful. He makes the mistaken assumption that brachycephaly will spread if it is dominant. An elementary analysis of populations shows that this is not so. He says (p. 198), "Apparently for the production of Boskopoid types the genes require to be carried by both parents." This would imply that the Boskop type is recessive to the Bush, but he found that a Boskop father has some Boskop and some Bush children by a Boskop wife. From the further analysis of head types, Dart concludes that an intimate admixture of the Bush and Brown (Mediterranean) races took place in east Africa, beginning in prehistoric times when the Bush race extended from the Cape of Good Hope to the Red Sea. He believes that this mixed type, by further mixture with the Negro, produced the present southern Bantu, and he estimates that the latter have 25–32.7 per cent of Bush ancestry.

As regards stature, Dart finds short, medium, and tall in both sexes of the Bushmen, the whole range being 145–160 cm. for males and 135–155 cm. for females. The Ituri Pygmies are definitely shorter, the means being about 146 cm. for males and 133.5 cm. for females. The male Bushmen are about 10 cm. taller than the females and the same difference exists between the Boskop and Bush elements. Both elements are virtually hairless except on the scalp. In the Boskop the peppercorn hair may be in larger, looser tufts, but in 22 per cent of Bushmen the hair is woolly. Body hair, when present, is ascribed to crosses with other races.

The spine is marked by an anterior displacement which Dart names *proptosis*, the abdomen being protuberant but the vertebral column vertical, thus making the buttocks prominent behind while the head is held forward over the chest. The Bushmen are incompletely erect, automatically take a crouched position, and sleep curled up in a foetal posture. There are Neanderthaloid features in the foot and the pelvis as well as in the vertebral column.

In a further analysis the well-known steatopygia is regarded as an expression of constitutional infantilism in posture and in fat distribution. The pelvis is also more anthropoid than in any other living race, and the sacrum is sometimes very simian. Three types of buttocks were first recognized by Drury and Drennan, the steatopygia being more obvious in females but also present in males. Angulated buttocks, with two to five gluteal folds, are regarded as characteristic of the Boskop and dominant over the other two types. The pancake (flattened posterior) is assigned to the Bush, while the rounded type is European. Among sixty-eight individuals, 50 per cent were of the extreme or angulated type, 13.2 per cent of the pancake type, and 36.8 per cent of rounded type. Sir Arthur Evans found at Knossos Neolithic figurines showing steatopygia, also in prehistoric Egypt and Malta and north of Greece.

A similar analysis of the mammae discloses three types of breasts: (1) Large and sphaeroid with a small areola. This is the Caucasian (Mediterranean) type and accounts for 22.2 per cent of the women. (2) Small and conoid with a large, globoid areola. This is the Boskop type, present in 40.8 per cent. (3) Pendulous, elliptical with a huge, dark non-elevated areola—the Bush type (37 per cent). It is well known that the external sex-organs of the Bushmen of both sexes differ markedly from all other races.

In the males the penis was horizontal in 16, diagonal in 7, the latter character being regarded as Boskop in origin. The scrotum was typically tight and contracted in all but four. In female Bushmen, Drury and Drennan (1926) described two types of labia minora: (1) the butterfly type with two finger-like pendants, at Sandfontein, southwest Africa; (2) the wattle type, resembling a turkey's wattles in shape, in the Cape Bush and the Masarwa of Bechuanaland. Where these races meet, both types were present. There is evidence that in some cases they formerly reached almost to the knee. Dart (1937) finds two additional types; (3) reversed butterfly, in which the finger-like pendant from the web is posterior instead of anterior; (4) reversed wattle. This lengthening of the labia at puberty occurs naturally and is absent in infants. Twenty-six adult or near-adult females were examined. In 9, which were of Boskopoid type, 2 had butterfly labia, 5 wattle (one reversed), and 2 European type. Seventeen were of Bush type, 5 of them having butterfly labia, 7 wattle, and 5 European. In Orford and Wells's data, four Boskopoid-Bantu females with marked steatopygia had butterfly labia and one with no steatopygia had European labia. Thirteen Bantu females with short Bush heads and facies included 12 with butterfly and 1 with European labia. In a total of 59, 4 had wattle labia and were of Negro or Mongoloid facies. Dart concludes that angulated steatopygia and butterfly labia are Boskop characters, but not linked in heredity, while pancake steatopygia and wattle labia are the corresponding unlinked characters in the Bush. There was some evidence of inheritance, one woman and two of her daughters having wattle labia, while a half-caste European had European labia. Her daughter by a Bushman had butterfly labia, suggesting that this type is recessive to the European type. Six women (23 per cent) had the Eur-Asiatic type of labia, indicating previous crosses with Europeans. The wattle type reached a maximum of 7 cm. in length, the butterfly 9 cm., the average length being 4–5 cm.

These remarkable secondary sexual characters of the Bushman race, as regards labia, mammae, and steatopygia require further genetic analysis, which should throw further light on the development and inheritance of these unique sexual conditions. Drury and Drennan interpret them as a persistence of infantile conditions in both sexes. The enlarged buttocks of the females are regarded as a fat reserve which enables the women to suckle their children through a drought, yet in central Australia, where the degree of aridity is much more extreme, this development has never taken place. When food is plentiful they grow to great size and in times of scarcity they become dull, flabby, and shrunken. They may thus be likened to the camel's hump. Drury and Drennan also note that in Bush women the perineum is wider than in other races. These peculiar conditions of the Bush race have been known in a general way since the eighteenth century.

Dart concludes that the Bushmen form a neatly balanced racial mixture in equal proportions, the Boskop being linked to the male sex and the Bush to the female. In addition, a certain amount of genetic basis has been derived from the Brown (Mediterranean or Hamite) race, as well as genes from the Mongoloids and Armenoids, but very little from the Negro. Further, the gerontomorphic Armenoid genes are linked with the Boskop type to the male sex, while the pedomorphic Mediterranean and Mongoloid genes are linked with the Bush type to the female sex. Dart also finds a close relation between the Bush and the Pygmies, but this is less convincing to the writer, who has considered the subject elsewhere (Gates, 1946). In another contribution, Dart (1940) has surveyed many problems of human history in South Africa.

Regarding the origin of the Bushman type, Drennan (1945) suggests that they have developed by a glandular adjustment to desert conditions, producing the pygmy form and the steatopygia as a "method of storing fat for the emergency of rearing children through a drought."

An anthropometric study of nearly a hundred Bantu women at Johannesburg by Orford and Wells (1937) showed the presence of some Bushman characters, presumably derived from ancestral crossing. For instance, 13 of 98 showed marked steatopygia and the labia minora were generally elongated. The women were moderately dolichocephalic and the facial features were classed as Negro (48), Bush (32), Mediterranean (15), Armenoid (1), Mongoloid (2). These were regarded as representing racial strains.

The important discoveries already referred to in Kenya remain to be considered and, if possible, related to the general story of man. We have seen that Boskop man, discovered in South Africa, was later found to have extended much further northward and to show relationships with north Africans and Cro-Magnon man. The Bushmen similarly appear to have had extra-African extensions in southern Europe.

At Oldoway, north of Lake Eyasi in northern Tanganyika, Professor H. Reck (1914) discovered a human skull and skeleton in 1913 in very hard tufa. It belonged to "modern" man (Homo) and had been buried in a contracted position. Reck and Mollison both concluded that it was not a recent burial, the latter placing it as contemporary with Magdalenian man in Europe, though not as old as the fossil-bearing strata in which it was buried. Five successive tufaceous layers were distinguished in the Oldoway gorge. The lowest contained few fossils, the second, in which the human skeleton was found, was rich in mammalian fossils, including two species of elephant different from the modern African elephant, and a hippopotamus. The third deposit contained bones of deer and elephant, the fourth abundant fish bones and elephant remains, the fifth (top) layer containing a steppe fauna of antelopes and gazelles.

In 1931, Reck joined a Leakey expedition to East Africa and pointed out the original site. Artifacts were found here belonging to the Upper Kenya Aurignacian and from consideration of all the evidence (Leakey *et al.*, 1933) it was concluded that bed v had been deposited after the burial of the skeleton in bed II. It was thus Upper Pleistocene, of the same period and type as the Gamble's cave skulls to be described later.

In 1931, in bed I at Oldoway Gorge, remains of a large Deinotherium had been found, associated with other animal remains and very crude human artifacts (Leakey, 1932). The artifacts later found at Oldoway made a complete culture sequence from pre-Chellean to advanced Acheulean. The chief interest of the Oldoway skeleton is that the cranium and face are (like the culture) of Gamble's cave type, but the mandible is of the Elmenteitan race—another example of dysharmony from crossing. The skull (Fig. 69 in Weidenreich, 1946) is clearly of the long, narrow-faced Nilotic type. Leakey (1933) pointed out that the Kanam and Kanjera remains were associated with two stages in the evolution of the *coup-de-poing* culture, a widely diffused industry the makers of which were now identified for the first time. He regarded Africa as the probable home of this culture and of light-browed "modern" man. The Kanjera skulls, of Lower Pleistocene age, were very thick, and agreed with the Piltdown skull in the structure of the diploe. Thus the genus Homo with very little change goes back at least half-a-million years.

Two of the most important finds described by Leakey (1935) were from Kanam and Kanjera, about four miles apart on the shores of the

Gulf of Kavirondo, Lake Victoria. In 1932, a fragment of a human mandible was found at Kanam West with a Deinotherium tooth and a human artifact. The Kanam mandible was dug out of a gully in a hard matrix. Its history was that it had been damaged and weathered before being washed down into an old Lower Pleistocene lake with other bones. Extremely rapid erosion in the newly formed gullies leading out of the lake cliff had nearly exposed these bones. Deinotherium and Mastodon, found here, were survivals into Lower Pleistocene from the Upper Pliocene. The series of old lake beds yielded a fauna and a culture sequence like that at Oldoway, but including one stage earlier, corresponding to Oldoway bed 1. The evidence from the fauna, geology, and culture stages shows that the Kanam mandible is of Lower Pleistocene age, an age as great as the Upper Pliocene of some areas. Thus man living here antedated Neanderthal man in Europe. These fossil beds, antedating the end of the Kamasian pluvial, were 160 feet thick. The Kanam series is near the base, the Kanjera series near the top. However, Boswell denies, on geological grounds, that the human bones are of this age.

Leakey named the mandible *Homo kanamensis*. Though massive, it belonged to the light-boned type of man, with a pronounced mental prominence, but it differs specifically from *Homo sapiens* in three features: (1) the premolars have no neck; (2) the roots of the premolars have a very large root-canal; (3) the shape of the root of the first molar differs. Leakey regards it as directly ancestral to *Homo sapiens*.

At Kanjera, Mr. D. G. MacInnes first found scattered fragments of a skull on the rapidly eroding surface. The pieces were heavily mineralized and had the same appearance and texture as the bones of extinct animals in the beds. Fragments of four skulls were found, some on the surface and some *in situ*. They were not from burials but had been broken before being incorporated in the beds, which are now under rapid erosion. A femur fragment was also picked up, coated with matrix and with a strongly developed pilaster area, as in the Cro-Magnon and Combe Capelle types. These remains were of Middle Pleistocene age, showing that in the highlands of Kenya near the equator man of the genus Homo had lived long before Palaeoanthropus (Neanderthal) in Europe.

These Kanjera skulls were ultradolichocephalic. Two of them could be reconstructed. They showed great thickness of bone, while in another the bones were less thick. All showed very poor development of the supra-orbital ridges, a feature which Leakey regards as both infantile and ancestral. Like Eoanthropus, of similar age, they have no frontal torus. From the endocranial cast Elliot Smith (in Leakey, 1935) found evidence that Kanjera man, like Boskop man, had some extremely primitive features combined with others distinctive of modern man. Like other early skulls, the Kanjera endocast shows exceptional symmetry of the occipital poles of the two hemispheres, with large protuberant bosses of the area striata. As in Sinanthropus, the sulcus lunatus coincides with the lambdoid suture, the prelunate area being depressed, as in other primitive brains. The fullness of the anterior ends of the cerebral hemispheres in the Kanjera cast is distinctive of modern man. Notwithstanding the great differences in skull structure of the heavy-browed and light-browed types of man, the brains of Sinanthropus and Kanjera man are found to be surprisingly alike. They seem to have reached about the same level of development. This is another indication that advance in brain size has been fundamental in human phylogeny, while skull shapes and "ornamentations" have been secondary.

It has usually been supposed that the heavy-boned type of man, with superciliary and occipital tori, represents an ancestral condition through which all men have passed. The accumulation of evidence that the heavy-browed and light-browed types were roughly contemporaneous, but the latter really older both in Europe and Africa, necessitates a fundamental revision of this concept. As pointed out in Chapter 5, the heavy-browed and light-browed types of man have evolved more or less independently and simultaneously, not one from the other; just as the orangutan and the gorilla have evolved in separate regions, probably from neighboring Dryopithecine ancestors, the one with a frontal torus, the other without. However, the parallel is not complete, for, as we have seen, *Homo australicus*, though descended from the heavy-browed Pithecanthropus, has now acquired relatively light superciliary ridges, though they remain heavier than in any other modern man. Also in Africa, while the light-browed Boskop type belongs to the independent line of descent, there has been some crossing with the heavy-browed Rhodesian type, as indicated, for instance, by the pseudo-Australoid individuals among the Koranas. There has also probably been some crossing between the heavy-browed and light-browed stocks in Europe and in Palestine, but in general they appear to have pursued their evolution independently on all three continents of the Old World. The fact that both these lines of descent show essentially the same series of culture development also implies contact between them from time to time, with interchange of culture elements, although the contacts were probably for the most part of a hostile nature. The situation in America will be considered in chapter 9.

Returning to Kenya, Galloway (1937b) regards the Kanam mandibular fragment as a probable variant of the Boskop mandible. Of the Kanjera remains, skull III is very long, the bones very thick and the shape chamaepentagonoid while skull I shows slight trigonocephaly. They both have a low forehead and a very low vault. The glabella and superciliary eminences are infantile, but there is evidence of the bilateral presence of the ophryonic groove. The relatively pure Boskop type must have persisted in northeast Africa over a long period.

At Gamble's caves, near Lake Elmenteita in the Great Rift Valley about a hundred miles east of the Gulf of Kavirondo, Leakey (1935) found heavily fossilized skulls of Late Pleistocene age in excavations between 1926 and 1929. Gamble's caves are really two rock shelters side by side, and the important findings were mostly in the second. This showed four successive occupation levels with unoccupied strata between. The upper levels revealed an Elmenteitan industry (described in Leakey, 1931) but no skeletons. After a long time interval, the second (lower) occupation level contained hearths and a Kenya Stillbay culture. A very short interval preceded the third occupation with a Late Upper Kenya Aurignacian culture. The fourth and lowest occupation contained implements of Upper Kenya Aurignacian. Bed x, in the second occupation level, at a depth of fourteen feet, had three skeletons buried against the wall, with their knees drawn up and facing the entrance. Large stones had been placed over the bodies, which were covered with red ocher. These remains were in a crumbling and fragmentary condition, but in the third occupation level two more skeletons were found, one of them nearly complete. The skull was of "modern" (Homo) type, length 192 mm., breadth 136 mm., cephalic index 70.8, the glabella rather marked, the superciliary ridges slight, the occiput bun-shaped. The face was orthognathous, of medium length, very wide, the nose very long (56 mm.)

but fairly wide, the chin slight. The other skull was mesocephalic (C.I. 76). Galloway regards these as raw unharmonious hybrids between the ancient Boskop and Caucasoid (Mediterranean) and Negroid types. The relatively long face is combined with a pentagonoid pedomorphic brain-case and low vault, the associated implements being Upper Aurignacian.

At Elmenteita, about 140 miles east of Kanam, Leakey investigated a remarkable collection of skulls and dispersed skeletons belonging to the Makalian wet phase when nearby Lake Nakuru rose to some 375 feet above its present level. Bodies which had been placed in crevices in the rock face and covered with large stones were washed out and the bones collected in eddies when the lake rose. At least 28 skeletons of all ages were involved. Some 17 skulls, some of them complete, 33 femora, and 24 tibiae yielded an extensive material of this Mesolithic people. The cephalic index of six skulls ranged from 67.45 to 80.4 (3 dolicho-, 2 meso-, 1 brachycephalic). Here, in Mesolithic times brachycephaly had already appeared, as had mesocephaly. The occurrence of mutations for head shape seems the most reasonable way to account for their origin. There is no evidence of an invading brachycephalic people, and no reason to make such an assumption. Galloway regards these Elmenteitans as the same mixed type, some having a few Boskop features. The long and relatively narrow noses allied them to the east African Hamites. They were quite distinct from the Boskop type, but the two must already have been in contact.

Leakey also excavated half a dozen skeletons of Neolithic age from old lake deposits on which shell-mounds were situated, near Kanam. They were dolichocephalic with no brow ridges, the cranial bones thick, the skulls large, mandibles massive, the faces small, and the teeth little worn even in the old. These shell-mound people showed resemblance to the Boskop type. The industry was Wilton, and Leakey regards them as belonging to the same stock as the Strandloopers. They also had the same method of subsistence on shellfish, which accounts for the unworn teeth. The refuse at Zitzikama is similarly compared with the middens near Abercorn on Lake Tanganyika. The Fish Hoek skull, of different type, like a large Bushman skull, is representative of the men probably responsible for the Stillbay culture at the Cape. This culture is widespread in Kenya, but hitherto only three teeth have been found of the men who made it.

At Eburru, not many miles from Elmenteita, three burial mounds at the foot of a kopje yielded three similar skulls from which the lower incisors had been removed in early life. The associated Neolithic Gumban A culture included a characteristic type of pottery with obsidian implements. One skull was 194 mm. long, 122 mm. wide, the cephalic index probably 65.7, allowing for distortion by earth pressure. In the others the C.I. was 67.5 and about 67.7. Another Neolithic site, at the great Ngorongoro Crater in Tanganyika, was excavated by the Germans in 1915, but the skeletons were lost. The culture was Gumban B, like that at Nakuru. Here Neolithic hut circles were surrounded by a rough stone wall. The burial mound of an important man was against the foot of a cliff, with beads and pottery. The body was folded up and covered with red ocher. Over it were piled stones and rubbish, discarded tools and the bones of animals. Among the stones were nine human skeletons, probably wives or slaves who had been slain to accompany their master. They were disintegrating, but the skull of the chief was in perfect preservation, C.I. 69.3. The face was very long, with long premaxilla, nose long, palate very deep, mandible not massive, chin well developed, angles of jaw everted. There are many more Neolithic burials in this part of Kenya.

The Njoroan Neolithic culture in Kenya consists of polished stone axes and full-length burial in cemeteries. It resembles the Gebel Moya Neolithic cemetery in the Sudan, but the skeletons were too friable to be saved. These Stone Age peoples of Kenya are not Negro, not having broad, flat noses. The modern population has the black skin and woolly hair of the Negro, but the skulls are generally as non-Negroid as the (presumably ancestral) Neolithic people. The dark skin was probably intensified in the modern Negro, but in these ancestral Kenyans the skin was probably at least brown. The woolly type of hair is probably at least as old as the ancestors of the Boskop race. The specialized features of the African Negro would appear to be of autochthonous origin in central Africa. His woolly hair is of much earlier origin than his race, which appears to be a relatively recent specialization. This is also supported by his high frequency of the B blood group.

The fundamental significance of Leakey's discoveries in Kenya is recognized when he says (p. 127), "Although our evidence from Kenya shows that true *Homo* and true *Homo sapiens* go back to the early part of the Pleistocene, yet this fact does not in any way exclude the possibility of the contemporary existence of other species and even other genera of man in the same region." He anticipates finding skulls of Rhodesian type in association with the Mousterian culture of Kenya.

Some of the numerous types mentioned in this chapter have been arranged in the scheme of human phylogeny drawn up in Figure 4, p. 161. In that scheme it is obvious that Eoanthropus and *Homo kanamensis* must have had a common ancestor. Homo from Kenya is most likely to be the type which spread northwards, so that his descendants probably hybridized with Neanderthal man at Mount Carmel. Meganthropus and the Eoanthropus-*Homo kanamensis* stock, being at opposite ends of the scale in their skull characters and widely separated geographically, there is no reason to suppose that they had a common ancestor. They probably originated independently at the Dryopithecine level, just as the Australopithecinae may have had a separate origin from other Hominidae.

Leakey (1946) and his wife have recently made another very interesting discovery at Olorgesailie Mountain near Lake Magadi, in Kenya. Lying on the sloping surfaces in heavily eroded deposits were hundreds of Acheulean hand-axes and cleavers. A number of ancient camp sites or "living floors" were found, the implements lying as they were abandoned when the Lake rose, probably 125,000 years ago. There is a succession of Acheulean cultures on successively flooded sites. These are the first Acheulean living sites ever discovered, although Acheulean hand-axes have been known since those found at Hoxne, Suffolk, in 1797 and at Gray's Inn Lane, London, in 1690. Since then, thousands of flint and stone tools of this type have been unearthed in England, France, southwestern Europe, and all over Africa and India. But the Acheuleans lived in the open, so their bones are very rare. Search will probably reveal them. Like some modern African tribes, the dead were not buried but left to the vultures and hyenas. That these people had no fire is shown by the absence of charcoal, ashes, or burnt bones. The bones of many gigantic animals were found, however: pigs, baboons, horses, and giraffes larger than the modern, as well as hippopotamuses and elephants. Another find of interest was the fourteen sets or groups of three round stones, showing that these Acheuleans, like the modern Patagonians, used the bolas—three stone balls connected by thongs—to throw at their prey and entangle their legs. Single stones of this character are already known from the Mousterian of Europe, and their use was suspected. The great age of this device is thus proved in East Africa.

REFERENCES

Allen, A. L. 1926. A report on the Australoid calvaria found at Mistkraal, C. P. S. *Afr. J. Sci.* 23:943–950. Figs. 3.

Arambourg, C., M. Boule, H. Vallois, and R. Verneau. 1934. Les Grottes Paléolithiques des Beni Segoual (Algérie). *Arch. de l'Inst. Paleont. Humaine.* Mém. 13. pp. 242. Pls. 22. Figs. 48.

Bauermeister, W. 1940. Neue paläolithische Funde aus dem ehemaligen Deutsch-Ostafrika. *Zeits. f. Morph. u. Anthrop.* 38:25–32. Pls. 4.

Baxter, H. C. 1944. Pangani: the trade centre of ancient history. *Tanganyika Notes and Records.* No. 17. Pp. 15–25.

Bell, E. H., and A. Hrdlička. 1935. A recent Indian skull of apparently low type from Nebraska. *Am. J. Phys. Anthrop.* 20:1–8. Pls. 2.

Bernstein, R. E. 1935. Fossil remains from Keurbooms River caves, C. P. S. *Afr. J. Sci.* 32:603–607. Figs. 2.

Berry, G. F. 1935. Human skeletal remains from Smitsdorp. S. *Afr. J. Sci.* 32:616–621. Fig. 1.

Bleek, D. F. 1931. Traces of former Bushman occupation in Tanganyika. S. *Afr. J. Sci.* 28:423–429. Pl. 1.

Boule, M., and H. Valois. 1932. L'homme fossile d'Asselar (Sahara). *Arch. de l'Inst. de Paleont. Humaine.* Mém. 9. pp. 90. Pls. 8. Figs. 33.

Broom, R. 1918. The evidence afforded by the Boskop skull of a new species of man (*Homo capensis*). *Anthrop. Papers, Am. Mus. Nat. Hist.* 23: Part II.

———. 1923. A contribution to the craniology of the yellow-skinned races of South Africa. *J. Roy. Anthrop. Inst.* 53:132–149. Pls. 2. Figs. 9.

———. 1929. The Transvaal fossil human skeleton. *Nature.* 123:415–416. Figs. 3.

———. 1941. Bushmen, Koranas and Hottentots. *Ann. Transv. Mus.* 20:217–251. Pls. 6. Figs. 15.

Burkitt, A. St. N., and J. I. Hunter. 1922. The description of a Neanderthaloid Australian skull, with remarks on the production of the facial characteristics of Australian skulls in general. *J. Anat.* 57:31–54. Pl. 1. Figs. 5.

Cassel, J., A. C. Harris, and D. F. Harris. 1943. A preliminary comparative study of some Bushmen and Strandloper cranial series. S. *Afr. J. Sci.* 39:235–246.

Caton-Thompson, G. 1931. *The Zimbabwe Culture.* Oxford University Press. pp. 299. Pls. 72.

Cooke, H. B. S., B. D. Malan, and L. H. Wells. 1945. Fossil man in the Lebomo Mountains, S. Africa: the "Border Cave," Ingwavuma District, Zululand. *Man.* 45:6–13. Figs. 3.

Coolidge, H. J., Jr. 1933. Pan paniscus, pigmy chimpanzee from south of the Congo River. *Am. J. Phys. Anthrop.* 18:1–59, Pls. 2. Figs. 4.

Dart, R. A. 1923. Boskop remains from the South-east African coast. *Nature.* 112:623–625. Figs. 4.

———. 1924. The Rooiberg cranium. *S. Afr. J. Sci.* 21:556–568. Pl. 1. Figs. 2.

———. 1937a. The hut distribution, genealogy and homogeneity of the Auni-Khomani Bushmen. *Bantu Studies.* 11:159–174.

———. 1937b. The physical characters of the Auni-Khomani Bushmen. *Bantu Studies.* 11:175–246. Pls. 94.

———. 1938. Fundamental human facial types in Africa. *S. Afr. J. Sci.* 35:341–348. Figs. 6.

———. 1939. A Chinese character as a wall motive in Rhodesia. *S. Afr. J. Sci.* 36:474–476. Fig. 1.

———. 1940. Recent discoveries bearing on human history in Southern Africa. *J. Roy. Anthrop. Inst.* 70:13–37.

Dart, R. A., and N. Del Grande. 1931. The ancient iron smelting cave at Mumbwa. *Trans. Roy. Soc. S. Afr.* 19:379–427. Pls. 4. Figs. 35.

Dearlove, A. R. 1935. Human skeletal material from the Lydenburg district. *S. Afr. J. Sci.* 32:635–641.

DeSaxe, H. 1935. A human skeleton from Naboomspruit. *S. Afr. J. Sci.* 32:632–635. Fig. 1.

Drennan, M. R. 1929. The dentition of a Bushman tribe. *Ann. S. Afr. Mus.* 24:61–87.

———. 1929a. Early man in Southern Africa. *J. Med. Assn. S. Afr.* Nov. 23. Figs. 7.

———. 1929b. An Australoid skull from the Cape Flats. *J. Roy. Anthrop. Inst.* 59:417–427. Pls. 3. Fig. 1.

———. 1931. Pedomorphism in the pre-Bushman skull. *Am. J. Phys. Anthrop.* 16:203–210. Figs. 3.

———. 1932. A report on human skeletal remains from a gold prospecting trench near the Matopo Hills. *S. Afr. J. Sci.* 29:651–654. Fig. 1.

———. 1937. The Florisbad skull and brain cast. *Trans. Roy. Soc. S. Afr.* 25:103–114. Figs. 4.

———. 1937a. Human growth and differentiation. *S. Afr. J. Sci.* 33:64–91.

———. 1938. Archeology of the Oakhurst shelter, George.

III. The cave-dwellers. *Trans. Roy. Soc. S. Afr.* 25:259–280. Figs. 6.
IV. The children of the cave dwellers. *Ibid.* 25:281–293. Figs. 6.
V. The pottery. *Ibid.* 25:295–301. (by J. F. Schofield)
VI. Stratified deposits and contents. *Ibid.* 25:303–320. Figs. 69.
VII. Summary and conclusions. *Ibid.* 25:321–324.

———. 1942. Report on the Likasi skeleton. *Trans. Roy. Soc. S. Afr.* 29:81–89. Fig. 1.

———. 1945. A macrodont Bushman skull in relationship to a Boskopoid skull with a similar dentition and large jaws. *S. Afr. Dental J.* 19:243–247. Fig. 1.

Dreyer, T. F. 1932. The Bushman-Hottentot-Strandlooper tangle. *Trans. Roy. Soc. S. Afr.* 20:79–92. Pl. 1. Fig. 4.

———. 1933. The archaeology of the Matjes River rock shelter. *Trans. Roy. Soc. S. Afr.* 21:187–209. Figs. 7.

———. 1935. A human skull from Florisbad, Orange Free State. *Proc. Roy. Acad. Sci. Amsterdam.* 38:119–128. Figs. 7.

———. 1936. The Florisbad skull in the light of the Steinheim discovery. *Zeits f. Rassenk.* 4:320–322. Fig. 1.

———. 1938. The fissuration of the frontal endocranial cast of the Florisbad skull compared with that of the Rhodesian skull. *Zeits. f. Rassenk.* 8:192–198. Figs. 6.

Dreyer, T. F., A. J. D. Meiring, and A. C. Hoffman. 1938. A comparison of the Boskop with other abnormal skull-forms from South Africa. *Zeits. f. Rassenk.* 7:289–296. Figs. 11.

Dru-Drury, E. G. 1920. An extreme case of microcephaly. *Trans. Roy. Soc. S. Afr.* 8:149–152. Pls. 2.

Drury, I., and M. R. Drennan. 1926. The pudendal parts of the South African Bush race. *Med. J. S. Afr.* 22:113–117. Figs. 3.

Elsdon-Dew, R. 1939. Blood groups in Africa. *Publ. S. Afr. Inst. Med. Res.* 44:33–94. Charts 17.

Fantham, H. B. 1936. Some race problems in South Africa. *Sci. Monthly.* 42:151–168.

Fischer, E. 1929. Zur Frage einer äthiopischen Rasse. *Zeits. f. Morph. u. Anthrop.* 27: 339–341. Pls. 5.

FitzSimons, F. W. 1926. Cliff dwellers of Zitzikama: results of recent excavations. *S. Afr. J. Sci.* 23:813–817. Pl. 1.

Fouché, L. 1937. *Mapungubwe*, ancient Bantu civilization on the Limpopo. Cambridge: Cambridge University Press. pp. 184.

Galloway, A. 1933. The Nebarara skull. *S. Afr. J. Sci.* 30:585–596.

———. 1936. Some prehistoric skeletal remains from the Natal Coast. *Trans. Roy. Soc. S. Afr.* 23:277–295. Pls. 3. Figs. 4.

———. 1937a. The nature and status of the Florisbad skull as revealed by its non-metrical features. *Am. J. Phys. Anthrop.* 23:1–16. Pl. 1.

———. 1937b. Man in Africa in the light of recent discoveries. *S. Afr. J. Sci.* 34:89–120.

———. 1937c. The characteristics of the skull of the Boskop physical type. *Am. J. Phys. Anthrop.* 23:31–46. Pl. 1.

Gates, R. R. 1929. *Heredity in Man.* London: Constable. pp. 385. Figs. 87.

Gear, H. S. 1925. The skeletal features of the Boskop race. *S. Afr. J. Sci.* 22:458–469. Figs. 4.

———. 1926a. A further report on the Boskopoid remains from Zitzikama. *S. Afr. J. Sci.* 23:923–934. Figs. 3.

———. 1926b. A Boskopoid skeleton from Kalomo, Northern Rhodesia. *Bantu Studies* 2:217–231. Figs. 5.

Goodwin, A. J. H. 1925. Capsian affinities of South African Later Stone Age Culture. *S. Afr. J. Sci.* 22:428–436. Figs. 5.

———. 1929a. A comparison between the Capsian and South African Stone cultures. *Ann. S. Afr. Mus.* 24:17–32. Figs. 16.

———. 1929b. Report on the stone implements found with Cape Flats skull. *J. Roy. Anthrop. Inst.* 59:429–438. Pl. 1. Figs. 3.

Goodwin, A. J. H., and C. van Riet Lowe. 1929. The Stone Age cultures of South Africa. *Ann. S. Afr. Mus.* 27:1–289. Pls. 45.

Graydon, J. J., and R. T. Simmons. 1945. Blood groups in the territory of Papua. *Med. J. Austral.* 32:77–80.

Haughton, S. H. 1917. Preliminary note on the ancient human skull remains from the Transvaal. *Trans. Roy. Soc. S. Afr.* 6:1–14. Pls. 10.

Hill, W. C. Osman. 1939. Observations on a giant Sumatran orang. *Am. J. Phys. Anthrop.* 24:449–505. Pls. 5.

Hooton, E. A. 1930. *The Indians of Pecos Pueblo*. New Haven: Yale University Press. Pp. 391. Pls. 10.

Keen, J. A. 1942. Report on a skeleton from the Fish Hoek cave. *S. Afr. J. Sci.* 38:301–309. Fig. 1.

———. 1943. Report on two "Korana" skeletons from burial sites near Upington. *S. Afr. J. Sci.* 39:247–255. Fig. 1.

Keith, Sir A. 1933. A descriptive account of the human skulls from Matjes River cave, Cape Province. *Trans. Roy. Soc. S. Afr.* 21:151–185. Figs. 33.

———. 1938. The Florisbad skull and its place in the sequence of South African human fossil remains. *J. Anat.* 72:620–621.

Klopper, A. I. I. 1943. A report on a collection of skulls from Kruidfontein, Prince Albert District, C. P. *S. Afr. J. Sci.* 40:240–247. Figs. 3.

Knoche, W. 1941. The Ice Age problem. *Smithson. Misc. Coll.* 99:No. 22. pp. 5.

Kohler, W. L. 1943. A survey of human skulls exhumed in the vicinity of Port Elizabeth. *S. Afr. J. Sci.* 39:227–234.

Kollmann, J. 1903. Die in der Höhle vom Dachsenbüel gefundenen Skelettreste des Menschen, *Neue Denkschr. allgem. schweiz. Ges. gesamt. Naturwiss.* 39:Pt. 1. pp. 37–126. Pls. 4. Figs. 11.

Laing, G. D. 1924. A preliminary report on some Strandlooper skulls found at Zitzikama. *S. Afr. J. Sci.* 24:528–541. Pl. 1. Figs. 4.

Laing, G. D., and H. S. Gear. 1929. A final report on the Strandlooper skulls found at Zitzikama. *S. Afr. J. Sci.* 26:575–601. Figs. 6.

Leakey, L. S. B. 1931. *The Stone Age Cultures of Kenya Colony*. Cambridge: Cambridge University Press.

———. 1932. The Oldoway human skeleton. *Nature.* 129:721–2.

———. 1933. The status of the Kanam mandible and the Kanjera skulls. *Man.* 33:200–201.

———. 1935. *The Stone Age Races of Kenya*. Oxford University Press. pp. 150. Pls. 37. Figs. 52.

Leakey, L. S. B., Reck, Boswell, Hopwood, and Solomon. 1933. The Oldoway human skeleton. *Nature.* 131:397–398.

———. 1946. A pre-historian's paradise in Africa: early Stone Age sites at Olorgesailie. *Illustr. London News.* Oct. 5.

Lewis, G. N. 1946. Thermodynamics of an Ice Age: the cause and sequence of glaciation. *Science.* 104:43–47. Figs. 2.

Lowe, C. van Riet. 1932. The prehistory of South Africa in relation to that of Western Europe. *S. Afr. J. Sci.* 29:756–764.

MacLennan, G. R. 1935. Supposed Bush skeletal remains from the Pietersburg district. *S. Afr. J. Sci.* 32:623–625.

Maingard, L. F. 1932. History and distribution of the bow and arrow in South Africa. *S. Afr. J. Sci.* 29:711–723. Figs. 7.

———. 1933. The Brikwa and the ethnic origins of the Bathlaping. *S. Afr. J. Sci.* 30:597–602.

Meiring, A. J. D. 1937. The frontal convolutions of the endocranial cast of the skull M. R. I. from the deepest levels of the Matjes River cave, C. P. *S. Afr. J. Sci.* 33:960–970. Figs. 4.

Moyne, Lord. 1936. *Walkabout.* London. pp. 366. Pls. 108.

Newman, M. T. and R. L. Eng. 1947. The Ryukyu people. *Am. J. Phys. Anthrop.* 5:113–157.

Nüesch, J. 1903. Der Dachsenbüel, eine Höhle aus frühneolithischer Zeit, bei Herblingen, Kanton Schaffhausen. *Neue Denkschr. allgem. schweiz. Ges. gesamt. Naturwiss.* 39: Pt. 1. pp. 1–31. Pls. 2. Figs. 3.

———. 1904. Das Kesslerloch, eine Höhle aus paläolithischer Zeit. *Ibid.* 39: Pt. 2. pp. 1–72. Pls. 30. Figs. 5.

Orford, Marg., and L. H. Wells. 1937. An anthropometric study of a series of South African Bantu females. *S. Afr. J. Sci.* 33:1010–1036. Figs. 22.

Paver, F. R. 1925. The Asiatics in South-East Africa. *S. Afr. J. Sci.* 22:516–522.

Péringuey, L. 1911. The Stone Age of South Africa as represented in the collection of the South African Museum. *Ann. S. Afr. Mus.* 8:1–177. Pls. 28. Figs. 26.

Phaup, A. E., and M. R. Drennan. 1939. *Trans. Rhodesia Sci. Assn.* 37:157–169. Pl. 1. Fig. 1.

Reck, H. 1914. Zweite vorläufige Mitteilung über fossile Tier- und Menschenfunde aus Oldoway in Zentralafrika. *Sitzungsber. der Gesells. naturforsch. Freunde zu Berlin.* pp. 305–318.

Schapera, I. 1926. A preliminary consideration of the relationship between the Hottentots and the Bushmen. *S. Afr. J. Sci.* 23:833–866.

——— (ed.). 1937. *The Bantu-speaking Tribes of South Africa.* London: Routledge. pp. 453.

Schepers, G. W. H. 1937. The hyoid bone of Negro and pre-Negro South African races. *S. Afr. J. Sci.* 34:328–350. Figs. 6.

———. 1941. The mandible of the Transvaal fossil human skeleton from Springbok Flats. *Ann. Transv. Mus.* 20:253–271. Figs. 7.

Schultze, L. 1928. Zur Kenntniss des Körpers der Hottentotten und Buschmänner. *Jenaische Denkschriften.* 17:148–227. Pls. 18. Figs. 16.

Schwarz, E. H. L. 1938. The Chinese connection with Africa. *J. Roy. Asiatic Soc. Bengal.* 4:175–193.

Selenka, E. 1896. Die Rassen und der Zahnwechsel des Orang-utan. *Sitzber. d. kgl. Akad. Wiss. Berlin.* 381–392.
Seligman, C. G., and B. Z. 1932. *Pagan tribes of the Nilotic Sudan.* London. pp. 565.
Shrubsall, F. C. 1907. Notes on some Bushmen crania and bones from the South African Museum, Cape Town. *Ann. S. Afr. Mus.* 5: 227–270. Figs. 3.
———. 1911. A note on craniology. *Ann. S. Afr. Mus.* 8:202–208. Figs. 4.
Slome, D. 1929. The osteology of a Bushman tribe. *Ann. S. Afr. Mus.* 24: 33–60. Pls. 4.
Smuts, Rt. Hon. J. C. 1932. Climate and Man in Africa. *S. Afr. J. Sci.* 29:98–131.
Smuts, J. C., Jr. 1938. Past climates and pre-Stellenbosch stone implements of Rietvlei (Pretoria) and Benoni. *Trans. Roy. Soc. S. Afr.* 25:367–388.
Steiner, Paul E. 1947. Okinawa and its people. *Sci. Monthly.* 64:233–241.
Stow, G. W. 1905. *The Native Races of South Africa.* London. pp. 618. Illus.
Toynbee, A. J. 1934–35. *A Study of History.* 3 Vols. London.
Verneau, R. 1924. La race de Neanderthal et la race de Grimaldi; leur rôle dans l'humanité. *J. Roy. Anthrop. Inst.* 54:211–230.
Weidenreich, F. 1939. The classification of fossil hominids and their relations to each other, with special reference to *Sinanthropus pekinensis. Proc. Internat. Anthrop. Congr.* Copenhagen, pp. 107–112.
Wells, L. H. 1929. Fossil Bushmen from the Zuurberg. *S. Afr. J. Sci.* 26:806–834. Pl. 1. Figs. 10.
———. 1934. A report on human skeletal remains from East London, C. P. *S. Afr. J. Sci.* 31:547–568. Figs. 7.
———. 1937. The status of the Bushman as revealed by a study of endocranial casts. *S. Afr. J. Sci.* 34:365–398. Figs. 6.
———. 1942. The Bayville "Australoid" skeleton. *S. Afr. J. Sci.* 38:310–318.
Wells, L. H. and J. H. Gear. 1931a. Skeletal material from early graves in the Riet River valley. *S. Afr. J. Sci.* 28:435–443. Pl. 1.
———. 1931b. Cave-dwellers of the Outeniqua Mountains. *S. Afr. J. Sci.* 28:444–469. Figs. 9.
Wilson, G. E. H. 1932. The ancient civilization of the Rift Valley. *Man.* 32:250–257. Figs. 3.
Wirth, A. 1898. Die eingeborene Stämme auf Formosa und den Liu-Kiu. *Peterm. Mitt.* 44:33–36.

8

Human Evolution in Europe

From the last chapter it is clear that the key to much human evolution is to be found in the African continent. There is also a closer relationship between the Hominidae of Africa and Europe than has generally been recognized. This applies not only to the similarities between Neanderthal man in Europe and the Rhodesian skull and Africanthropus in Africa. It is equally evident in the relation between Eoanthropus in Britain and *Homo kanamensis* in eastern Africa. The close relation between Boskop man in Africa and Cro-Magnon man in Europe has also been emphasized, as has the presence of a pygmy type in France, Switzerland, and Italy in the Neolithic, which may belong to the same race as the remnant of modern South African Bushmen. The Mediterranean was a lake until the pillars of Hercules were broken through. A broad land connection across it formerly existed via Malta and Sicily. Neither the Aegean nor the Sea of Marmora existed. The northern part of the North Sea was only submerged at the second (Mindel) glacial advance and in the second interglacial the English Channel and much of the North Sea were dry land, so that the British Isles were still connected with Europe. These geographical connections increase still further the significance of the human intercourse between northern Africa and Europe.

The evidence has already led us to the conclusion that there have been two main lines of descent in the Hominidae, with occasional crosses between them. One of these lines we call the gorilloid, beginning with heavy occipital and supra-orbital ridges. In the other, which we designate the orangoid, these skull appendages have been lacking from the beginning, but they were present in the third possible ancestral line, the Australopithecinae. As regards the lower extremities, it is necessary to assume that all three of these lines of descent, orangoid, gorilloid, and Australopithecine, had already achieved essentially erect bipedal gait at the beginning of their evolutionary careers. The origin of the bipedal posture therefore constitutes an earlier episode in human evolution and there is at present no fossil evidence as to when the arboreal phase of man's ancestry came to an end, but it must have been early. Smith Woodward's (1944) guess that man originated from ground-apes in Turkestan can now be superceded by the actual knowledge of the Dartians of South Africa.

Let us consider first what I have called the orangoid line of descent in Europe. This begins with *Eoanthropus dawsoni* in England. Mr. Charles Dawson, an amateur geologist, first received from the hands of workmen a portion of a very thick human parietal bone. It belonged to a skull which the men had apparently mistaken for a petrified "coconut" and broken to pieces. The material came from a gravel bed on a farm at Piltdown, Sussex, which was being used in road making. Some years later, in 1911, Mr. Dawson picked up from spoil heaps of the gravel pit a larger piece of the frontal bone of the same skull. He took these bones to the British Museum (Natural History) where Dr. Smith Woodward was impressed with their importance. They then began excavation and sifting of all the gravel left undisturbed by the workmen. The gravel bed was three to five feet thick, finely stratified, and cemented together by iron-oxide. It was covered with only a few inches of surface soil beneath which three layers were recognized. Layer two from the top consisted of undisturbed gravel varying in thickness from a few centimeters to a meter, pale yellow in color with some darker patches. The third layer was stained dark red with iron-oxide and was about fifty centimeters thick. Layers two and three both contained rolled flints, and all the fossils except a deer, as well as the human skull, came from the dark layer three. The fourth layer, about twenty-five centimeters thick, contained larger unworked flints but no fossils or implements. All the fossil bones and teeth were found in the darker stratum of gravel, just above Cretaceous bedrock. Here Dawson found the right half of a mandible *in situ* in undisturbed gravel, apparently on the identical spot where the skull was first found. Smith Woodward dug up a portion of an occipital within a yard of where the jaw was found, and at exactly the same level. The animal bones found were a Pliocene elephant tooth, the molar of a Mastodon, two Hippopotamus teeth, and two molars of a Pleistocene beaver. In the spoil heaps was the metatarsal of a deer, split longitudinally probably by man. All these bones were highly mineralized and red with iron-oxide. Several flint implements and numerous eoliths were also obtained.

Dawson and Smith Woodward (1913) concluded that the stratified gravel was Pleistocene in age but containing in its lower stratum animal remains secondarily derived from some Pliocene deposit not far away. These were mixed with early Pleistocene mammals in a better state of preservation, both being associated with the human skull and mandible. A Chellean flint was found in the same bed, in a very slightly higher stratum, and others in the spoil heaps. It was concluded that the skull probably belonged to the warm cycle of the Lower Pleistocene.

The remains finally recovered included nine fragments of a skull which could be joined together into four pieces, a pair of nasal bones, half of the mandible, and a canine tooth. Later, Dawson found two fragments of another skull and a molar tooth (Smith Woodward, 1917) showing the same ferruginous color and mineralization. These were obtained by Dawson in 1915 from a field about two miles from the original site, where the stones had been raked from the surface of the soil into heaps. Although the Piltdown gravel was apparently a shingle beach with its characteristic brown flints widely distributed in the district, persistent search yielded no other fossils, and the fossiliferous bed appears to have been quite local.

The fragments of the second skull included the supra-orbital region of the right frontal bone near the midline, the middle portion of the occipital, and another molar tooth. The occipital fragment showed little cerebellar asymmetry, unlike the first skull, but indicated that the neck muscles extended farther up the occiput than is usual in modern man. The frontal fragment showed that the superciliary ridge was small, fading away at the glabella. But the pieces were in exactly the same mineralized condition and of the same thickness as the first skull. The lower first molar tooth agrees closely with the previous one, but is more obliquely worn by mastication. It closely resembled a modern Melanesian molar, and while resembling a chimpanzee tooth in some respects was much less brachyodont. Elliot Smith found in the endocranial cast certain features of resemblance to Neanderthal man and to anthropoids. Original photographs of these fragments were published by Weinert (1933), who concluded that the first skull and mandible could not be separated. He regarded the Piltdown riddle, set by nature, as one of the most interesting in human paleontology, but too difficult to understand at present.

Two difficulties at once arose with the main skull. One was its correct restoration, the other concerned the great contrast between cranium and mandible. They probably belonged to a woman between thirty and fifty years of age. The most striking feature of the cranium was the great thickness of the bones—about twice that in modern man, although it was otherwise essentially "modern" in type, in other words, not at all Neanderthaloid. Nevertheless, the mandible was "almost precisely that of an ape" (Smith Woodward). This contrast was so marked that the jaw has actually been described as belonging to a new species of chimpanzee (Miller, 1918), while Weidenreich (1943) regards it as the jaw of an orangutan. However, the chances against such an accidental association are almost infinitely high, and no fossil anthropoids have ever been found in Britain. Moreover, the molar teeth, two of which were in the jaw, are typically human.

Keith (1939), in a meticulous restudy of the skull, having proved his ability to make a correct restoration of similar fragments of a modern skull, is convinced that the cranium and mandible belong to the same individual. He finds the length of the skull to be 194 mm., the cephalic index 77.3 (mesaticephalic), the capacity about 1358 cc. The nasal bones are extraordinarily thick, resembling in this respect Negro and Melanesian skulls. In a critical study of the original specimens, Hrdlička (1930) while recognizing the contrast between the massive cranium and the relatively light, slender, and chinless jaw, yet finds jaws of Australian aborigines in which there is "a very close approach to the condition such as seen in the Piltdown mandible." The simian shelf is like that found in most chimpanzees and there are only traces of it in other ancient human jaws. The canine is also larger than in modern man. In these and certain other respects the jaw is nearer ape than man. But the height of the jaw is beyond the range of variation in chimpanzees, the size of the three molar alveoli in the Piltdown jaw is close to that in early and recent macrodont human jaws, and the height of the molar crowns is 8.5 mm. (hypsodont) whereas in chimpanzees it is only 5.5–7 mm. (chamaedont). The molars are 13 × 11 mm., nearly like those of the Ehringsdorf skull (12 × 11 mm.), whereas in chimpanzees they only measure 11.4 × 10.4 mm. The ramus of the jaw is also more human than chimpanzoid. Hrdlička concludes (p. 86) that "it is no longer possible to regard the jaw as that of a chimpanzee or any other anthropoid ape," yet he also finds that the two crania do not belong to the jaw but to "possibly chronologically younger human individuals." This is because of the "advanced" character of the cranium and the primitive character of the jaw. We have already seen in earlier chapters that such independent variation of different characters and organs is to be expected on the basis of modern genetical conceptions. The original view of the skull as having a cranium of essentially modern and a mandible of very primitive type thus appears to be fully justified.

In an earlier study of the Piltdown molars, Hrdlička (1923) finds them remarkably like those of *Dryopithecus rhenanus,* the latter differing only in the detailed sculpturing of the cusps, which is not human. He finds that the *D. rhenanus* molars could be ancestral to those of man. They are as close to the Piltdown molars as the latter are to those of the chimpanzee. Closely similar to the Ehringsdorf molars (see below), they fall within the range of early man and not within that of living apes.

As regards the brain, Keith (*Antiquity of Man*), from the endocast, finds it "primitive in some respects . . . but in all its characters directly comparable with that of modern man," whereas Elliot Smith (1913) regarded it as "the most primitive and most simian human brain so far recorded." Keith (1939) recognizes a greater degree of asymmetry in the Piltdown cranium than is anywhere found in modern man. Smith Woodward (1944) emphasized that the Piltdown skull resembles Sinanthropus in its very broad base and the thick, spongy brain-case. These would be primitive characters held in common in the two separate (gorilloid and orangoid) lines of human descent.

Several other skulls found in England, as well as *Homo kanamensis* in Africa, serve to strengthen the view that in one line of human evolution heavy skull appendages never developed. Of these, perhaps, the Swanscombe skull, found twenty-four feet deep in the gravels of the one hundred-foot (Lower Pleistocene) terrace of the lower

Thames valley, is the most completely authenticated. Portions of a parietal and an occipital were found in the same seam of gravel eight yards apart in the Barnfield Pit. These fluviatile deposits are the infilling of a stranded river 1400 feet wide. The skull bones were at the precise level of the Middle Gravel Erosion Stage, between the upper and lower gravel deposits. There was an interval of years between the two finds, but their sutures fitted together perfectly. Many Acheulian *coup-de-poing* or hand-axe implements were associated with the skull. A few Clactonian flake implements were found, but none in the vicinity of the skull. It is thus probable that the Acheulian implements were made by Swanscombe man. The implements associated with Neanderthal man are always Mousterian. It is probably not without significance that the industries associated with Oldoway man in Africa were also Acheulian (as well as Chellean). The committee which investigated the Swanscombe remains were in complete agreement [1] as to the indigenous nature of this fossil in the Thames gravels. Although the incompleteness of the skull-remains leaves the possibility that there were heavy brow ridges, yet the similarities to Eoanthropus make it more probable that this skull was not of Neanderthal type, a type which has never been found in Britain. The skull bones are thick but these fragments otherwise show little to distinguish them from a modern skull. Keith calculated the length of the skull to be 185 mm., breadth 144 mm., cranial capacity (estimated) 1350 cc. The estimate of LeGros Clark and Morant was 1325 cc. Other estimates are as small as 1065 cc. The mammalian fauna in these deposits indicate a temperate interglacial climate between the great Eastern glaciation of East Anglia and the cold period which followed.

The Galley Hill skeleton, originally found at a depth of eight feet in undisturbed strata in the one-hundred-foot terrace of the Thames not far from Swanscombe in 1888, was long regarded as of doubtful provenance because the skull was "modern" and not Neanderthaloid. The associated implements were Abbevillian. As Keith (1939) points out, the Bury St. Edmunds cranial fragments and the London skull ("the lady of Lloyds") found in 1925 in excavations fifty feet deep in the London clay [2] (on the third terrace of the

---

[1] *J. Roy. Anthrop. Inst.* 68:17–98. 1938.

[2] In the original description, Young (1938) finds this skull most like those of Upper Paleolithic Solutré, of modern type. It was associated with the woolly rhinoceros and Aurignacian implements.

Thames), having a cranial capacity of not more than 1250 cc., probably belong to the same period and race as the Swanscombe skull. He concludes that the Swanscombe skull is definitely not of the Neanderthal type but is probably a mid-Pleistocene descendant of the early Pleistocene Eoanthropus.

Together, these remains make a strong case for the presence of an orangoid race of man in southern England in Lower Pleistocene times, earlier than Neanderthal (gorilloid) man in central Europe. The Steinheim skull, belonging to Riss I/Riss II, is also chronologically older than Neanderthal but morphologically of more "modern" type. *Homo kanamensis* in East Africa appears to belong to the same evolutionary level in the orangoid series as Galley Hill and Swanscombe man in England. This type was therefore as widespread—from England to Africa—in early Pleistocene, as Neanderthal man of the gorilloid series.

Coming now to the Continental records of early man, we have first the Heidelberg man or Mauer mandible, discovered in 1907 by two workmen in the face of a large gravel pit near Heidelberg. The jaw was nearly eighty feet from the surface and less than three feet from the floor of the excavation. Since the whole deposit ranges from Tertiary to Recent, the lower strata must be early Pleistocene. Schoetensack (1908) published a monograph on the jaw and its surroundings. There are no human artifacts in the Mauer sands, but the animal bones found at the same level as the jaw included *Elephas antiquus, Rhinoceros etruscus,* and a lion (*Felis leo fossilis*). The Heidelberg jaw is large and chinless (Plate XXII, a, b) and is unique in being arched upward at the base in front. There is no lingual shelf and the incisors are not shovel-shaped. The teeth are of normal size, small for such a massive jaw, and Hrdlička (1930), in a detailed study with numerous measurements, shows that the second molar (M_2) is larger than M_1 or M_3. The jaw is much larger than that of a modern European and has a number of primitive features, but is equaled in some of its measurements by Eskimo jaws. The crowns of four teeth were broken off by the shovel of the workman, but otherwise the mandible (Plate XXII, a, b) is in practically perfect preservation. It is presumably ancestral to Neanderthal man, but of this there is no certainty.

NEANDERTHAL MAN

The numerous remains of Neanderthal man (the gorilloid line of descent) have already been referred to in Chapter 4, but certain further points need to be considered here. Many of these skeletons were found in the nineteenth century, beginning with the Gibraltar skull (Plate VII, a) in 1848. Hrdlička (1930) has made a valuable comparative study of them all, from personal observations of the originals. He also gives full references to the original descriptions, and numerous photographic illustrations. Boule (1923) also gives full descriptions, with the conditions in which they were found. In an earlier review of the Pleistocene, Osborn (1915) gives many details, particularly of the European fauna at different periods of the Ice Age, in relation to changing climates.

The Gibraltar skull was described in 1865. The median incisors were absent and had probably been removed in youth—a custom of various African and other tribes. It is the most complete of all the Neanderthal specimens, but has the smallest brain and is probably a female. In 1910 another cave at Gibraltar yielded Mousterian, Aurignacian, Solutrean, and Magdalenian stone implements, and in 1926 Miss Dorothy Garrod, by further excavations, discovered the skull of a baby, associated with Mousterian implements, embedded in hard rock.

The Neanderthal skeleton proper, from which the whole tribe is named, was found by quarrymen in a cave in the Devonian limestone sixty feet above the Neander river in western Germany in 1857. The bones included the skull cap, femora, humeri, ulnae, the right radius, part of the left pelvic bone, part of the right scapula, and the right clavicle and five rib pieces. The supra-orbital torus is the heaviest in any Neanderthal skull. The cephalic index is 72. The skeleton is more primitive than in modern man and the long bones show that the stature was submedian. A bear "tusk" was found later in mud in a branch of the cave, at the same level.

In the cave of Spy, near Namur in Belgium, which is now sixty feet above the stream, in 1879 were found an ivory bead, decorated bones, a hyena tooth pierced (for a pendant), and Mousterian implements. Further excavations in 1886 uncovered flint blades, needles, awls, beads, and pendants of Magdalenian age, with two human skeletons and Mousterian artifacts below. The Spy skeletons are of Neanderthal type, but the skull vault of No. II is higher and is regarded as far in advance of Neanderthal and nearer to modern man. The chin is erect, not receding. These were probably burials of Upper Mousterian or even later age. The cephalic index of No. I is 72.1 and of No. II, 77.4 (mesaticephalic), but they have never yet been fully studied and their vicissitudes in the recent war are unknown.

In the Krapina rock-shelter in Croatia were found the remains of twenty skeletons of all ages, in deposits twenty-six feet thick showing nine Mousterian culture layers, probably belonging to one long warm interglacial period. The associated animals were *Rhinoceros Merckii, Ursus spelaeus, Bos primigenius, Castor fiber,* and *Equus caballus,* representing a cold fauna, but there was no mammoth or wooly rhinoceros. Gorjanovič-Kramberger collected over 1000 Mousterian implements. Many of the skeletons were fragmentary and showed evidence of cannibal feasts. The supra-orbital arch and sixteen mandibles were massive with only a trace of a chin. The teeth were large and the upper incisors generally shovel-shaped. The skulls, though Neanderthaloid, showed progressive tendencies, some having higher foreheads, and they showed the first disposition towards broader and shorter heads, a condition never found in the western Neanderthals. The femora were also more modern than in the Spy and Neanderthal skeletons, lacking the forward arch. *Palaeoanthropus krapinensis* thus represented a single race of man with certain specific distinctions, such as a lighter build, more slender bones, and smaller head, somewhat in advance of the typical Neanderthals although probably somewhat older chronologically.

At Ehringsdorf, three kilometers from Weimar, Germany, blasting operations in a limestone quarry in 1914 uncovered a lower jaw with worked flints and evidence of fire. It was at a depth of 11.9 meters below the surface, in travertine overlying the loess, and was later described by Schwalbe. The mandible is nearly complete, having all the teeth except two incisors. It is chinless and shows pronounced alveolar prognathism. Two years afterwards a child's skeleton, including a lower jaw and nine ribs, was found in solid rock. Later a femur and stone implements were found, and in 1925 further blasting revealed an adult human skull which was described by Weidenreich (1928). This skull was in rock fifty-five feet below the surface. The lower travertine layers where the bones were found belonged to the last

(Riss-Würm) interglacial (hence earlier than Neanderthal) and the "Weimar stone culture" was of mid-Paleolithic age. In the mandible the third was smaller than the first and second molar—an advanced character. There were shovel-shaped incisors in the child's jaw. The adult skull resembles Spy II, somewhat advanced, with a cephalic index of 74, but having a complete heavy torus and a primitive protruding occiput as in Neanderthal. However, the forehead and vault were higher, the upper face and forehead large, the jaws also showing peculiarities, thus characterizing a separate race. The fauna resembled that at Krapina and may represent a warm interlude. Besides Mousterian implements there were drills, double points, and other tools of more advanced character.

The Steinheim skull, still earlier than Ehringsdorf, was discovered in 1933 in a gravel pit at Steinheim in Wurtemburg and is described by Weinert (1936). It was 7 meters from the surface, the upper 2 meters being loess, then 5 meters gravel, and below the skull 9 meters more of gravel. The skull (Plate XXII, c) is believed to be female, the occiput rounded as in modern man, but the supra-orbital torus is exceptionally heavy even for Neanderthal man, showing that these two features vary independently and are separately inherited. The skull is dolichocephalic, C.I. 72, but the endocranial index is 77. It is chamaecephalic, but higher than in Pithecanthropus, leptene, chamaeconch, and orthognathous. The *antiquus* fauna indicates an early stage of the Riss-Würm interglacial. Mammoth remains were found in the gravel layers above the skull.

At Taubach, near Weimar, excavations for sand and stone have been going on since 1874 in a terrace of calcareous tufa alternating with sand. The fauna included *Elephas antiquus* and *Rhinoceros Merckii*, representing the warm Riss-Würm interglacial. Early Mousterian implements were found and in 1892 two human teeth were found seventeen feet below the surface.

Excavations in a cave on the island of Jersey, chiefly by Marrett, unearthed thirteen human teeth, all probably from the same person. They were larger than modern, taurodont (the roots stout and fused), and the first premolars were larger than the second. The animal bones included *Elephas primigenius* and *Rhinoceros tichorhinus* (the woolly rhinoceros), and over 15,000 flints were found, including 155 which were typical Mousterian, others foreshadowing the Aurignacian.

The old man of La Chapelle-aux-Saints is one of the more typical Neanderthal skeletons. It was found as a burial in a cave in southern France in 1908 and described by Boule in 1910. One of the oldest known burials, it was lying on its back with the limbs flexed. The teeth were lost in old age and the skull sutures nearly closed. There were abundant flints in the excavation—points and scrapers—with evidence of fire and bones broken for their marrow. The accompanying animals, including *Rhinoceros tichorhinus* and *Rangifer tarandus*, represent a cold fauna of the last glacial period. The skull had a length of 208 mm., cephalic index 75. It had a very large palate, the mandible was chinless and there was a trace of a lingual shelf, that is to say, a swelling between the lingual border of the anterior portion (symphysis) of the jaw and the genial tubercles. The brain, as shown by the endocast, has many primitive features, but the cranial capacity is very large, 1600–1626 cc. In the skeleton the femora were curved forward as in the Neanderthal and Spy skeletons, but there was no platymery. The spinal column and tibia were short, and Boule estimated the height at 155 cm. The hands were large and strong, the right humerus and radius stronger than the left. A marked feature was the low height of the vertebrae and especially the nearly horizontal spines of the cervical vertebrae—a simian character.

The rock-shelter at La Ferrasie, Dordogne, France, was excavated in 1909. There was a long succession of occupations, as shown by the Acheulian, Mousterian, Lower, Middle, and Upper Aurignacian implements, above which was twelve feet of humus and gravel. Two adult skeletons, probably male and female, were found as shallow burials, the bodies flexed, and three infants were also found of which few bones remained. The male skeleton was lying under three flat stones. The skull was as large as La Chapelle-aux-Saints and had the beginning of a chin. The two skeletons are important, having many Neanderthal characters but some that are more modern. An incised bone accompanying the skeleton further suggests that these were perhaps hybrids of Mousterian and Aurignacian man. The astragalus (ankle bone) and calcaneum (heel bone) as well as the fibula show primitive characters relating the foot to that of the lower monkeys rather than the anthropoid apes (Boule, 1923).

At La Quina, Dept. de Charente, France, there were early excavations. Vast quantities of cultural and faunal remains included bones of horses, bovidae, and deer. In the valley of the Voultron there were formerly many rock-shelters which became filled with middens and rock falls. At La Quina in 1911 a skeleton was found which was horizontal, but may have been a burial. M. Henri Martin has published many papers on his later excavations. The skeleton was only two feet six inches deep in clayey sand. It resembles Spy 1, but the teeth were larger than Spy or the Mauer jaw, and taurodont. The mandible was very primitive, with no chin, and the molars do not decrease in size from front to back. The cephalic index was 73.8, the forehead narrow, low, and sloping and there was an occipital torus. A second mandible has the beginning of a chin. Nineteen other parts found are believed to belong to nine skeletons. Two Mousterian astragali found in 1908 differ from the modern in shape and are larger. The implements show a very long continuous occupation, with marked improvement in type. Perforated bones were used as pendants. The fauna indicates cold conditions, but there was no evidence of fire, breaking of bones, or cannibalism. A child's skull was also found, in which the supra-orbital arch was complete, but the forehead was as high as in moderns, the shape ovoid. The cephalic index was 76.7 and the incisors were shovel-shaped. This skull approached the modern in size, form of the vault, and shape of the occiput. The adult skull differs from the type of Neanderthal in its narrowness, the stout jaws and teeth, and the shallow glenoid fossa.

At Le Moustier (Dordogne) in 1909 a skeleton was found extended on its side, three feet deep in a terrace facing the rock-shelter. It was removed by Klaatsch and Hauser and has sometimes been known as *Homo mousteriensis Hauseri*. The skull has a complete frontal torus, a massive mandible with no chin, and some features like the Krapina skulls. It is incomplete and has been reconstructed four times. It is less markedly Neanderthaloid because it is that of a youth, probably about sixteen years of age. The length is found to be 196 mm., cephalic index 76.5, capacity (estimated) 1564 cc. About the skeleton were numerous worked flints, a charred bone of *Bos primigenius*, and other fragments.

The Galilee skull, the first Neanderthal skull found outside of Europe, was excavated by F. Turville-Petre in 1925 and described by Keith in 1927. A limestone cave high in the cliff of a stream had a floor in which the upper four feet had accumulated through use as a goat stable; below this was three and one half feet of human occupation, from Paleolithic to Neolithic or Bronze Age, and beneath this an undisturbed layer. At a depth of six and one half feet, near the lower limit of the Paleolithic, four fragments of a skull were found together—the frontal, the right zygoma, and two fragments of the sphenoid. The skull has biarched supra-orbital tori and the vault is of moderate height. It is of advanced Neanderthal type in the shape of the forehead, height of vault, form of orbits, and general brain features. The accompanying implements show Mousterian affinities, but there are also some blades and later types. The animal bones show no evidence of a cold fauna, but rather a hippopotamus and a rhinoceros with African types like *Hyaena crocuta* and a river hog, as well as the Palaearctic brown bear.

At Saccopastore, near Rome, in 1929 laborers found in a breccia-filled cave a Neanderthal skull which Sergi described as a woman of thirty years. The cranial capacity is not more than 1200 cc., the vault low with facial prognathism, but M_3 is smaller than M_1 and M_2 which are of equal size. The accompanying animal bones included *Elephas antiquus, Hippopotamus major, Rhinoceros Merckii, Cervus elephas,* and *Bos primigenius,* probably indicating the Riss-Würm interglacial period.

Other remains of Neanderthal man include the La Naulette jaw and ulna found in Belgium in 1866 at a depth of four and one half meters under four separate stalagmitic layers, accompanied by a cold fauna. The Sipka jaw is a fragment of a child's mandible found in Moravia in 1880, in a cave at a depth of four and one half feet in hardened ashes. The mandible was primitive, with large teeth. There were eight layers of paleolithic occupation in the cave, with quantities of animal bones indicating a cold fauna. The Malarnaud jaw was found in a cave in Ariège, southwestern France, in 1889 at a depth of seven feet below a layer of stalagmite, associated with *Rhinoceros tichorhinus,* the cave lion, cave bear, and cave hyena. In 1887 the Bañolas mandible was discovered in northeastern Spain (Catalonia) in a block of calcareous tufa, sixteen and one half feet below the surface. An account of it was published in 1909. The third molar (M_3) was larger than M_1 and M_2. Other remains of Neanderthal man, notably at Mount Carmel in Palestine and the Monte Circeo skull in Italy, have been considered in Chapter 4.

Considering the whole record of Neanderthal man in Europe, the skull was of moderate to large size, dolichocephalic to mesaticephalic, the bones generally thicker than in modern man. The supra-orbitals were always biarched but forming a complete torus; the forehead was low and sloping to high, the occiput broad and flattened, with a transverse torus. The molar teeth were large, frequently taurodont, and the upper incisors often shovel-shaped; there is no caries, but a tendency for the third molar to diminish in size. The mandible is rather heavy, the chin receding, the nose stout, the mouth large. The brain was of moderate to large size but morphologically inferior, having simpler convolutions. In stature these people were short, rarely medium in height, the neck was short and thick, the skeleton was stout, especially the ribs, giving a different shape to the thorax. The muscles were heavy, the ends of the long bones larger, and the gait was probably not erect. The tibia was relatively short, its head more inclined backwards, and the astragalus showed peculiarities.

On the basis of these and other morphological features relating to many parts of the skeleton, there is no doubt that McCown and Keith were right in placing Neanderthal man in a different genus (Palaeoanthropus) from Homo. If these fossils belonged to horses or elephants no paleontologist would question for a moment their generic distinctness. Nothing could show more clearly how the dogma of *Homo sapiens* as including all living men (and all fossil men who could be supposed to cross with them) has muddled all thinking on the subject of species and evolution in mankind. As a recent example, Angel (1946) speaks of modern and Neanderthal man as "distinct subspecies," presumably because there is, as we have seen, some evidence that they could intercross. To such straits are biologists reduced when they try to adhere to the absurd conception of interspecific sterility as a sole criterion of specific distinction. Innumerable attempts have been made to found the conception of species on a single criterion. They have all broken down, and none more hopelessly than that of interspecific sterility.

In Neanderthal man, Hrdlička (1930) attempts to "grade" the skull, jaws, teeth, and bones separately, showing their relative independence in evolution. He says (p. 341), "In one and the same skeleton are found parts and features that are very primitive . . . with parts and features that are practically modern; and every skeleton is found to differ in these respects." This morphological instability was found also in the brain, endocasts of the skull showing a low type in some, but a relatively high type in the Weimar and Galilee skulls as well as in Spy II and the Gibraltar child. Keith concluded that the brain of Galilee man was at least at the level of the Australian aborigines in mass and markings. The characteristic Neanderthal occiput is not found in all, but on the other hand some recent skulls nearly approach the same condition. Hrdlička shows the skull of a modern Piegan Indian having the Neanderthal supra-orbital torus. This torus is again not equally developed in all Neanderthalians, but on the other hand is found partly developed in the Podkoumok, Brüx, Brno, Predmost, Obercassel, and other Aurignacian skulls of later age. The Krapina skulls show more variability than would be found in a modern race in any one locality. The Neanderthal skeleton No. II is "so superior in size, shape, height of the vault, and height of the forehead" that the difference between Neanderthal Nos. I and II is greater than between No. II and the Aurignacian Brüx or Brünn skulls, the latter being transitional to the Aurignacian type. Neanderthal skulls thus show considerable variation.

As regards implements, the Chellean, Acheulean, and Mousterian types form an evolutionary series. Acheulean man had the same food habits, as a hunter and trapper, as the Mousterians. Although there is at present no certainty, the association with Acheulean implements makes it not improbable that they were made by the "orangoid" type of man represented by Swanscombe, Galley Hill,[3] and other skulls in the south of England. No bones of Acheulean man have been generally accepted as such on the continent of Europe, but in east Africa Leakey found implements all the way from pre-Chellean to advanced Acheulean, some of which at least were made by Oldoway and Kanam man. From the African evidence, it seems likely that this developmental series of implements originally evolved there and spread into Europe by culture contact. Without attempting to enter into the complicated question of culture successions, we may refer to Burkitt's (1938) conclusion that in the Lower Paleolithic there were at least two distinct civilizations in Europe, each with a number of cultures. One civilization with a large number of cultures was based on flaking. In eastern France and England these industries are associated with a

[3] For the original full account of the Galley Hill skeleton see Newton (1895).

core (*coup de poing*) civilization. Although the flaking industries are all allied, it is not satisfactory to class them all together as Clactonian, Levalloisian, and so on.

After the warm Chellean (Abbevillian) period in Europe, man began using shelters. Hrdlička shows how this habit grew with increasing severity of the climate. Some 360 Paleolithic sites are known in Europe, from the pre-Chellean to the Neolithic. Neanderthal man with Mousterian culture extended from the latter part of the last (Riss-Würm) interglacial to deep in the Würm glaciation. Some Swiss caverns at an elevation of 5000–8000 feet were occupied. This could obviously have been only during the interglacial. Of the sites for Mousterian man, 34 per cent were in the open and 66 per cent in rock-shelters or caves. For Aurignacian man only 18 per cent were open sites, and for Magdalenian man only 10 per cent. In some Neanderthal caves there were as many as six successive occupations. Hrdlička points out that the Aurignacians took over Neanderthal rock-shelter and cave sites and lived in them in exactly the same way as their predecessors, except for technical differences in flint chipping. There is frequently a direct sequence, Acheulean→Mousterian →Aurignacian.

Many have assumed that Aurignacian (Cro-Magnon) man came from Africa as invaders of Europe and exterminators of the Neanderthals. Hrdlička's objection is that during the Würm glaciation Europe was cold and inhospitable while northern Africa, including the Sahara, was then a most favorable region for man. More complete evidence shows, however, that Aurignacian man arrived in Europe during the warmer Laufen interstadial between Würm I and Würm II. It is possible, of course, that other reasons, such as over-population, might have driven some elements of Cro-Magnon man northwards across the Mediterranean. That he had already evolved physically to a higher level than Neanderthal man is certain. Since some Neanderthalers showed more advanced characters than others, these have quite possibly been derived, at least in part, from an early period of hybridization, when Aurignacians were few and Mousterians many. We may suppose that later, as the Aurignacians multiplied and began to outnumber the Mousterians, the process of exterminating the Neanderthalers went on more rapidly, so that finally only the Aurignacians were left; but some of these had a certain amount of Neanderthal ancestry, chiefly from women who had been spared in the process of extermination.

This is much the same as happened when the Indians in eastern America were exterminated by the invading Europeans. In the latter case the cultural differences were greater but the physical differences less. There is also the alternative possibility that the descendants of our orangoids in the south of England played a part in the process of finally exterminating the Neanderthalers; but the evidence we know seems to substantiate the view that whatever the reasons, Aurignacian man probably came to Europe from northern Africa. His art, which seems to have developed slowly from humble beginnings, finally flowered into artistic productions which modern man might envy. All this developed in caves, some of which had previously been occupied by Neanderthal man.

As regards flint implements, there are now many uncertainties and complications, compared with the simple succession of types of implements which was formerly supposed to exist in western Europe. The subject has been broadly discussed by Miss Garrod (1936), who finds that at the beginning of the Upper Paleolithic the first blade industry (Chatelperronian) was already present not only in France but in Palestine and East Africa, having had an Asiatic origin. She believes that the Capsian may have originated from it in East Africa and then crossed the Sahara to North Africa, belonging to the extreme end of the Upper Paleolithic.

Hrdlička believes that Neanderthal man evolved directly into *Homo sapiens* in western Europe, without any extraneous element coming in. There are insuperable difficulties with this view. It would imply a rate of evolution out of all proportion to what we know, and the assumption is unnecessary since we know that man in Africa and England had already reached the Homo level of evolution. Probably, as Hrdlička suggests, with increasing severity of the climate natural selection would play a part in killing off those of less vigorous physique, and he finds evidence of a large reduction in population. A reduction in food supply would seem inevitable with the oncoming ice. But there is no evidence that climatic selection favored Aurignacian as against Mousterian man. In withstanding hardships there is every reason to suppose that lower Paleolithic Neanderthal man was quite the equal of the Upper Paleolithic Aurignacians. We suspect that the superior mentality of the former led to his supplanting Neanderthal man in Europe. The advantage derived from the bow, which the Mousterians did not have, would be in itself sufficient to account for the rapid extermination of Neanderthal man, but it is not quite certain that Aurignacian man had the bow. The period of the Upper Paleolithic Aurignacians (especially the Magdalenian) is frequently called in France the Reindeer Age.

If we again compare the case of the American Indians, exterminated in eastern America by the Europeans, we see that the former was at least the equal of the latter in physique, but the bow and the tomahawk were no match for the shotgun.

Hrdlička, Weidenreich, and others have assumed that if Neanderthal man did evolve directly into modern man they must therefore both belong to the same species. This is one of the biological absurdities which has been widely current, and I hope to show in Chapter 11 that it is based on mixed thinking and is necessarily without foundation. If it were true, the whole animal kingdom would belong to one species, unless we revert to special creation, for evolution assumes that one species has been derived from another by continuous breeding. How mistaken even an eminent anthropologist can be in his first impressions regarding a new species or genus of man, is shown by the remarks of Hrdlička (1930) regarding Sinanthropus. Davidson Black had sent him photographs of a skull cap and portions of a mandible, which he reproduced in an appendix, with the remark (p. 368), "The skull is clearly neanderthaloid. It appears to represent no distinct genus, species or even a pronounced variety!"

As we have seen, the evidence strongly indicates that Cro-Magnon man or a related ancestral type entered western Europe from Africa and that he afterwards moved eastwards during the Upper Paleolithic.[4] We need not refer here to the many books which have been written on Upper Paleolithic or Aurignacian man, his arts and his implements, but it is necessary to refer to some of the main records of this remarkable people. As indicated above, although there was some hybridization between Mousterian and Aurignacian man, yet this was not sufficient to bridge the gap completely between the genera Palaeoanthropus or Neanderthal man and Homo or modern man. There re-

---

[4] Verneau, however, finds that in Spain the more recent Cro-Magnon burials are in the South, indicating a southward movement.

mained a hiatus between Lower and Upper Paleolithic, although in some places the culture transition appears to have been a gradual one.

Coon (1939) states that over one hundred Upper Paleolithic skulls have been disinterred in the last century, but only sixty odd have been measured and published. They have come mostly from France, but some from England, Spain, northern Italy, Germany, Czechoslovakia, Poland, and Russia. Morant (1930) made a biometric study of twenty-seven of these skulls and marshaled the published data on twenty-five others. He gives full references to the original descriptions. Seventy per cent of the skeletons belong to the Aurignacian, others to the later Solutrean and Magdalenian culture periods. Morant found that although these remains from western and central Europe extend over a period of perhaps ten thousand years, yet the total range of variability of the skulls is "surprisingly small," being no greater than some modern European series considered racially homogeneous. Although the bulk of Aurignacians in Europe were regarded as belonging to the Cro-Magnon race, yet the Grimaldi skeletons of Mentone were considered to be Negroid and the Chancelade skull, of Magdalenian age, Eskimoid.

Although Morant (1930), in his survey of Upper Paleolithic skulls, concluded that there was no statistical reason for detaching the Chancelade from other Upper Paleolithic skulls, yet in an earlier paper (Morant, 1926) he had shown as clearly as biometric methods can that this skull agreed essentially with the modern Eskimos and differed in race from the modern population of western Europe. The Eskimo skull is generally recognized as dolichocephalic with a distinct sagittal crest, low nasal index, high orbital index, large cranial capacity, with superciliary ridges and mastoids feebly developed. The Chancelade skeleton, representing a man only 1.5 meters (5 ft. 2 in.) in height, was found in 1888 lying on bedrock in a tightly flexed position on its left side at a depth of 1.6 meters in a rock-shelter at Raymonden near Chancelade, seven kilometers northwest of Perigueux, Dordogne. Above it were three distinct Magdalenian hearths separated by sterile layers. The body had been powdered with red haematite. The artifacts were flints and incised bones of reindeer as well as fragments of the Greenland seal. The skull was 194 mm. in length, the cephalic index 69.3 (Boule says 72), the nose long and narrow. Morant compared it biometrically with the numerous available measurements of Eskimo skulls, which incidentally show that the Greenland Eskimo form a "perfectly homogeneous" population. The mean C.I. for eastern Eskimo males is about 71.4 and for western (Alaskan) Eskimo about 76. The conclusion reached was that "judging from the given cranial measurements only, the Chancelade appears to be an Eskimo skull differing from the mean type of that population in very few essential ways." No measured character will clearly distinguish the Chancelade skull from the mean type of the Eskimo.[5]

This Magdalenian skull thus appears to represent an Eskimo race and culture in Europe near the end of the last glacial advance. However, a caveat must be entered. Sir Arthur Keith discovered an old photograph of the Chancelade skull (*New Discoveries*, p. 395), before the nasal bones were lost, in which they are shown to be highly arched and quite unlike the modern Eskimo nose. Of course, nasal shapes often vary greatly within a race, so this does not prove anything. But it does show a feature in which this skull differed from the modern Eskimo. Vallois (1939) pointed out that the torus mandibularis, present in the Chancelade mandible, argues for its relation to the Eskimo, since it is very characteristic of the Mongolian race, including the Eskimo, being found in prehistoric as well as modern Chinese and in Sinanthropus. He finds, however, that the form of the face, nose, and orbits is much less different from other Upper Paleolithic skulls than has generally been supposed.

Pittard (1941) points out that the Chancelade skeleton was short. He believes that the type continued through the Mesolithic, especially in France, and in Brittany as the Magdalenian race. According to Boule, the cranial capacity was 1710 cc.—a very large size, but Keith's estimate is 1530 cc.

---

[5] It is desirable here to direct attention to the fact that stone implements belonging to late Paleolithic man have been found in northern Norway (Brögger, 1937). This is contrary to the general belief that man only advanced into northern Europe in Mesolithic times, after the ice had disappeared and the tundra had been replaced by forest. These men were living under conditions very similar to those of the Eskimo in Greenland today, in open sites on the seashore and at the head of fiords. Sixty such sites are now known, but under the conditions no skeletal remains are to be expected. Their culture appears to have centered mainly around the seal and the reindeer. A variety of primitive implements, including scrapers for the preparation of hides, were made from loose stones in the glacial moraines. Some were of Clactonian, Levalloisian, and Mousterian types, while others (blades and flake-knives) show Aurignacian technique. There were also microliths, and arrowheads of Lyngby type. These Finmark finds are believed to belong to the last glaciation, about 12,000–15,000 years ago, when glaciers still existed in the fiords and severe arctic conditions prevailed.

If a part of the Chancelade race came to depend on the reindeer, either wild or domesticated, for sustenance, then as the glaciers disappeared and post-glacial conditions became warmer he would naturally retreat northwards and (because of the late survival of the glaciers in Scandinavia) eastwards, either following or taking with him the reindeer herds. This retreat might well carry him to Siberia and the Asiatic coast.

In a recent full study, Vallois (1946) concludes that the resemblance of Chancelade man to the Eskimo can be explained without recourse to the theory of an eastward migration in pursuit of the reindeer. He emphasizes the extreme leptorrhiny (as in the Eskimo), N.I. probably 42.6; but the lower border of the nasal aperture is of the modern European type. The torus mandibularis is present in no other Upper Paleolithic jaw found in France. A detailed study is made of the skeletal bones, including humerus, radius, cubitus, femur, tibia, clavicle, hand, and foot. He concludes that there are profound differences between Chancelade man and the Eskimo. He finds a close resemblance to Mesolithic man in Brittany and accounts for similarities to the Eskimo by the fact that Paleolithic man was ancestral to the Magdalenians and the Chancelade type, both having certain characters which persisted in the Eskimo. But he finds the Chancelade skull nearer the old Eskimo than the modern race. Vallois also compares Chancelade man with the so-called Eskimoid female skull from the upper cave at Choukoutien. In a table, he shows near-identity in a number of indices, but the Choukoutien skull is more prognathous. He therefore concludes that the Chancelade race in Upper Paleolithic times extended from France to northern China.

The results regarding the negroid character of the Grimaldi "race" are perhaps less clear. Verneau in 1910 concluded that the two lowest skeletons in the Grotte des Enfants [6] showed Negroid affinities. Morant (1930) finds that the male and female "Negroid" skulls both show posthumous deformation, which narrowed the nose, and reduced the breadth of the male skull. As regards the female skull, he concludes that its facial characters (especially the alveolar prognathism), which have been taken to indicate its Negroid nature, "may all have been influenced to some degree by earth pressure." Keith and Elliot Smith both concluded that these skeletons do not show Negroid characters.

[6] For a full account of these (and other) excavations, see Hooton (1946).

Later measurements, however, show that, as in the Negro, the distal segments of the limbs are relatively long and the arms are long in relation to the legs; also the long heels are characteristically Negroid (Bonin, 1935a).

The possible relations of these Grimaldi skeletons to Asselar man in the Sahara have already been referred to. Verneau (1924), in an extended discussion of the Grimaldi race, finds this Negroid type represented by certain Neolithic skeletons from Brittany to Switzerland and the Balkans. The finding of a number of Aurignacian statuettes and engravings of the human figure gives clear evidence of a people with a number of Bushman characters in France at this time. One statuette from the Grimaldi caves is typically steatopygous, and a head is Negroid. Boule (1923) describes them in detail. Whether the Grimaldi skulls are Negroid or not, there is little doubt that the Boskop race with many Bushman characters had spread into southern Europe at this time. This hypothesis fits the facts much better than that the Boskop race originated in Europe and spread southwards, an idea which appears to be untenable.

Mendes Correa (1919) studied about two hundred human skeletons from the kitchen middens of Muge in the Tagus valley of Portugal, with Tardenoisian artifacts. They are mostly very early dolichocephalic representatives of the Mediterranean race, but a few are said to be brachycephalic although this was mostly a result of post-mortem deformation. Correa finds in them certain Negroid traits. The dolichocephalic skulls were smaller than in modern Mediterraneans, they were mesorrhine, prognathous, occasionally with a sloping forehead, and of low stature. The brachycephalic skulls had different head shapes. The brachycephali of Muge and of Ofnet in eastern Bavaria are the earliest in Europe. The former, at any rate, could not have come from the east and must have developed autochthonously.

Neolithic remains abound in Portugal. Some of the Neolithic peoples were probably descended from the Muge type and contributed to the modern Portuguese. One of the dolichocephalic types resembled the Baumes-Chaudes in France, whom Bonin concluded were Nordics, at least in skull measurements. They constructed dolmens and castros and are believed to form a large element in the modern population, while the protobrachycephalics are supposed to have disappeared. In a discussion of the origin of blond Europeans, Bloch (1911) concluded that in the Neolithic a large part of Europe from Holland to Russia was occupied by a blond race from which all the blond peoples of history are descended.

Morant's conclusion that the Upper Paleolithic population of Europe (including the Chancelade skull) showed no more biometric variation than a single race, was supported by Bonin (1935a), both as regards skull characters and long bones. He confirms that the Upper Paleolithic type is surprisingly uniform as measured by statistical constants. He also shows that the males and females differ greatly in stature, the mean stature for males being 173 cm. and for females 155 cm. This gives a sex-ratio of 1:1.116, whereas in ten modern races it is only 1:1.0813. The skulls are also easier to sex than in modern populations. In an extensive study of the Pleistocene in European Russia, Golomshtok (1938) finds that while the Mindel, Riss, and Würm are present, the (earlier) Gunz glaciation is very weak or scarcely recognizable. Some 113 sites are described, 10 of which produced skeletal material. No Chellean or Acheulean sites were recognized. The Podkumok skull was found in 1918 at Piatigorsk in digging sewers. It probably belongs to the last glaciation in north Caucasus. The presence of a continuous supra-orbital torus relates it to Neanderthal man, but it is nearer modern man than other Neanderthaloids. A new Neanderthal character is recognized in the presence of deeper supra-orbital depressions than in modern man. In the Kiik-Koba cave in the Crimea, two culture layers, Lower and Middle Paleolithic, were found, separated by a long time interval. The skeleton was a burial, of Neanderthal type but somewhat more primitive.

Cro-Magnon Man

We may now refer to some of the chief remains of Cro-Magnon man. In 1868, five fragmentary skeletons were discovered in the Cro-Magnon cave at Les Eyzies, France, and described by Broca. Four were adults and one a child. No. I is an almost complete skeleton, known as "le vieillard"; the length of the skull is 202.5 mm., the C.I. 73.8. No. II is a female, described in detail by Quatrefages and Hamy (1882), C.I. = 71.7 (?). The skull cap, No. III, has a C.I. 74.8.

At Solutré (Saone-et-Loire), one of the most famous prehistoric sites in Europe, the first specimens were obtained before 1868. Excavations in 1923 discovered an adult female, two adult males, two children (all burials), and a foetus. The following year two more adult skeletons, male and female, were found. Morant (1930) gives photographs of the five adult skulls, of which no full account has been published. Some showed posthumous distortion by pressure. Superciliary and occipital ridges were well developed in two, not prominent in a third, and feeble in a female skull. In No. III the third molar was smaller than M_1 and M_2; in No. V (a female) M_2 was smaller than M_1, the C.I. being 81 (brachycephalic). In two other skulls the C.I. was 76 and 78.7(?) respectively. Solutré No. II is a male skull, brachycephalic. The C.I. is 85.2, but Morant believes that the height and the cephalic index were considerably increased by earth pressure. The occurrence of brachycephaly will be referred to later.

At Combe-Capelle (Perigord) a Lower Aurignacian skeleton was discovered in a rock-shelter in 1909. It was excavated by Klaatsch and Hauser and sometimes goes under the name *Homo aurignaciensis Hauseri*. The skull was very large, length 202 mm., C.I. 66.3.

Near Mentone, in the Riviera, there is a series of nine Grimaldi caves. One of these is the Grotte des Enfants, already referred to. Two nearly complete children's skeletons were discovered here in 1874 and 1875, at a depth of 2.7 meters, but the skulls were broken into fragments and are of little use. In 1900 further excavations unearthed four adult skeletons, two male and two female. They were all Aurignacian but at depths from 1.9 to 7.75 meters, so there must be a long interval between them. The two lowest of these were the "Negroid" skeletons referred to on p. 256. In the fourth Grimaldi cave Rivière found a male skeleton in 1872 at a depth of 6.55 meters. The skull was badly broken but was very dolichocephalic. In the fifth (Barma Grande) cave an early skull has a C.I. of 72.2. In 1884 a complete male skeleton was found by a collector, but the skull was afterwards broken. In 1892 three more skeletons were discovered buried together, one skull having a C.I. estimated at 63. Two years later another male skeleton was found, the C.I. 71.4(?); and a very incomplete skeleton was afterwards found. These are badly preserved in glass cases in the cave. In the sixth cave Rivière found very fragmentary skeletons of two adults and a child, but these appear to have been lost.

Obercassel is an open station near Bonn, Germany, in which two Magdalenian skeletons (male and female) were unearthed in 1914. The C.I. is given as 73.6 (?) and 71.2. The male skull has heavy brow ridges (like Predmost) but no occipital torus (Weidenreich, 1946, p. 42).

Several skeletons have been found in a rock-shelter at Laugerie-Basse, near Les Eyzies, but only one is generally admitted to be Paleolithic. It was excavated in 1872 at a depth of four meters. The man was lying on his side and had evidently been killed by a large rock falling from the roof. The skull was broken and covered with stalagmite. Hardy recorded the C.I. as 73.2. The accompanying artifacts were Magdalenian. La Madelaine is the type station of the Magdalenian culture in a rock-shelter near Les Eyzies. In 1864 fragments of a skeleton were found, the skull including only the frontal bone and part of the mandible.

In Le Placard cave near Vilhonneur (Charente) a Magdalenian skull was discovered in 1881. The following year de Maret found nine other skull vaults or fragments, seven of them Magdalenian and two Solutrean. Two nearly complete skulls had a C.I. of 76.2 and 78.4 respectively.

In Duruthy rock-shelter at Sorde (Landes), in 1874, fragments of a Magdalenian skeleton were found. At a higher level in the same shelter thirty-three Neolithic skeletons had previously been excavated.

At La Vallée du Roc (Charente) Dr. Henri Martin uncovered in a rock-shelter, in 1923, three skeletons belonging to the end of the Solutrean or the beginning of the Magdalenian period. In the adult female skull, which has a sagittal crest, Martin found the C.I. to be 80 and in the adult male, with a low sagittal ridge, the C.I. was 75.9. Thus we have another instance of female brachycephaly in the Solutrean. The sagittal crest is characteristic of the Eskimo and the Chancelade skull. Its presence in these Solutrean skulls may represent a single genic difference being transmitted in the population. The Magdalenian cave at Bruniquel (Tarn-et-Garonne) yielded fragments of two mandibles before 1863. Richard Owen (1870) visited the cave in 1864 and gives a very interesting and full account of it. The cavern is in the face of a cliff forty feet above the bed of the river Aveyron. The two human crania were found *in situ* in the wall of the cave, having been exposed by a fallen rock. There were other remains on the floor of the cave, with flint and bone implements, the bones of animals, and antlers. Some of the mammals are now extinct and some confined to the extreme north of Europe. The animals included *Rhinoceros tichorhinus*,

Hyaena spelaea, and *Elephas primigenius*, as well as remains of thirty horses (*Equus spelaeus*). The artistic productions included incised heads of reindeer and horse, on bone.

Excavations at Les Hoteaux in France in 1894 produced another Magdalenian skeleton, the skull having a C.I. of 77.3.

In England, a Magdalenian skeleton was found in Gough's cave at Cheddar (Somerset) in 1903, and described by Seligman and Parsons in 1914. The skull has a covering of stalagmite which prevents accurate measurements. An Aurignacian skeleton without a skull was found in Glamorganshire. Another at Halling, Kent, was excavated at a depth of 1.8 meters in 1912. Paleolithic man has not been found in Ireland.

At Eguisheim near Colmar, Alsace, in 1865, Late Paleolithic remains were found, consisting of a frontal and right parietal.

The eastern branch of the Aurignacian in Poland, and especially in Moravia, has yielded results of great interest. Cranial fragments were obtained from the Ojcow cave in Poland in 1902. Predmost, in Moravia, is, however, the most important Paleolithic site in central Europe. It is of Solutrean age. Excavations were begun there in 1880 and numerous human remains have been uncovered since. Before 1884 an incomplete mandible was described. Maška in 1894 exposed a cemetery with twenty skeletons (fourteen complete, six fragmentary). Ten skulls were partly reconstructed. In the following year Křiž found the skull of a child of twelve, numerous fragments of skulls, and long bones belonging to at least six individuals. Morant (1930) was able to measure four of these skulls. No. III has unusually prominent superciliary ridges (though not so marked as in Neanderthal man). The temporal lines, which are feebly developed in Neanderthal man, are strongly marked here, and the mastoid processes are also massive, and prominent. The C.I. of this skull is given as 71.2 (?). No. IV also has distinct superciliary and occipital transverse ridges, C.I. 74.9. No. IX is a young adult with distinct superciliary ridges, C.I. 73.7 (?). In No. X the superciliary ridges are much less marked, the C.I. 75.9. This Solutrean find was as remarkable as that of Tutankhamen's tomb. These people hunted the mammoth, built sepulchers of mammoth jaws and shoulder blades, and made many objects of bone and ivory ornaments with geometrical designs. Predmost represents the earliest neanthropic man in central Europe. Although their superciliary ridges were rather prominent, they were of short or medium height and were representatives of the Caucasian race.

In the Moravian cave at Lautsch, in 1881, two incomplete skulls were found, fragments of a child's skeleton, and two maxillae. In 1903 further digging yielded the front half of a cranium, and the following year an adult skull, a cranium, a calvaria, and other skeletal fragments, all of Aurignacian age. These ranged in C.I. from 71.5 to 72.5, two of them having prominent brow ridges. Brno, in Moravia, is another important site. In 1891 a skeleton was discovered here accidentally, at a depth of 4.5 meters, in the loess on which Brno is built. The skull (male) had prominent brow ridges with a retreating frontal, but had undergone posthumous deformation. Another skeleton was described in 1927 by Absolon and Matiegka. In the calotte found by Absolon in 1925 at Věstonice in Moravia the brow ridges are only prominent over the nasion and the inner half of the orbits, the C.I. near 73. Matiegka (1929) has since described the Brno III skull, and Bonin (1935) the Cap-Blanc remains, an almost complete Magdalenian skeleton of a young woman.

These Upper Paleolithic people of eastern Europe are thus dolichocephalic to mesaticephalic, whereas in western Europe there was a small brachycephalic element. The C.I. of all Upper Paleolithic skulls ranges in males from 65.7 to 78.4 (mean 72.6) and in females from 68.3 to 81.0 (mean 75.3). Possibly as they moved eastwards the brachycephalic element was left behind. In the eastern group, including the Russian skulls, the C.I. ranges only from 64 to 76, mean 71. The difference between Cro-Magnon man proper, with its center in France, and the Moravian group which is more exclusively dolichocephalic, is similar to that between the western and eastern Eskimo, although none of the Eskimos are brachycephalic. Coon (1939) points out resemblances of Predmost No. III to Skhul No. V from Mount Carmel, regarding Predmost III as intermediate between Combe-Capelle and the Neanderthals. The resemblance to Skhul V reminds us that Upper Paleolithic man may have also entered Europe from the east. We have been accustomed to think of brachycephaly coming from the east, but in Cro-Magnon man it must either have arisen through mutations or crossing in western Europe or have been brought in with the race from northern Africa. Since the Alfalou people in northern Africa had a brachycephalic element, it is probable that this was at any rate the chief source of brachycephaly in Cro-Magnon man.

Coon (1939, p. 38) points out that it is the brow ridges which make all the Neanderthals dolichocephalic. If their cranial measurements are taken from the ophyron, a point on the frontal bone behind the brow ridges, then five out of eight are brachycranial or brachencephalic. He reasonably suggests that the brachycephaly in Cro-Magnon man may thus have arisen in the descendants of crosses with Neanderthal man. Similarly, the brow ridges found in some Upper Paleolithic men and not in others may have arisen as an intermediate heterozygous condition in crosses between Neanderthal man and others without brow ridges. However, if brachycephaly arose in Upper Paleolithic man from crossing with Neanderthals in western Europe, it is difficult to see why it did not arise in the same way either in eastern Europe, or earlier on Mount Carmel. On the whole, the hypothesis of mutations from dolichocephaly to mesaticephaly and from mesaticephaly to brachycephaly seems more probable.

The brow ridges in many Upper Paleolithic skulls, and especially in the Predmost people, are rather conspicuous. This has led some writers to speak of them as Australoid, a misleading appellation because, as shown in connection with the Korana in South Africa, it is a parallel development and not derived from an Australian source.

Morant finds the mean length of all Upper Paleolithic skulls to be 198.1 mm., greater than in any modern race (except the Ona). In the French Neolithic it was 195 mm., the British Neolithic 193.7, and Anglo-Saxons, 190.6. The head of Upper Paleolithic man was larger in several measurements than that of modern man. Only the orbital height is less. The Barma Grande skull, with a length of 211.5 mm., is the longest non-pathological skull on record. The Rhodesian skull approached it, with a length of 208.5 mm., but this includes the enormous bony frontal torus.

The tallness and strength of Upper Paleolithic man in well known, but early estimates of 190 cm. in height were exaggerated. Bonin (1935a) finds the stature to range in males from 161.1–181.8 cm., the mean being 173 cm. for males and 155 cm. for females. This is another example of the very marked difference in the sexes. Another feature of these people is that in the upper extremities the proportions are like those of modern primitive races, while the lower extremities are perhaps less stable in the proportions of femur and tibia than in modern races. The femur was platymeric and pilastric, the tibia platycnemic, and the fibula deeply fluted. These latter conditions still persist.

Bonin (1935b) describes in detail the Magdalenian Cap-Blanc skeleton found in a cave in the upper valley of the Tarn in southwestern France. This cave was ornamented with sculptures on the walls and a fine frieze of animals, and there were numerous chipped implements on the floor. The skeleton is now in the Field Museum at Chicago. Bonin shows that the biometric measurements of the Beaumes-Chaudes Neolithic people are closely similar to those of the British Neolithic, the Norwegian Iron Age people, and the Anglo-Saxons. Since these three peoples are said to be all Nordic, the Beaumes-Chaudes skeleton should also represent "Nordics" in the Neolithic period, but their pigmentation is unknown. He suggests that the Nordics arose from hybridization between Upper Paleolithic man surviving in France and an invading Neolithic race in western Europe.[7] Although we do not know when the blue eyes and fair hair of the Nordics arose, it was probably in Cro-Magnon man himself, as there is an autochthonous Nordic element in north Africa. These negative mutations probably began to appear long before the Neolithic. We may think of them as arising in Cro-Magnon man during the Paleolithic and accumulating in certain areas of north Africa and France through repeated mutations and sexual selection. The relations between the Cro-Magnons of France and the orangoid descendants of Swanscombe and Galley Hill man in England remain obscure. There is no reason to suppose that this early English race ever mutated from dark eyes and hair. They may be the ancestors of the Brythons.

[7] Coon (1939, p. 83) takes the view that the Nordics were simply derived by depigmentation from hypsicephalic Mediterraneans, arising in the area north of the Black Sea and the Caucasus, the Mediterraneans being descended from Galley Hill stock and related to Combe-Capelle and Alfalou types. Morant (1926a) found the population of the late Neolithic long barrows in England and Scotland to be homogeneous, the skulls long (C.I. 71.3) and of medium height. The brachycephalic Bronze Age beaker-folk invaders (in the round barrows) were also homogeneous, indicating that they drove their dolichocephalic predecessors as remnants into inaccessible parts of the country, although some hybridization also took place. Thus among 151 Bronze Age skulls Morant found the C.I. to show the enormous range, 64.5 to 92.5, the brachycephals being most numerous (91); but 30 were of pure Neolithic type and 30 hybrids. The Iron Age invaders were again dolichocephalic, and of low cranial height. Finally, the Anglo-Saxons of the fifth to tenth centuries were again homogeneous and with skulls like the Iron Age men except that they were hypsicephalic. These two populations apparently lived side by side for some centuries without intercrossing.

Coon (1946, p. 109) believes that the British Neolithic people who introduced the megalithic culture came to England by sea and were of large (western) Mediterranean type, perhaps from north Africa. Bonin suggests that the Nordics drifted from England to northern Europe, whence they spread to Germany and France, finally returning to England as the Anglo-Saxons.

According to the Scandinavian chronology based on varves, the end of the Upper Paleolithic and the beginning of the Mesolithic in Europe can be dated 11,800 B.C. Relatively little attention has been paid to the relation of the reindeer to civilization in the Magdalenian and later times. Unlike the horse, which has a long paleontological ancestry, the reindeer appears suddenly in Europe in the Middle Pleistocene, just as he is today. It is possible that he was the first of the larger mammals to be domesticated, long before Neolithic times, but the antlers do not show evidence of castration, which is an indication of domestication. This reindeer culture survives with the Lapps in Scandinavia and the Tungus in Siberia. Leroi-Gourhan (1936) believes that Cro-Magnon man played a considerable, perhaps an essential, role in the Upper Paleolithic history of the reindeer.

We may now consider certain other aspects of Cro-Magnon man— the original *Homo sapiens* in nomenclature but not in age. His relations to Boskop man and *Homo kanamensis* in Africa were referred to in the last chapter. It now seems clear that he originated in north Africa, and his relationship to Alfalou man in Algeria and Boskop man (*Homo capensis*) were previously noted. The subject of Cro-Magnon man has also been discussed elsewhere (Gates, 1946). Hooton (1925) was the original exponent of the view that Cro-Magnon man originated through hybridization between a short-faced dolichocephalic and a broad-faced brachycephalic type. This of course implies the previous origin of brachycephaly elsewhere. His views are developed in his classical study of skulls in the Canary Islands. In the museum at Santa Cruz, on Teneriffe, he measured 454 skulls. The Guanche mummies in the caves of Teneriffe have long been famous. Three or four hundred in a cave were shown to Thomas Nichols in 1526, and there are supposed to be over twenty caves containing bodies. In 1770 one very large cave had over a thousand mummies. In medieval Europe pieces of mummies were an important ingredient of medicine. Only the upper classes were embalmed, but a catacomb full of mummies was found at La Palma in 1758 and others were well known on Gran Canaria.

As regards the coloring of the modern population, a Florentine visiting the Canaries in 1341 refers to the long, light hair of four young men, although the majority were swarthy. Records of eye color in 39 persons, in 1893, show that 12.8 per cent had blue eyes and 2.6

per cent grey-blue. There is evidence that some of the north African tribes have changed from blond to brunet within historical times, as a result of "mixture with strangers." Is this the same process that Dart finds in the history of Egypt, or does it rather result from the fact that light eyes and hair, if simple recessives, will only reappear in a quarter of the offspring from crosses with dark eyes and hair, if in the original population one of each pair of parents were dark and one light? It is, of course, recognized that a dominant character does not spread in a population through crossing. But if numerous matings occur between two races one of which has a dominant character and the other the corresponding recessive, then the frequency of the recessive will change from 50 per cent in the beginning to none in the F_1 and 25 per cent in the F_2. This is, in effect, a large increase in the dominant character. Whereas in Egypt selection in some form was at work, as shown by changes in skull shapes, in the present case the result could be purely genetical. There was a large blond element in northern Africa which probably began in the Paleolithic. The blondness in the Canaries was probably an extension of this. Verneau (1887) concluded after five years' investigations that the Guanches were predominantly tall, blond Cro-Magnons in Teneriffe, less so in the other islands. They only became extinct in the sixteenth century, and their language is related to that of the Berbers. Hooton estimated their height at 166 cm. We are here involved in the relation between the north African branch of the Boskop race on the one hand and the Alfalou and Cro-Magnons on the other. There is not yet sufficient evidence to determine what these relations were, but the elements of the problem can be further analyzed.

From cranial measurements of the Guanches, Hooton found the mean C.I. 75.99 for males and 77.57 for females, about 26 per cent being dolichocephalic and 7 per cent brachycephalic. The skulls of Teneriffe were most like the seventeenth century Londoners. Their cubic capacity was 1451 cc. according to Shrubsall and 1672 cc. from results of Verneau. These included some very small and some very large crania. In compiling the measurements of fourteen "Cro-Magnon" crania from Teneriffe, Hooton found the C.I. ranging from 63 to 79 and the N.I. (nasal index) from 37 to 62. He concluded that skulls with such a great range of variability could not be regarded as belonging to one uniform race. Their sole resemblance to typical

Guanche crania was in their large size (selected), chamaecephaly, and leptorrhiny. The Cro-Magnon type of skull is dysharmonic—a long, narrow cranium with a very broad face in front of it. The low, wide (chamaeconch) orbits are another dysharmonic feature.

Hooton was further able to show that in a modern family in which the father was dolichocephalic with a very broad face and a narrow straight nose, the mother brachycephalic with very prominent malars and a short fleshy nose, the daughter inherited the dolichocephaly and leptorrhiny of the father with the prominent molars and zygomatic arches of the mother and the great facial breadth of both. He concluded that the Cro-Magnon cranial type arose as a hybrid between brachycephalic and dolichocephalic races, the tallness being a result of heterosis or hybrid vigor.

As regards the Canary Island skulls, Hooton classed them as nearly half Guanche, the remainder Mediterranean-Nordic, near-Nordic, Alpine, and Negroid. Besides the dysharmony of a narrow skull and broad face, there is also the opposite dysharmonious combination of a short, broad skull with a long, narrow face. Some of the Canary skulls showed the association of (a) mesocephaly with a leptene face (upper face long or narrow), or (b) brachycephaly with a euryene face (upper face short or broad). Such types may be expected to occur wherever races having a long head and long face mix with those having a short face and short head. It was concluded that the first inhabitants of the Canaries were probably in the Neolithic, these people being dolichocephalic, mesorhine, short brunets of the Mediterranean race with some Negroid mixture. From a study of the measurements of one thousand Canary Island skulls by Hooton and others, Temagnini (1931) concluded that they were clearly a heterogeneous population, including brachycephaly and Negroids, the dysharmonic types having their origin in crossing. The dysharmonic elements show a tendency to chamaerrhiny, but the percentage showing Cro-Magnon affinity is believed to be insufficient to support the theory of the survival of this type in the Canaries.

There is another important feature, which has been pointed out by Dart (1939, p. 132). In Chapter 5 (p. 133) we found that Boskop man is chamaecephalic whereas Nordic man is hypsicephalic. Cro-Magnon man was also chamaecephalic, and to that extent Boskopoid. His skull was also obtusely pentagonal (i.e., foetal) in vertical view. If the

European Cro-Magnons gave rise to the Nordic type then there must somewhere have been a mutational change from chamaecephaly to hypsicephaly. Dart believes that the Boskopoid Cro-Magnons emerged from crossing between the hypsicephalic, leptoprosopic (narrow faced), leptorrhine, ellipsoidal, Nordic race (believed already existing) and the chamaecephalic, chamaeconch (with low orbits), chamaeprosopic, platyrrhine, pentagonoid Boskop race. We have already seen that Nordics existed in the Neolithic in the Baumes-Chaudes people. How much earlier they can be postulated is not at present clear. The evolutionary origin of hypsicephaly requires further investigation. Hooton (1925) regards the Cro-Magnon type as the product of crossing between the ancient dolichocephalic stock represented by Galley Hill and Swanscombe and the brachycephalic strain represented by Ofnet. This would account for the long, narrow head combined with a broad face, so characteristic of Cro-Magnon crania. Since, as we have already seen, there is a small brachycephalic element present in the early Upper Paleolithic in western Europe, this cross could probably have arisen earlier, because the Ofnet strain are Mesolithic.

Perret (1937) finds the Cro-Magnons surviving in modern Hesse (Germany) as the Falic race; they were also in the stone cysts of the Neolithic in Hesse, the Bronze Age, and the early and late Middle Ages. He believes that the Nordics were derived from Cro-Magnon in Late Paleolithic by mutations in this region (see also Gates, 1946, p. 1370). French anthropologists find the Cro-Magnon type persisting in parts of France, especially the Dordogne.

The find at the big Ofnet cave in Bavaria is one of the most remarkable in all human paleontology (Scheidt, 1923). Two nests containing in all 33 skulls were found. These skulls were laid together and sprinkled with red ocher, and were probably of Azilian age in the late Mesolithic, or perhaps late Magdalenian. Four skulls belonged to men, 10 to women, and 19 to children. All had been beheaded and the bodies are missing. This appears to have been a case of "women and children first," with dim indications of a "political" or racial background. A detached male skull found at Kaufertsberg, a few miles away, has already been referred to (p. 257). Only 2 of these 34 skulls were dolichocephalic. One of them was of Cro-Magnon type with very broad face, heavy brow ridges, a retreating forehead, very low orbits, a strong mandible, and a cranial capacity of over 1600 cc. The other was similar but less massive. The bulk of the skulls were mesaticephalic, but 4 (two of each sex) were brachycephali, the highest cephalic index (a female) being 87. They all have broad faces and belong to the Cro-Magnon type. Further details regarding them are given in Coon (1939, p. 67). So early is the beginning of brachycephaly in Bavaria.

It is not our purpose to follow further the racial history of Europe, but it may be pointed out that brachycephaly has continued to increase, especially in central Europe, since that early time.

Boule (1923) shows that brachycephaly continued to increase gradually throughout Neolithic and later times in many parts of Europe. For instance, of 688 Neolithic skulls found in France up to 1896, 58 per cent were dolichocephalic, 21 per cent brachycephalic, and 21 per cent intermediate. In England, as is well known, the Neolithic long barrows contained long skulls of the Mediterranean race. The later round barrows contained brachycephalic people who were late in arriving in Britain from western Europe. They appear to have been almost eliminated in the modern population of Britain. It is customary to ascribe them to the Alpine race, believed to be an intrusion into Europe from the east. But the early distribution of brachycephali among dolichocephali, and their gradual increase in frequency over a long period in western Europe seems to show not a new intruding race but a mutation in a single character—head shape—gradually increasing in the population.

In Switzerland on the contrary, according to Boule, the early Neolithic people were nearly all brachycephalic, the dolichocephali and mesocephali becoming much more numerous in mid-Neolithic, dolichocephali predominating in the Bronze Age. Towards the end of the Neolithic the Nordics arrived, with long skulls, narrow noses, and tall stature. Neolithic Germany was dolichocephalic, of the long-faced Nordic type, brachycephali and Mediterraneans coming in later. In Poland and the Ukraine tall dolichocephali were dominant in the Neolithic kurgans as well as farther north and east and around Lake Ladoga. The origin of the Neolithic brachycephali of Switzerland, the so-called Alpine race, remains to be determined. Was it possibly autochthonous, the result of a high mutation rate to brachycephaly combined with other changes? Coon (1939, p. 119) discusses the possibilities and points out that the Neolithic Alpines differed from the Mediterraneans in more than brachycephaly. They were often taller, with larger vaults, lower orbits, shorter faces, and wider noses. Coon agrees that the Alpine broad-heads have not come from the east as a Neolithic invasion, nor from the south (Africa). Since brachycephaly has appeared independently in many parts of the world, repeated mutations appear to be the only way to account for its spread in these different areas. But of course each such case of spread would be accompanied by concomitant racial changes. If mutations are to account for the basic material of human evolution it is difficult to see how else the mutation theory can be applied.

A brief reference must be made to the Mediterranean race (another branch of the Caucasian) because they are now recognized as extending far eastwards beyond the Mediterranean area into India. Elliot Smith suggested that Brown race was a more appropriate name. In recognizing this racial type, Sergi (1914, p. 31) says, "I hope to show . . . that there was really a centre of dispersion of the Mediterranean stock, which . . . anterior to all tradition, occupied the regions which surround this great basin, and that the various peoples derived from this stock have possessed the most ancient native civilization in the countries, islands and peninsulas they occupied." They inhabited Asia Minor, Syria, Egypt, Libya, north Africa, Greece, Italy, and the Iberian peninsula—a rather small dolichocephalic, brunet race, derived from neither white nor black. Their branches included the Iberians, Ligurians, Pelasgians, Hittites, Libyans, and we may add the Hindus. Sergi derives them from Hamitic-speaking peoples in north Africa, probably including Somaliland and the sources of the Nile in the Great Lake region. As the ancient Libyans were blonds (Procopius speaks of a population with white skin and fair hair, and Callimachus noted fair Libyan women in Cyrenaica) it is possible that blondness originated through mutations in the Libyans, independently of the blond Nordics. In Egypt, the mother of Amenhotep IV, about 1700 B.C. (18th Dynasty), was blond, with blue eyes and a rosy skin; and in the 19th Dynasty (*ca.* 1400 B.C.) Egypt was invaded by nomads from the west with blue eyes and fair hair. Under Rameses II the blond invaders were established in Egypt and furnished troops to the king. These Libyan invaders had no European customs or dress. Much turns upon whether they were really of Mediterranean or Nordic stock. The former is orthocephalic, the latter hypsicephalic.

REFERENCES

Angel, J. L. 1946. Race, type, and ethnic group in ancient Greece. *Human Biol.* 18:1–32.
Bloch, A. 1911. Origine et évolution des blonds européens. *Bull. et Mem. Soc. Anthrop. Paris.* 2:442–450. Figs. 11.
Bonin, G. von. 1935a. European races of the Upper Palaeolithic. *Human Biol.* 7:196–221.
———. 1935b. The Magdalenian skeleton from Cap-Blanc. University of Illinois Press. pp. 76. Pls. 9.
Boule, M. 1923. *Fossil Men.* Edinburgh, London. pp. 504. Figs. 248. Trans. Ritchie.
Brögger, A. W. 1937. Late Palaeolithic man in northwestern Norway. In *Early Man* (ed. G. G. McCurdy). pp. 53–60.
Burkitt, M. C. 1938. The middle Palaeolithic. *Rept. Brit. Assn.* p. 435.
Coon, C. S. 1939. *The Races of Europe.* New York: Macmillan.
Dawson, C., and A. Smith Woodward. 1913. On the discovery of a palaeolithic skull and mandible in a flint-bearing gravel overlying the Wealden (Hastings Beds) at Piltdown, Fletching (Sussex). *Quart. J. Geol. Soc.* 69:117–151. Pls. 4. Figs. 11.
Garrod, D. A. E. 1936. The Upper Palaeolithic in the light of recent discovery. *Rept. Brit. Assn.* (Blackpool). pp. 155–172.
Golomshtok, E. A. 1938. The Old Stone Age in European Russia. *Trans. Am. Phil. Soc.* 19:191–468. Pls. 37. Figs. 100.
Hooton, E. A. 1925. The ancient inhabitants of the Canary Islands. *Harvard African Studies.* 7:xxi, 401.
———. 1946. *Up from the Ape.* 2nd Edition. New York: Macmillan. pp. 788.
Hrdlička, A. 1923. Dimensions of the first and second lower molars with their bearing on the Piltdown jaw and on man's phylogeny. *Am. J. Phys. Anthrop.* 6:195–216.
———. 1930. The skeletal remains of early man. *Smithson. Misc. Coll.* 83:1–379. Pls. 93. Figs. 39.
Keith, Sir A. 1939. A resurvey of the anatomical features of the Piltdown skull with some observations on the recently discovered Swanscombe skull. *J. Anat.* 73:155–185, 234–254. Pl. 1. Figs. 32.
Leroi-Gourhan, André. 1936. *La Civilization du Renne.* Paris. pp. 178. Pls. 32.
Matiegka, J. 1929. The skull of the fossil man Brno III and the cast of its interior. *Anthropologie* (Prague). 7:90–107. Pls. 4. Figs. 7.
Mendes Corrêa, A. A. 1919. Origins of the Portuguese. *Am. J. Phys. Anthrop.* 2:117–145.

Miller, G. S. 1918. The Piltdown jaw. *Am. J. Phys. Anthrop.* 1:25–52. Pls. 4.

Morant, G. M. 1926. The Chancelade skull and its relation to the modern Eskimo skull. *Ann. Eugen.* 1:257–276. Pls. 10. Figs. 3.

———. 1926a. A first study of the craniology of England and Scotland . . . with special reference to the Anglo-Saxon skulls in British Museums. *Biometrika.* 18:56–98. Figs. 6.

———. 1930. A biometric study of the Upper Palaeolithic skulls of Europe and of their relationships to earlier and later types. *Ann. Eugen.* 4:109–214. Pls. 12. Figs. 63.

Newton, E. T. 1895. On a human skull and limb-bones found in the Palaeolithic terrace-gravel at Galley Hill, Kent. *Quart. J. Geol. Soc.* 51:505–527. Pl. 1. Figs. 2.

Osborn, H. F. 1915. Review of the Pleistocene of Europe, Asia and North Africa. *Ann. N. Y. Acad. Sci.* 26:215–315. Figs. 20.

Owen, R. 1870. Description of the cavern of Bruniquel and its organic contents. *Phil. Trans. Roy. Soc.* 159:517–557. Pls. 4. Figs. 9.

Perret, G. 1937. Cro-Magnon typen von Neolithicum bis heute. *Zeits. f. Morph. u. Anthrop.* 37:1–101. Pls. 29. Figs. 6.

Pittard, E. 1941. Les origines de l'humanité et les bases préhistoriques de la civilization. *Actes Soc. Helvét. de Sci. Nat.* 121:27–37.

Quatrefages, A. de and E. T. Hamy. 1882. *Crania Ethnica.* Paris. pp. 528. Pls. 100. Figs. 486.

Scheidt, W. 1923. Die eiszeitlichen Schädelfunde aus der Grossen Ofnethöhle und von Kaufertsberg bei Nördlingen. Munich: Lehmann. pp. 112. Pls. 8. Figs. 7.

Schoetensack, O. 1908. Der Unterkiefer des *Homo heidelbergensis,* aus dem Sanden von Mauer bei Heidelberg. Leipzig. pp. 67. Pls. 13.

Sergi, G. 1914. *The Mediterranean Race.* London. pp. 320.

Smith Woodward, A. 1917. Fourth note on the Piltdown gravel, with evidence of a second skull of *Eoanthropus dawsoni. Quart. J. Geol. Soc.* 73:1–10. Pl. 1. Figs. 2.

Smith Woodward, Sir A. 1944. The geographical distribution of ancestral man. *Geol. Mag.* 81:49–57. Figs. 3.

Temagnini, E. 1931. Estado actual dos nossos conhecimentes acêrca da antiga populaçao des Canaries. *Internat. Congr. Anthrop.* (Portugal) 1930. 1:241–245.

Vallois, H. V. 1939. Nouvelles recherches sur l'homme fossile de Chancelade. *Proc. Internat. Anthrop. Congr. Copenhagen.* pp. 112–114.

———. 1946. Nouvelles recherches sur le squelette de Chancelade. *L'Anthrop.* 50:65–202. Figs. 10.

Verneau, R. 1887. Rapport sur les resultats anthropologiques de la mission dans l'archipel des Canaries. *Arch. des Miss. Sci. et Litt. Paris.* Ser. 3, vol. 13. pp. 557–817. Pls. 4. Figs. 44.

———. 1924. La race de Neanderthal et la race Grimaldi: leur rôle dans l'humanité. *J. Roy. Anthrop. Inst.* 54:211–230.

Weinert, H. 1933. Das Problem des "Eoanthropus" von Piltdown. *Zeits. f. Morph. u. Anthrop.* 32:1–76. Pls. 7. Figs. 36.

———. 1936. Der Urmenschenschädel von Steinheim. *Zeits. f. Morph. u. Anthrop.* 35:463–518. Pls. 6. Figs. 18.

Young, Matthew. 1938. The London skull. *Biometrika.* 29:277–321. Pls. 3. Figs. 7.

9

From Sinanthropus to the American Indians

In previous chapters we have traced a line of human evolution from Pithecanthropus or Meganthropus to the Australian aborigines in Sundaland and Australasia. On the African continent, beginning with Africanthropus and the Rhodesian and Florisbad types we found abundant evidence for another line of evolution through Boskop man to the South African Bushmen. In Europe the Pithecanthropus level has never been found, but beginning with Eoanthropus in the orangoid series and Neanderthal man in the gorilloid series we have traced descent to modern Homo, partly through crossing of different human species and genera, partly through influx of human types from other continents, and partly through mutational changes in particular races and species.

As regards the peopling of North and South America, agreement is practically universal that it has occurred by migrations from northeastern Asia, the same route by which, in earlier geological ages, large numbers of species of reptiles and mammals had passed into and out of the western continent. It is evident that southern Asia formerly had a widespread Australoid population, descended from Pithecanthropus. In China, we have seen that Sinanthropus was a source from which later human types have descended, and that certain features of the Sinanthropus skulls have been transmitted to the modern Chinese and the American Indians. That the Amerind is allied to the Mongolian has long been recognized, but the relationship is not a simple one. It happens that the blood groups are a definite help in elucidating these problems of relationship.

The geographical position of Sinanthropus in northern China, as well as the various morphological points of resemblance between Sinanthropus and skulls of Mongoloids and American Indians, makes it probable that the latter are mainly descended from Sinanthropus; but there is a big gap between them, just as there is between Rhodesian man and the Negro or Bushman. In the Red Indian that gap is partly filled by the finding of occasional skulls of "Neanderthaloid" character, such as must have occurred in the evolution from Sinanthropus to the modern type of man. The term Neanderthaloid, like Australoid, is only allowable in the sense that it represents an intermediate stage of evolution, in this case independent of the European Neanderthal but belonging to the level of the genus Palaeoanthropus.

Little is known archaeologically of the development of Mongoloid characters, but a brief reference may be made to the origins of the Chinese although it comes within proto-historic times. Williams (1918) has discussed the possibilities. It appears that the nucleus of the Chous (later the Chinese) was autochthonous in the valley of the lower Yellow River. They gave up nomadism, became agricultural, and absorbed many allied tribes. Legendary emperors existed in the third millennium B.C. Long before 1200 B.C. a civilization had developed in northeast Honan. This involved the driving southwards of the Burmese Karens, who were originally near Tungting Lake, A.D. 778. Racial movements from north to south have continued, the Miao being driven into Yunnan. The Lolos have been regarded as "Aryans," with white skin, hook nose, brown hair, and blue or gray eyes not almond shaped. Père Richard states that the Chinese of Szechuan are of mixed Mongol and Aryan descent, many having blue or gray eyes and brown hair.

There is another theory of Chinese origins, based partly on resemblances between the Chinese and Sumerian languages, both using idiographs and vertical writing. These languages probably had a common origin, but the Chinese symbol for the eye as early as 1500 B.C. shows a type of eye with the Mongolian eyefold. Although there is thus evidence of early contacts with the west, the Chinese race and civilization appear to be largely of native origin in central China. In any case, the sources of the American Indians have to be sought for in a more northerly region and a much earlier time.

One of the main problems regarding man in America is this: Did the earliest passages of man across Bering Strait occur during the

Ice Age or only at the end of it? A limiting factor of much importance regarding the immediate ancestry of the American Indian and the time of his passage into North America is furnished by the blood groups. So far as now known, all the Mongoloid peoples now living in northeastern Asia have a high frequency of the B blood group. It is well recognized that the A, B, and O are all inherited as simple Mendelian units (see Gates, 1946, p. 679). In the Manchus the frequency is about 35 per cent, Buriats 37.5–40 per cent. In the Gilyaks of Saghalien, who resemble the Eskimos in some respects, it is only 14.5 per cent; but the Ainu of northern Japan, who are not Mongoloid, nevertheless range from 32–38 per cent. Now the Indians of North America are devoid of B so far as known, except when acquired from Europeans by crossing. They could therefore obviously not be derived from a population like that of the modern Mongolians, apart from the physical differences. It was formerly pointed out, however (Gates and Darby, 1934), that some of the isolated primitive tribes in Formosa not only have Indian-like features but are low in the B blood group and high in O. The Tsou, for example, in some localities in the mountains, have only 6 or 7 per cent of B, and the Paiwan, who run over 30 per cent in some places, drop to 7.5 per cent in one locality. It was therefore suggested that some of these tribes represent remnants of an early population, who were driven to isolation on coastal islands such as Formosa, while others crossed Bering Strait to America. Hrdlička (1926) also pointed out that some Tibetans have features strongly resembling the Indians, as well as straight black hair. Hrdlička says (p. 61), "True American Indian types are also to be found, so true that if they were transplanted to America nobody could possibly take them for anything but Indian. They—men, women and children—resemble the Indians in behavior, in dress, and even in the intonations of their language."

A Tibetan woman in Darjeeling, in the foothills of the Himalayas, is shown in Plate XXIII (a). By comparison, Plate XXIII (b) represents three old slave Indians who came to hospital at Fort Simpson on the Mackenzie River in 1928. These thus appear to be survivors in eastern and central Asia of the early groups which peopled much of North America. It is clear that the B blood group must have developed in the modern populations of Asia adjacent to Bering Strait *since* the ancestors of the Indians emigrated to America, and since that channel for emigration was finally blocked by the Eskimos settling astride of the Strait. The possibility of *late* immigration of Indians into America is thus nullified not only by the Eskimos occupying the Alaskan coast, but also by the absence of the B blood group from the modern Indians of North America.

The best archaeological evidence from caves in western America and from other sources indicates that man was present in western North America during the last interglacial, but further waves of immigration probably occurred in post-glacial times. Some of this evidence must now be briefly recapitulated; it has become quite substantial in recent years but only a few of the investigations bearing on this problem can be referred to here. Having stated the conclusion, we may proceed to develop the evidence. Hrdlička (1942) maintained that there is "no evidence of man more than a few thousand years old in any of the parts of northeast Asia from which man could have reached America"; but this negative evidence, in a region which has been but little explored archaeologically, can not be regarded as final. He states, however, that in Siberia the Neolithic and Upper Paleolithic have been found, but nothing older. If man first entered America in the last interglacial period nothing older is required.

A few words may be interposed here regarding the divisions of the Pleistocene in North America. According to accepted results of geologists (see Hay, 1923), there were five different advances of the ice sheets in the northern States, with four interglacial periods between them. The whole series of nine periods, beginning with the earliest (the interglacial periods being in italics), are as follows: Nebraskan, *Aftonian,* Kansan, *Yarmouth,* Illinoian, *Sangamon,* Iowan, *Peorian,* Wisconsin. The various ice advances took different directions, so that while they partly overlapped they by no means coincided. Many fossils of mammoths, elephants, horses, tapirs, camels, peccaries, deer, bison, musk ox, and other animals have been unearthed belonging to these different stages of glaciation, but man's relation to them is imperfectly known.

As regards correlation with the glacial periods in Europe, the Gunz and Mindel advances, separated by a warm wet interglacial, would correspond to the Nebraskan-Kansan advances with the interglacial Aftonian between. The long Yarmouth interglacial, representing possibly a third of Pleistocene time, and believed to have been a period of cold, dry climate, will correspond with the long Mindel-Riss interglacial in Europe. The later Illinoian-Wisconsin advances apparently correspond with the Riss and Würm, the short, warm, wet Sangamon and Peorian probably corresponding roughly with interstadial periods in Europe, the Iowan being a substage of the Wisconsin advance. Sauer (1944) suggests that in high northern latitudes the Nebraskan-Kansan merged into one, without an intervening warm period, and similarly the Illinoian-Wisconsin. There is also the important difference that whereas in Europe the main glacial advances were all from the Scandinavian ice-fields, in North America each main glaciation spread from a different center, from Labrador in the east to an area west of Hudson Bay. Five substages or lobes of the Wisconsin advance are recognized. The glacial conditions in western Canada and Alaska, and their relation to Rocky Mountain uplift, have yet to be determined.

Recent observations show that recession of the glaciers is taking place in Greenland, Iceland, and Norway, and there is also reduction in the Arctic pack ice north of Siberia. We may be still emerging from the last glaciation into a new and warmer interglacial period.

Regarding the particular route taken by man into America, many writers have pointed out different possibilities. As far as the Aleutian chain of islands is concerned, the evidence indicates that they received their population mainly from Alaska and not from Asia. Blood grouping of the Aleuts would no doubt throw light on this problem, even though they interbred to some extent with Russians in the eighteenth and nineteenth centuries. The most probable view appears to be that of Kirk Bryan (1941) and others that man crossed Bering Strait either on a land bridge or on ice north of the Strait, or that some of them possibly crossed in boats. The lowered sea level formed a land bridge in glacial times. Entering Alaska would undoubtedly be easier during a glacial period, when the sea level was low. It is now recognized that central Alaska and much of the Mackenzie valley were free from ice throughout the Pleistocene. Such immigrants might then accumulate in unglaciated areas of central Alaska until warmer interglacial conditions permitted their entrance further into the continent. Various writers have pointed out that during the last interglacial period (Peorian), preceding the Wisconsin glaciation, the Mackenzie River valley was an unglaciated corridor between the Alaskan and Rocky Mountain glaciation on the west and the Keewatin glaciation on the east. Man may then have entered Alaska from northeastern Siberia during the penultimate glaciation, passing across to the Mackenzie and down the corridor during the following interglacial, having multiplied largely in the meantime. If, on the contrary, a dribble of small groups from Siberia continued for a long time, then the unity of the American race would be less than in fact it appears to be.

In early Pleistocene a host of animals crossed the isthmus into America. Jochelson (1926) states that the Siberian sheep spread into America during an interglacial period and afterwards moved back into Asia. The reindeer, elk, musk ox, black bear, and other mammals were hunted by man in northeastern Asia, and probably were followed across the isthmus plateau into Alaska. Peoples such as the Koryaks and Chukchi, who afterwards may have recrossed into Asia, Jochelson calls Americanoids.

Hrdlička (1932) did not admit any interglacial migration but, from his excavations in the Aleutians and travels in Alaska, concluded that man's post-glacial movements from Asia were in driblets by boat across Bering Strait, thus bringing groups already having differences in physical type, language, and culture. Some of these passed eastwards along the northern coast of Alaska, others southwards down the western coast, ultimately to reach British Columbia and California.

The most likely conclusion is that early invaders with a primitive hunting culture passed southwards over the Canadian plains when these plains were free from ice but flanked with glaciers in the Rockies and west of Hudson Bay. Bryan concludes that the preponderance of evidence implies a single migration route to the empty New World—a one-way road where the attractions of abundant game would lead the hunters onwards. He says, "To the relatively advanced and well-organized tribes developed in Asia in the past 2000 to 4000 years, many of the difficulties would be small." But these are the tribes that did not enter America, as shown by the absence of the B blood group. We must then assume that these later peoples, in whom the B blood group had developed, were blocked from entering Alaska; and the Eskimos, who were already tied to an ice culture, were quite sufficient to constitute such a block.

The length of time that the Eskimos have occupied the Alaskan coast will be considered later, but we may conclude that for several thousand years no immigration from Asia, except Eskimos, has taken place. Incidentally, the fact that the B blood group is absent, not only from American Indians but also from the Australian aborigines and (except for a low percentage through contact with other peoples) the Bushmen in South Africa and Lapps in northern Europe, shows that the rise or spread of B took place late in human evolution, after the Australians were isolated in the island continent and the Amerinds in America. The post-glacial Indians entering Alaska also must have been early enough to have been uncontaminated by the B blood group of the later Asiatics.

If only three or four main immigrations into America occurred, owing to the exigencies of changing glacial climates, then the fascinating possibility emerges that some comparison may be made between the rate of evolution or differentiation of man and the rate of development of his languages. Harrington (1940) assumes a single wave of migration from Asia some 20,000 years ago, having one language which later differentiated into all the Indian tongues. These number some fifty in North America alone, with many more in Mexico, Central, and South America. He points out that in six hundred years Chaucer has become unintelligible to modern ears, although the process of language evolution has been retarded by education and writing, and in the last five centuries by printing. Vedic Sanscrit, perhaps the oldest of the Indo-European languages, dates only from about 1500 B.C. or 3400 years ago, a matter of over 100 human generations.

Harrington believes that all the American Indian languages had a common origin, since the first person singular pronoun, *na,* is found in all of them throughout North and South America. They are generally supposed by Sapir and others to belong to a number of independent stocks between which no similarities can be detected. Harrington assumes that the first immigration wave moved southwards along the American coast, but there are serious difficulties with this view. He believes that further immigration from Asia to Alaska was blocked by the Chukchean peoples of Kamchatka, whom he regards as having been unmoved in 20,000 years. This appears to be a considerable overestimate of time. Harrington applies the term Athapascawan to a bloc of related languages—Eyak, Tlingit, Haida, and Athapascan, in northwestern America and the Great Plains. The Athapascans include the Carrier, Sarcee, and Chipewyan Indians. From this stock it is well recognized that two prongs or migration routes extended southwards: one, west of the Rockies into Washington, Oregon, and northern California; the other through the Great Plains. This included the Navaho and Apache in the southwest. These latter migrations were relatively very recent, and it is suggested that the Navaho probably arrived in their present area as late as A.D. 1300.

Bonin and Morant (1938) made a survey of the detailed cranial measurements of 1167 undeformed male skulls of United States Indians, from the studies of Hrdlička and Gifford. They were divided into sixteen series, and as many as thirty-one characters were available for comparison in some skulls. In accord with von Eickstedt, they found marked divergences between various tribes in the United States. An unexpected close relationship was found between the skulls of Florida and central California (Eickstedt's marginal group). His sylvan group was broken into at least two: (1) the prairie Indians (Sioux and Arikara), (2) the eastern Algonkian of the northeastern forests. Eickstedt's central group, which resemble Peruvian skulls, were supposed to be brachycephalic, but may be represented by the dolichocephalic Old Zuni and the mesaticephalic Basketmakers. Four Algonkian series were all different, indicating that a linguistic grouping has little ethnic significance. The Pecos Pueblo skulls were found to be heterogeneous, in agreement with Hooton. Applying the coefficient of racial likeness, which is of doubtful significance, they found a close relationship between western Eskimo and some United States types. Also the Chukchi were closely allied to some North American but not to Asiatic types.

In a more recent study, Hoijer (1946) finds (p. 9) that "there is no evidence whatever that these languages [of North and South American Indians] are historically interrelated. Indeed, quite the contrary appears to be true: we find in the Americas a region of greater linguistic diversity than any other in the world." Such remarkable diversity is difficult to account for even if some of the original stocks were from different sources. The diversity is found to be in grammatical structure and phonology as well as in historical origin, nor is any American linguistic stock found to be related to idioms outside America. It seems certain that language evolution is a much more rapid process than the physical differentiation of races, as might be expected.

That more than one movement into America took place is indicated, for instance, by the occurrence of low-headedness (chamaecephaly). Stewart (1942) finds that this condition is limited to the western portion of North America, many low-headed groups having moved southwards in relatively recent times. In the Aleutians the low-headed type has been superposed on the high. The evidence is therefore that the low-heads belonged to a later influx into America than high heads. Stewart finds evidence of a reservoir of low-heads in northeastern Siberia in Neolithic and Recent times, but here again the blood groups would be a limiting factor as regards time of emigration. Brachycephali were also later than dolichocephali.

In a later study, Stewart (1943) finds that a mean height index (which relates head height to the mean of length and breadth) below 83 is unknown in Indians east of the Mississippi, while in the west it is restricted to Siouan, Caddoan, and Athapascan tribes. These low-heads are believed to have arrived late on the migration route from Asia. The mean head height index ranges in North America from 88.2 to 76.9 and in South America, from 89.1 to 76.5. But the low-heads in South America are mainly confined to regions near the north coast, indicating again a late immigration. In the modern Mayas this index is 76.9, so they would serve to complete a line of chamaecephali from western North America to the South. However, there is a striking exception in the case of the Tierra del Fuegians, who are near the border of low-headedness (81.9–83.1) yet are generally regarded as early arrivals, pushed to the inhospitable tip of the continent by their successors. Some of them (the Yaghan) were formerly represented as very high in B, the Ona having no B. It has recently been shown that the B was derived from crossing. The Fuegians may be a remnant of the first Amerind wave.

In Figure 8 an Eskimo and an Aleut skull are compared. They have the same cephalic index and approximately the same nasal index, but the Eskimo is hypsicephalic, the Aleut chamaecephalic. The absolute height from basion to bregma is 121 mm. in the Aleut and

145 mm. in the Eskimo, the mean height index being 74.6 for the former and 88.4 for the latter.

A digression is necessary here, to deal with another method which has been used in comparing skulls. Dixon (1923) assumed that the early migrants to America from Asia would be pushed to the periphery of the continents and into refuge areas by later arrivals—a view which various later writers have accepted. The earlier dolichocephalic Indians are in fact found mainly on the margins of North America, while the brachycephali form a solid mass occupying the

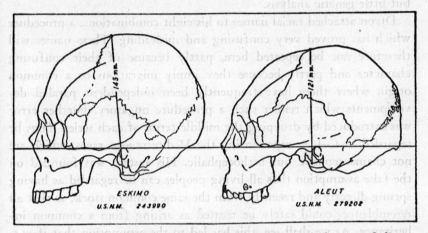

FIGURE 8. Comparison of hypsicephalic Eskimo and chamaecephalic Aleut skull.

center of the continent. Dixon concluded that by the sixteenth century the brachycephalization of North America as well as the other continents was "almost complete," which is a considerable overstatement. Dixon took three sets of contrasting characters, (1) dolicho-, meso-, and brachycephaly, (2) hypsi-, ortho-, and chamaecephaly (height in relation to length of skull), (3) lepto-, meso-, and platyrrhiny. These nine characters could occur in 27 possible combinations. He then assumed that the middle term in each was a blend of the extremes. There would thus be 8 fundamental and 19 blended types. Using the initial letters (for example, D for dolichocephalic), he derived the following eight types:

D–H–L	D–C–L	D–H–P	D–C–P
B–H–L	B–C–L	B–H–P	B–C–P

Thus he had a formula for combining head length and head height with nasal width. We have already seen in Chapters 5 and 7 how head length and head height can be usefully combined in a racial analysis of head shapes. It is not so clear that nose width, although an independent variable, can be used with equal advantage in the same way. Many consider that nasal width is affected by climate, broad noses developing in the tropics and narrow noses under Arctic conditions. Apart from this, nose shapes show great variability in Amerinds, Europeans, and other peoples—a variability which has had but little genetic analysis.

Dixon attached racial names to his eight combinations, a procedure which has proved very confusing and misleading. These names will therefore not be repeated here, partly because of their confusing character and partly because they imply migrations or a common origin where there have frequently been independent parallel developments which render such a procedure nugatory. Further error was introduced by dropping the middle terms of each series. Thus he regarded D–C–L as representing the Mediterranean races, which are not chamaecephalic but orthocephalic. His system was founded on the false assumption that all living peoples can be regarded as having sprung directly and recently from the same common stock, so that all resemblances could safely be treated as arising from a common inheritance. As we shall see, this has led to the assumption that if certain skulls in South America have Australoid characters they must necessarily belong to the Australian race, whereas it is highly probable that they represent a parallel stage in skull development of a widely different race. Although Dixon's method of racial analysis is not genetically sound or permissible, it was nevertheless a remarkable tour de force in its time.

In a further study of head shape in North America, Stewart (1940) finds a wedge of brachycephalic leptorrhine Indians with its base from British Columbia to the Gulf of St. Lawrence and running south between the Great Lakes and the Rockies through the western plains almost to Mexico. These are believed to have driven the brachycephalic, platyrrhine Shoshonean and Athabascan tribes west of the Rockies and into the southeast. Low-heads (below 83) occur on the Pacific coast, from lower California to northern California and in Piegan, Apache, Sioux, Aleuts, and certain others. Stewart concludes that the older Indians were dolichocephalic, hypsicephalic, with noses at least somewhat broader than in more recent types. Cranial deformation developed later in three or four different centers. There seems no reason to doubt that head width and head height are equally stable and similarly under genetic control.

If man came to America during the last interglacial, then we may reckon his advent as at least 10,000 and perhaps 25,000 years ago or more. So long as he continued in the north he must have remained a nomadic hunter, roaming over great areas. It was only further south that the domestication of plants and the beginnings of agriculture could take place. This is shown by the geographical distribution of the plants he cultivated. Whether maize originated in South or Central America is still a matter of dispute, but its many varieties, specialized culture methods in Indian hands, and the spread of maize culture even into parts of Canada shows that a long period must have elapsed since the domestication of the plant began. The numerous plants developed for culinary purposes, especially in the tropics and sub-tropics, again indicates a period of time comparable with the passage from Paleolithic to Neolithic in the Old World. The scattered publications in ethnobotany of different Indians tribes (for example, Densmore, 1928, Rousseau and Raymond, 1945) show how they had gradually accumulated extensive knowledge of the uses, for food, medicine, textiles, and so on, to which the plants in many local floras could be put. Much of this knowledge was of direct practical value and has been lost, but some of their uses of plants as medicine were no doubt erroneous. Their knowledge, such as it was, could only have been acquired by a long and intimate acquaintance with nature.

If we except the turkey (and the dog, a variety of which was used for wool by the coastal tribes of British Columbia), man found no animals which could be domesticated until he reached South America, and even here only certain surviving members of the Camelidae were available. The agricultural achievements of the Indians must therefore be one of the best comparative measures of his age in America. Pottery also may have been invented independently in America, since the nomadic hunters of the north would have lost the art of making it, even if their Siberian ancestors possessed it. If man came down through the interglacial corridor he would be too early to have pottery anyway, and the possibility of its introduction later by culture contact over such a long and unfavorable route is very doubtful. The type of pottery of the Mound-builders, who were formerly referred to as pre-Indian, has since been linked to the Indians of the historical period, reaching into the eighteenth century. Spinden's (1942) estimate that Folsom flints are not earlier than 4000 B.C. would therefore appear to be a serious underestimate. He concludes that Mayan astronomy was most flourishing in the period A.D. 400–500, the Dresden Codex covering over a thousand years, to A.D. 950.

There is now convincing evidence that man co-existed with extinct Pleistocene mammals both in North and South America. The difficulty is to determine whether these animals were contemporaneous with the ice, or whether they had survived into post-Pleistocene times. We may now examine some of the numerous records bearing on this problem. One of the earliest concerns Koch's (1841) "Missourium," which was found twenty-two miles south of Saint Louis in 1839. Albert Koch was a German collector of fossils, who made a profitable business by excavating large fossil mammals, exhibiting them in various places in America and Europe, and finally selling them to museums. He had no training in natural history and many of his claims will not bear examination. Yet it is possible that he incidentally made the first observations in America of human artifacts associated with extinct Pleistocene mammals. He also apparently found what are now known as Folsom points, belonging to what Roberts (1940) calls the Paleo-Indian. It would be unprofitable here to attempt to determine whether Koch's evidence of the association of man with the remains of a Mastodon and a Mylodon (ground sloth) is trustworthy where so many of his other statements are obviously without foundation. Some of his statements (Koch, 1857) may here be recorded. The skeleton of "Missourium" from Gasconade County, Missouri, in 1839, was stated to have been mired in gray clay at a depth of eight to nine feet with the legs upright and the head partly burnt. This is the famous specimen now in the British Museum (Natural History). Mingled with the bones and ashes were pieces of rock from the riverbank below, believed to have been thrown to kill the animal. Several arrowheads and stone axes were found. Another entire skeleton was excavated in Benton County the following year.

Montagu and Peterson (1944) try to make out a case for Koch, but do not state some of the strongest points against him. His "Missourium" was excavated before 1839 and afterwards remounted by Owen in the British Museum under the name *Mastodon giganteus*. It still shows clear signs of fire. Koch's statement that the skeleton was under twenty feet of sand, clay, and gravel was a pure fiction, as shown by Hay (1871), who visited the site in 1840. Andrews (1875) also visited the spot and talked with the men who assisted Koch. The skeleton was said to be close to the surface and covered with muck. As the place was not drained, the men dug out the bones, often working up to their waists in water. These statements were published after Koch's death, apparently in order to prevent his false claims being accepted as true.

One can only conclude that Koch's own statements about his finds are thoroughly unreliable and sometimes fantastic. Simpson (1943) in his history of the beginnings of vertebrate paleontology in America, condemns him completely so far as reliability is concerned, and Eiseley (1946) does likewise, pointing out the absurdity of his story of "large pieces of skin" and "sinews and arteries . . . on the earth and rocks," small pieces of which he says he "preserved in spirit." These seem to be the statements of a showman anxious to stimulate the public imagination but with little or no regard for scientific truth.

There were, in fact, as Simpson shows, collections of mastodon bones even a century earlier. The Baron de Longueuil found them on the Ohio River in 1739. These were taken from New Orleans to France and later studied by Cuvier. Another important collection of "elephant" tusks and molars, by Croghan in the Ohio region in 1766, was sent to London, partly to the British Museum and partly to Benjamin Franklin. Even in Missouri, mastodons had been found in 1806 or earlier, and skeletons had been assembled and exhibited long before Koch. Mastodon bones collected by the Indians in Tlascala, Mexico, were shown to the army of Cortes in 1519. It is evident then that the bones of mastodons and other large mammals were rather easily accessible in many localities before the disturbances of man cleared them away.

An early record with certain features of unique interest, which has received little attention, comes from Louisiana. On the island of Petite Anse there was a salt mine and during the Civil War several pits were sunk to a depth of ten to fifteen feet, through sand and mud, to obtain the pure rock salt. Leidy and Clew (1866) describe how "elephant" bones were found in several of these pits, and "beneath them, within a few inches of the rock salt, abundance of matting," which was plaited diagonally, of tough split cane, well preserved. There was some evidence of stone implements too. At the sides of one pit were "elephant" bones, and beneath them pieces of matting could still be seen. Joseph Leidy was a well-known anatomist and paleontologist, so we may rely upon the accuracy of these observations. Two questions arise in connection with the matting. Was it perhaps preserved by the salt? Its survival for even a few centuries would otherwise seem impossible. If pieces of matting were directly beneath mastodon bones, is it possible that the natives had made a pitfall covered with matting, into which the animals fell or were driven?

In a survey of remains attributed to early man in North America, Hrdlička (1907) cites a number of skulls and skeletons, with interesting comments on their morphology. He is possibly too rigorous in saying that none of them have any great age, but it is significant that many of them are described as "Neanderthaloid" or having other peculiar features. The New Orleans skeleton, excavated at a depth of sixteen feet in 1844, was found in four layers of soil and river mud, with cypress stumps and burnt wood. The skeleton crumbled, except the skull. At Soda Creek, in what is now central Colorado (Berthoud, 1886), two miners opening a claim at the foot of Soda Hill reached a depth of twenty-two feet. At this depth and three yards from the foot of the hill slope they found a skeleton lying on its face embedded in auriferous gravel, sand, and rock fragments. The skull and larger bones were almost entire. Two feet lower they came to the trunk, limbs, and roots of a pine tree similar to the red pine. They concluded that before these enormous beds of glacial gravel, sand, and boulders had covered this area man roved here and timber grew. Apparently no competent geologist checked the possibilities of landslides or other explanations of the burial.

During excavations with dynamite the "man of Peñon" was found (Barcena, 1885) near Mexico City. The skeleton was embedded in very hard rock (silicified calcareous tufa). The cranium was recovered, with the mandible and fragments of vertebrae, ribs, and long bones. There were no animal remains or artifacts, but animal bones were found in similar rocks three miles away. Vulcanism had begun before the skeleton was deposited, and continued afterwards. The canines are said to be the same shape as the incisors, as in very ancient Toltec graves. Perhaps this means that the incisors were shovel-shaped. In the southeastern part of the Valley of Mexico, ceramics were found in pumice tufa under a basaltic lava formation which was two meters thick in some places. Observations in connection with the volcano Paricutin of recent origin in Mexico indicate that the best conditions for fossil preservation are a thick deposit of ash followed later by a lava flow sealing it in.

Webster (1889) described the excavation of three Indian mounds near Floyd, Iowa. They were about 220 yards from the Cedar River. The largest circular mound was thirty feet in diameter and two feet high. It contained five well-preserved skeletons, all placed in a sitting posture, facing north, in a bowl-shaped excavation nearly four feet below the surrounding surface, lined with gravel and fragments of limestone. They were covered with a thin layer of earth, then nine inches of earth and ashes containing charcoal. Above this was four feet of earth which was red and very hard from the continued action of fire. The skeletons were four adults, two of each sex, and a baby with a thick skull. Three crania were of Neanderthaloid type. The figure of one indicates that the superciliary ridges were almost as heavy as in Neanderthal man, the forehead extremely low (chamaecephalic), the vault lower than in the latter. At Charles City, six miles below, thirty-one mounds were excavated, but the skeletons were mostly decomposed. As already indicated, the Mound-builders were of no great age, such mounds being found especially in Illinois, Iowa, Wisconsin, and Dakota. Hrdlička (1907) states that there were thirty-six Indian skulls in the National Museum having pronounced superciliary ridges and a low forehead. All were also dolichocephalic or mesocephalic, but many other Indian skulls are brachycephalic. This indicates that the Neanderthaloid type was a definite one, differing in several respects from other Indians but no doubt intercrossing with them.

At Rock Bluff, Illinois, a skeleton was found like those of the Illinois River mounds. The forehead was low and the brow ridges were greatly developed, not in arcs as in Neanderthal man but in the median three-fifths only. Here is further evidence, if it were needed, that the American Neanderthaloids (perhaps better pseudo-Australoids) were a parallel to the European Neanderthal, and unrelated.

The Nebraska loess man was first found in a mound in 1894 near Omaha (Barbour and Ward, 1906). Further excavations in 1906 unearthed seven more skulls with stone tools scattered through the mound and evidences of ancient fire. Five skulls from the lower level (below a layer of burned clay) were more primitive and probably belonged to an earlier period. One of these skulls was low and Neanderthaloid with a marked frontal torus, but it was brachycephalic and narrow at the temples. The length was 182 mm., C.I. 87.9, the femora were also massive and with certain peculiarities. In the Gilder mound (Barbour, 1907) about a dozen skeletons were found, old and young of both sexes. The upper layer contained mound-builder remains, and beneath a layer of fire-hardened clay were eight skulls and many bones in the loess deposit. The upper skeletons had been buried but the lower ones were scattered and disarticulated, the jaw of a youth being found four feet down in the loess. This loess rests on the Kansan glacial drift and is as young as the later Wisconsin sheet or younger, but early post-glacial. Another skeleton was found at Lansing, Kansas, in digging a tunnel. It was at a depth of twenty feet, but whether in loess or more recent is disputed. The skull had moderate brow ridges and was dolichocephalic, C.I. 73.75. Two other skulls with very low vault from near Trenton, New Jersey, were not of Indian type. Hrdlička believed that they represent a coastal type in Holland and in Germany to the Elbe—early Colonial burials. Thirty skulls of the same type, of the ninth or tenth century, were found under Bremen cathedral.

One of the earliest records of human remains apparently associated with the bones of extinct mammals in America is found in a note by Dickeson (1846). At Natchez, Mississippi, a fossil innominate bone of man was found in blue clay two feet below skeletons of the ground sloth, Megalonyx, and other extinct quadrupeds. The bones were *in situ*, not a re-deposit, and the human bone resembled in color, density, and fossilization the Megalonyx and associated bones.

To show that every form of caution is necessary in drawing conclusions from these early records of fossil excavations, we may cite the exhibit by Holmes (1859) of a collection found near Charleston, South Carolina, in digging a ditch to reclaim a swamp. A mastodon tooth and tusk were found, and alongside the latter a fragment of Indian pottery. Bones of deer, raccoon, opossum, and other living species, the musk-rat and beaver (then extinct in South Carolina) as well as peccaries, capybara, and tapir now confined to South America, and the extinct Mastodon, Megatherium, and Mylodon, made a large collection at a depth of only three feet. In marine beds underneath, 150 species of Mollusca occurred, all recent in the area except one species now found on the coast of Florida and Cuba and another species now in the Gulf of Mexico. In addition there were exposed on the river bank bones of the horse, ox, sheep, hog, and dog, for which Louis Agassiz ventured the explanation that these domestic animals had been "called into existence" in America before the white man arrived!

In Florida, early indications of the association of man with extinct Pleistocene mammals have been followed by fuller confirmatory investigations. The whole peninsula is of Tertiary age, being gradually reclaimed from the sea in a southwards direction, Oligocene, Miocene, Pliocene, and post-Pliocene occurring in succession from north to south. Heilprin (1887) described the fossilized remains of a human skeleton at Sarasota Bay which had been entirely converted into limonite. The skeleton was near sea level and had been exposed by the wash of the sea. It was embedded in ferruginous sandstone and most of the parts had been removed by curiosity-seekers, but two vertebrae which had been extracted from their matrix were pronounced by Leidy to be probably the last thoracic and the first lumbar. The bone had been completely replaced by limonite. Heilprin concluded (p. 67), "Man's great antiquity on the peninsula is established beyond a doubt."

At Vero, on the west coast of central Florida, in digging a drainage canal in 1913, numerous animal bones were thrown out (Sellards, 1916a). They included the elephant, mastodon, camel, horse, bison, tapir, and sloth, all extinct. In 1915, human bones were exposed where the canal bank had caved in, and more were found in the following year. Two skeletons were exposed at different levels, but without the skull in either case. Further excavations showed the presence of three layers: 1. Underlying Pleistocene marine shell-marl, which is extensive in Florida and is considered of Nebraskan age— the earliest Pleistocene. 2. Above this, three to five feet of cross-bedded sands, in which human remains were found. 3. On the surface an alluvial deposit of loose sand and muck two to six feet thick. There was no break between layers 1 and 2, but a change from marine to fresh-water shells. In the sandy stratum 2, the human bones were about ten inches above the base. The overlying alluvial beds were stratified and conformed to the irregularities of the underlying formation. There was thus an erosional break between layers 2 and 3. In the latter, the second skeleton was found with bone and flint implements, pottery, arrowheads, and ornaments. These were not intentional burials, the strata being undisturbed. Sellards regarded this as demonstrating that man lived in Florida contemporaneously with extinct Pleistocene mammals, but one skeleton was of considerably greater geological age than the other. In a more complete account (Sellards, 1916b), the topography and geology were described, as well as the human and animal remains. *Elephas columbi* and *Equus leidyi* were found at an equal or higher level on either side of human remains. The first incomplete skeleton was probably a female. It was in layer 2, two feet from the original surface with two feet of overlying hard marl; and the bones were not eroded or worn, indicating that they had never been displaced far from their original resting place, though they were somewhat scattered. In immediate association were deer bones (common in this bed) as well as *Megalonyx jeffersoni* and mastodon teeth. The human bones were mineralized as much as those of the Pleistocene animals. The skeletal remains in bed 3 were also of great antiquity because the bones were mineralized like those of the accompanying extinct mammals, and there were plant remains at a higher level above the human bones. Sellards (1937) has since given a connected account of these finds at Vero. The skull bones were scattered, but could be fitted securely together by their sutures.

Hay (1917) strongly supported Sellards' conclusions on geological grounds. The human remains in bed 2 were regarded as early Pleistocene, as shown by the character of the species and the fact that 74 per cent of the animals were extinct forms. In bed 3, about 44 per cent of the animals belonged to extinct species and it was considered mid-Pleistocene (Illinoisan stage). He cites various cases, some already mentioned, of arrows and other artifacts buried deep in loess in the Middle West, from which he concludes that man was present in the succeeding (Sangamon) interglacial period. He also argues that since horses, cattle, bears, dogs, and so on, reached their modern physical development in early Pleistocene there is no reason why man should not have done the same.

One of the citations of Hay refers to observations by Witter (1891) at Muscatine, Iowa, which is on the bed of a glacial lake. In the yellow-brown loess overlying the drift was found a spear point at a depth of twelve feet, which left an impress when removed from the bank. An arrowhead at a depth of twenty-five feet in loess, another in a bed of blue clay eight feet from the surface of the loess, fragments of an elephant tooth, and many flint chips in a gravel bed overlaid by ten feet of loess are also recorded.

An expedition to Florida sponsored by Amherst College and the Smithsonian Institution, made further excavations in 1922 at Melbourne, thirty miles north of Vero Beach. Exactly the same three strata were found, and Gidley (1916) confirmed the conclusions reached by Sellards, that the human remains belong in the level where they were found and are not intrusions through burial or otherwise. Gidley and Loomis (1926) described in detail the findings at Melbourne. As no part of the coast is more than twenty feet above sea level, the country was originally swampy and the streams sluggish. Above the underlying coquina shell-layer were six to nine feet of deposits in two layers. Human artifacts and remains were, as before, mostly near the bottom of layer 3 or at the contact plane between 2 and 3. From an examination of the fossil plants, mostly in the top layer (3), Berry concluded that they were late Pleistocene, immediately following the retreat of the Wisconsin glaciation. Human remains were found in three localities near Melbourne. On the golf course, the top bed consisted of less than two feet of stratified deposits containing much vegetable matter. The middle bed, five feet thick, contained fossilized bones of Pleistocene mammals, and a crushed human skull was found, with finger, arm, and leg bones, apparently near the top of this bed. Over the skull lay an undisturbed layer; within a foot of it was a fossil horse tooth, and twenty-five feet away was the jaw of a tapir. On another site a mile away, artifacts, an arrowhead, flint chips, and broken pottery were near the top of the middle bed. At the top of this layer in the third locality a fine arrowhead, a human rib, and charcoal occurred with teeth of

Mylodon, Megalonyx, and so on. The Mylodon tooth was lying almost in contact with the arrow. Several square feet here were thick with bones. The top layer, three or four feet thick, had been removed for agricultural purposes, leaving only a few inches of soil.

It was concluded that man reached Florida before the mammoth and mastodon became extinct, since their bones were in the same layer as the human bones and artifacts, but Hrdlička concluded that they are physically indistinguishable from the modern Indian. A recent reconstruction of the skull fragments by Stewart (1946) shows that it was not brachycephalic, as Hrdlička thought, but dolichocephalic like the Vero skull, hence conforming to the Paleo-Indian type. The age was regarded as early post-Pleistocene, antedating the Mound-builders. Further digging in 1926 (Gidley, 1927) followed a different plan. Hrdlička and others had objected that it was uncertain whether the crushed skull belonged in the upper six inches of the Pleistocene fossil-bearing deposit or possibly in the dividing zone between this bed and the overlying layer. Gidley therefore continued this excavation, but instead of taking off horizontal layers from the top downwards a vertical face was made, so that a cross section of the three layers could be seen. The bone-bearing layer was thus seen to be plainly distinguished from the overlying bed which rested on it unconformably, showing an interval of erosion between the two periods of deposition. From further observations it was clear that the skull was originally in the middle or bone-layer, though near its top.

The inevitable conclusion from all the evidence appears to be that man in Florida was contemporary with Pleistocene mammals, especially the mammoth, mastodon, and ground-sloth, although these may have survived in that region into post-Pleistocene time. However, if they persisted here it seems probable that they would have survived even longer in more northerly latitudes, where the climatic conditions would have departed much less from the cold temperatures to which they must have been adapted. It is quite possible that these early Indians played a considerable part in the extermination of the large Pleistocene mammals, just as they did afterwards in the slaughter of the bison.

We turn now to the other side of the continent. The gold mining days of California led to some remarkable finds of human and mastodon bones, but the exact location of few if any of them is sufficiently authenticated. Captain David B. Akey relates with many details (Winslow, 1873) how a complete human skeleton was found in Table Mountain, in 1855 or 1856. When running a drift or tunnel into the mountain about two hundred feet from the brow, the skeleton was removed from gravel in another tunnel fifty feet away. The bones were in excellent preservation and were found at a point which was stated to be one hundred and eighty feet from the mouth of the tunnel and at a depth which Akey thinks was two hundred feet. The spot was pointed out to him, and nearby was a "petrified pine tree" sixty to eighty feet long and two to three feet in diameter, *in situ*. He states that many animal bones were taken from other tunnels in the same mountain, including mastodon's teeth and smaller animals, and petrified wood was common. Some of the overlying strata were lava. As the skeleton was not preserved nothing further can be learned about it.

The famous Calaveras skull still exists in the Peabody Museum of Harvard University. In his well-known report on the gold-bearing gravels of the Sierra Nevada, Whitney (1880) deals with the skull and its history at length. He does not hesitate to say that it affords "clear and unequivocal proof, beyond any possibility of doubt or cavil, of the contemporaneous existence of man with the mastodon, fossil elephant, and other extinct species, at a very remote epoch as compared with anything recorded in history." He found a large body of evidence that man existed in California before the cessation of volcanic activity in the Sierra Nevadas, at the time of maximum extension of the glaciers in that region and before the erosion of the present river canyons and valleys, when the topography as well as the flora and fauna were different from now. He assumed that man extended back unchanged to the Pliocene, but of course little was known of Pleistocene geology at that time.

The Calaveras skull belonged to a very old man. All the teeth in the maxilla except one molar had been lost and the alveoli more or less absorbed. The mandible was missing and the skull was so encrusted with earth and stony material that it was not at first recognized as a skull. Its base was embedded in a "mass of ferruginous earth, water worn pebbles of much altered volcanic rock, calcareous tufa, and fragments of bones." It consists mainly of the frontal bone, showing heavy superciliary ridges, the face and maxilla; most of the parietal and occipital are missing. In the material covering the face were found two metatarsals, the lower end of a fibula, and fragments of the ulna and sternum. There was also a portion of a tibia which was too small to belong to the same person.

The skull was found by Mr. Mattison in the shaft of a mine which he had sunk on Bald Hill, near Altaville. This hill and others in the vicinity consists of layers of more or less consolidated volcanic ashes alternating with gravel beds. Whitney, in his geological analysis in a shaft 153 feet deep, measured ten alternating layers of lava and gravel, the skull being found in the eighth layer from the top, just above a stratum of red lava. It was taken out by Mr. Mattison with his own hands, in February 1866, "one hundred and thirty feet from the surface, and beneath the lava, in the cement, and in close proximity to a completely petrified oak." He placed it in a bag with some pieces of fossil wood and gave it to Mr. Scribner, the Agent of Wells, Fargo Express Company. The latter cleaned off some of the encrusting material and, finding it to be a human skull, gave it to Dr. W. Jones, a well-known local medical practitioner, who was making a fossil collection. The latter wrote on June 18, 1866, to the Geological Survey that he had the skull in his possession, and gave the skull soon afterwards to Mr. Whitney when the latter visited the locality to determine the geological conditions under which it was found. It is difficult to see how any such find could be much better authenticated, and Whitney is very emphatic on this point.

After the discovery was announced, religious papers took up the matter to discredit it and Bret Harte wrote his famous poem on "The Pliocene Skull." Apart from the fact that the skull is Pleistocene, and not Pliocene, this would appear to be one of the outstanding records of early man in America. From his geological reconnaissance, Whitney concluded that the lava flow of Table Mountain filled the valley after running forty miles down the Sierra slopes and forming a ridge 2000 feet high. He estimated (probably incorrectly) that the subsequent denudation had amounted to 3000–4000 feet. Whatever geological work may have been done more recently in this area, it seems clear that the skull was found beneath 130 feet of lava and gravel in seven alternating layers, unless there is some mistake regarding the source. Early man, therefore, witnessed volcanic eruptions here and was engulfed along with samples of the local vegetation and animals. Hrdlička (1907), in his remarks on this skull, says that the supra-orbitals are strong but not more so than in some recent Indians, but they "extend along the whole superior border" of the orbit and the glabella is little less prominent than the ridges. He refers to the two other crania from caves in Calaveras County, one of which is similar to the Calaveras skull. He concluded that both came from the cave, because they have the same calcareous coating. The third cranium had "very heavy supra-orbital ridges"; but is considered not entirely normal, for reasons which are not stated.

Whitney shows that this skull, although it got all the publicity, was by no means the only record of early man and Ice Age animals in California. Among other records which he cites are the following, mainly statements from miners. At Horse-Shoe Bend on the Merced River (Mariposa County) in 1869, bones of the mastodon were found twelve feet below the surface. In the immediate vicinity were numerous human bones and stone implements. One which was preserved was an obsidian spear or lance head five inches long. At Hornitos and No. 1 Gulch and in several other localities were reported in 1864 stone implements, chiefly "mortars" (metates?), with elephant and horse bones at a depth of fifteen feet. At Dry Creek (Stanislaus County) in 1870 a tusk and molar teeth of an elephant were found thirty-seven feet below the surface. Tuolumne County was prolific in human remains and "works of art." Whitney concludes (p. 263), "It is hardly possible to escape the inference that the human race existed before the disappearance of these animals from the region which was once so thickly inhabited by them." At Gold Springs a wagon-load of mastodon and other animal bones was destroyed by a fire.

Under Table Mountain, at a depth of sixty to seventy-five feet and three hundred feet from the mouth of the tunnel, was found a stone hatchet with a hole for hafting, together with stone mortars, animal bones, and fossil wood. In 1857 small fragments of a human skull were found in Valentine shaft (which was vertical and boarded to the top), in a nine-foot seam of auriferous gravel one hundred and eighty feet below the surface. From the same shaft at a depth of one hundred and twenty-five feet a mastodon tooth and a white marble "bead" were taken. At other places in Calaveras County were stone mortars under volcanic ashes, gravel, and sand at depths of one hundred to one hundred and fifty feet. El Dorado County was prolific in human relics and implements. In 1853, near Placerville, Dr. H. H. Boyce, as a miner, found a scapula, clavicle, and parts of three human ribs in clay with over thirty feet of sand and gravel capped by eight feet of basaltic lava. This by no means completes the list of finds throughout the Sierras, but there were none in the Coast Range. Apart from the supposed Pliocene age of these deposits, and the view that the whole of the Tertiary was represented in the Sierra detrital beds, Whitney appears to have been correct in the conclusions he drew regarding the association of man with extinct mammals in California.

In later times, Merriam (1909) investigated various limestone caverns in California. Human remains of considerable antiquity were found in Mercer's cave (Calaveras County) and in Stone Man cave (Shasta County). In the former, human skeletal remains were in close proximity to a Quaternary ground-sloth. Both were encrusted in stalagmite, but it formed a thicker layer on the sloth. In a remote gallery of Stone Man Cave, human bones were found embedded in stalagmite one-eighth inch or more in thickness. The cavities of a vertebra were filled with calcite crystals.

The Hawver cave near Auburn, California (Eldorado County), was entered from the top by means of a rope (Furlong, 1907). The bones of many animals which had fallen in included Megalonyx, *Equus occidentalis*, species of Felis, and rodents. In 1908, at the entrance to the lower cave, human bones were found under twelve feet of fallen rocks and soil at the lower end of a passageway which had long been closed. The skull did not differ appreciably from that of modern California Indians. It might have been centuries or millennia old. Bones of extinct animals were found near. This cave differs from the Shasta caves in the absence of the goats, Euceratherium and Preptoceras, and of deer. Split bones were also numerous in Shasta caves, few here. Merriam (1934), in a summary of work with a full list of references on early man in America, agrees that there are a "considerable number of apparently authentic occurrences of human remains in the older gold-bearing gravels of California." The Calaveras skull was believed to be from a cave deposit near Angel's Camp. When mining operations changed from tunneling to hydraulic sluicing, numerous artifacts in the auriferous gravels were washed in by the mining operations. He refers to more recent discoveries in western deposits and caves, which we may consider next, as they constitute a climax in the reconstruction of early Indian prehistory.

The new developments began with the announcement by Cook (1925) of the finding at Lone Wolf Creek, near the Colorado River in southwestern Texas, of the complete articulated skeleton of a large species of bison associated with beautifully made arrow or lance points, and a Pleistocene fauna in undisturbed deposits. Three points of fine workmanship and distinct design, since known as Folsom points, were closely associated with the bison skeleton, one under the cervical vertebrae, another under the femur, and the third (afterwards lost) also by the skeleton. Another bison, with a Folsom point embedded between two ribs, has since been mounted in the Colorado Natural History Museum at Denver. Hundreds of the cruder arrowheads of later Indians were also found here. With the bison were fossilized remains of Elephas, Equus, and camels, in solidly cemented gravels overlaid by five to seven feet of undisturbed Pleistocene sands and gravels, above which were several feet of sand, silt, and soil. These extinct animals were thus shown to be contemporaneous with man.

At Folsom, New Mexico, in 1926 a deposit was studied near the head of a gully, containing bison bones under six to eight feet of tough, hard clays on both banks of the gully. The associated points were like those from Colorado, Texas, but more pointed and of finer workmanship. One point was associated with an extinct bison. A second was broken, but the tip was found inserted between a bison's ribs. The bison from Colorado, Texas, was later found (Figgins, 1927) to have a complete arrow or spearhead inserted between the fifth and sixth cervical vertebrae. These points were of thin, grayish flint, with no notch, and very different from those on the surface. They show skillful secondary chipping around the margin. The bison was named *B. taylori* and considered of Pleistocene age.

On the face of a gravel pit one hundred and fifty yards long and over twenty feet high, at Frederick, Oklahoma (Cook, 1927, 1928, Hay, 1928, 1929), were found more Folsom points in the lower levels, together with quantities of fossils. The base of this sand pit lies directly on Permian red beds. In 1928, Mr. Holloman, owner of the pit, found another arrow in the gravel thirteen feet from the surface (Gould, 1929). It was photographed *in situ* before removal from the matrix. There was evidence that the whole region had been eroded away one hundred feet or more, and Cook suggested that the age of this river gravel was 365,000 years, which seems excessive. Of these three discoveries within two years, all of which are believed to be of Pleistocene age, Cook (1927) considers the Oklahoma deposit to be the oldest, next the Texas beds, those from Folsom, New Mexico, being youngest. The basal stratum at Frederick contained Mylodon, Equus, Elephas, and Trilophodon. Hay suggested that interglacial man came from Asia, having already the art of chipping flint, which was perfected in the Folsom blades.

At the Lindenmeier site near Fort Collins in northern Colorado, eighty-three Folsom points were collected on the surface (Roberts, 1935). Variations of the Folsom point are now known from the Rockies to the Atlantic and from southern Canada and even Alaska to the Gulf of Mexico, mostly found on the surface. They were collected in various parts of Virginia in 1934, but are more abundant in the West than the East. Their significance as belonging to the Paleo-Indian was only recognized in 1927, but they were collected as early as 1897. In a cave in the Guadeloupe Mountains (southeast New Mexico) Folsom points have been found associated with the musk ox (indicating a cold climate) in a stratum beneath Basketmaker materials (the oldest Pueblo culture). Near Clovis, New Mexico, is another site where Folsom and other chipped implements have been found with extinct animals. At Dent, Colorado, a Folsomoid point was found with mammoth bones, and several sites in Nebraska and Kansas have yielded similar results. The true Folsom point of best workmanship is found in the High Plains east of the Rocky Mountains, but those of less developed type are widely distributed in the eastern states, notably the Finger Lakes of New York and in Ohio, Tennessee, and Virginia. The Yuma type, from Yuma, Colorado, is a related form named and described by Renaud, which is found with Folsom points in some places. One form of the latter is fluted, that is to say, with a longitudinal groove on each side. Yuma points have been found associated with mammoth bones in Alaska, also in Alberta, Saskatchewan, and Minnesota. They have been compared with the Solutrean blades of Europe, but are of finer workmanship and apparently earlier. Sauer (1944) cites more primitive pre-Folsom artifacts from Texas (Abilene) and from the beach of glacial Lake Algonkin in Ontario. He suggests that the extinction of the large post-Pleistocene mammals of the western plains was gradually accomplished by fire-drives organized by these early people. He also believes that the treeless condition of the grassy plains was another effect of fires, although ecologists have other theories of the origin of the prairies.

Further excavations at Lindenmeier (Roberts, 1936, 1942) yielded 750 artifacts, but no human bones have yet been found in association with Folsom points. In a bison pit at least nine skeletons were found, a vertebra of one having a Folsom point embedded in the foramen for the spinal cord. Bryan and Ray (1940) conclude that the Lindenmeier hunters camped on the edge of a meadow on a vestigial valley bottom when the streams were at the level of the present twenty-foot terrace. Solifluxion phenomena in the terrace gravel show that the climate was almost Arctic at times, the plains dry and cold. The site was probably occupied only in summer, the occupants moving southwards with the herds of bison in winter. There was also the mammoth and a big camel, and the evidence of glaciers still in the mountains. As the Wisconsin glaciation is now recognized as having three main stages, separated by two warmer interstadial periods, the Lindenmeier site would belong to one of the colder periods, with a climate colder than the present. A series of papers edited by Howard (1936) discusses early man in the southeastern states.

Two finds of human bones near Los Angeles need to be mentioned. In digging a sewer trench, human remains were found at a depth of nineteen feet, three miles south of the famous asphalt pits at Rancho La Brea (believed of Aftonian or first interglacial age) where great numbers of extinct and recent animals were trapped in the viscous pitch. At least six human skeletons are represented in an area of only twelve square feet (Stock, 1924). They were apparently mired in a bog and afterwards covered with sands and clays. They represent the Indian type and appear to be Recent rather than Pleistocene, but their study is incomplete. In more recent excavations for an extensive drain near Los Angeles, Bowden and Lopatin (1936) describe the finding of a skull and other human bones thirteen feet below the surface. The skeleton was in undisturbed river deposits, in the next to the bottom of five strata of sand and clay, the fourth from the bottom containing boulders (presumably glacial). Only the posterior part of the skull is preserved. It is small and resembles the Basketmaker type, but the bones of the braincase average 7 mm. thick. About 1000 feet away mammoth bones were found in the *same* stratum. Clements (1938) has determined the age of these river beds as Late Pleistocene, the fauna being comparable with that of Rancho La Brea. Human bones were also found in pit 10 at the latter locality. From a comparison of the fossil avifauna of pit 10 with other parts of Rancho La Brea which are regarded as Pleistocene, Howard and Miller (1939) conclude that this pit is somewhat more recent, having a larger percentage of living species.

In a study of twenty skulls and some skeletons from caves in southwestern Texas, Stewart (1935) found the hair like that of modern Indians, the skulls rather small but hyperdolichocranic, the face and orbits rather low and the nose moderately broad. They differed from the chamaecephalic California Indians in having a good height of vault, and they were not identical with the Basketmakers of southern Utah and adjacent Arizona. This early hyperdolichocephalic Texan type has often been compared with Melanesian skulls, but is clearly a parallel development, probably representing an early arrival from Asia.

A good discussion of the age of man in America is in a later paper by Roberts (1940), who refers to many skeletal remains not mentioned here. He assumes that man came down the interglacial corridor through Canada, penetrating as far south as Mexico by the beginning of Recent time. The early migrations are believed to have been in small groups, the larger movements taking place only at the beginning of Recent time 9,000–10,000 years ago. The opening of the corridor east of the Rockies is placed at 15,000–20,000 years ago, with an earlier temporary break 35,000–40,000 years ago. There is a gap between the Paleo-Indian and his earliest modern counterpart, the Basketmakers, and we know nothing definite as yet of his physical anthropology. He evidently belonged to the genus Homo and was probably the progenitor, at least in part, of modern Indians. One of the skeletons recovered from the upper cave at Choukoutien (Weidenreich, 1939), associated with Upper Paleolithic artifacts and a Pleistocene fauna, had a skull like that of the long-head Indians of Texas. The associated artifacts (Pei, 1939) included a bone needle, a polished deer antler, scrapers, a chopper, stone beads, and perforated pendants made from the teeth of the deer, fox, and badger.

The Gypsum cave of South Nevada is another site where early man is associated with extinct mammals. It is twenty miles east of Las Vegas at an elevation of 2000 feet. Stock (1931) describes the big cavern with several chambers. The largest is sixty-five feet below the entrance and has at least fourteen feet of deposits on its floor. Limestone fragments from the ceiling alternate with bedded sands deposited under water, probably from glacial Lake Lahontan. Below a gypsiferous layer is one of dung of the sloth, Nothrotherium, some of it trampled and some burned or charred. This layer was twenty-six inches thick and contained also claws and a hind foot of the sloth. The dung of a mountain sheep was found in one chamber. Over this was six feet of limestone detritus, then a layer containing Pueblo and Basketmaker cultures. In another chamber, camel and horse bones were found at a depth of ten feet. It is uncertain whether man was contemporaneous with the sloth or came shortly after. The age of man in this cave has been estimated by Roberts (1940) at 8500 years.

Another important excavation of this period was reported by W. A. Bryan (1929) in a limestone cavern near El Paso, Texas. The small entrance led to a floor of wind-blown sand (loess) eight feet below and twenty feet in diameter. It was excavated to a depth of thirty feet without reaching rock bottom. At a depth of ten feet a human skull-cap and the phalanges of a ground-sloth were found associated. A hard horizontal diaphragm of stalagmite two to four inches thick was found at a depth of twenty feet from the surface. Below this were more skull fragments, as well as bones of extinct horse, cave bear, a very large camel and a nearly complete ground-sloth skeleton, proving that man was a contemporary of the sloth.

A definite succession of Paleo-Indian cultures has been established in one case. Hibben (1941, 1942) shows that in a cave in the Sandia mountains of New Mexico, the Sandia culture is older than Folsom. In this cave the upper layer is recent, containing Pueblo potsherds which are dated A.D. 1400–1600. Below this is a hard crust of stalagmite formed under moister conditions, and sealing in the lower beds. Folsom artifacts (points, gravers, and scrapers) are followed by a zone of yellow ocher, which represents a wet period. Kirk Bryan concludes that the sterile ocher deposits belong to the last Wisconsin ice advance (W 3), about 25,000 years ago. Below the ocher are the deposits containing Sandia artifacts, and below this a layer of sterile gray clay. The Sandia period would then be just before, and the Folsom period just after the last glacial advance. This basal cave layer contains hearths and charcoal as well as Sandia points, which are larger than Folsom points but crudely chipped and sidenotched. They occur also from southeast Colorado to central Texas. These Upper Paleolithic artifacts may bear some resemblance to eastern Siberian types, but should not be expected to bear a close likeness to those of Europe, which must have evolved separately to a large extent. Nevertheless, some of the artifacts collected by Renaud (1942) in Wyoming and elsewhere bear a striking resemblance to Lower and Middle Paleolithic artifacts in western Europe. But they lie on the surface and their age is unknown. In 1935 over a thousand implements were collected from nine sites on four river terraces in the Black's Fork Basin.

The Sandia cave was only a hunting station and contains no burials. The animals in the Folsom layer are the horse, camel, Nothrotherium, mammoth, wolf, and a bison like *B. taylori* but smaller. The sloth remains were found only in the uppermost recent layer. The Sandia layer contained a horse (*Equus excelsus*), bison (*B. antiquus*) as well as camel, mastodon, and mammoth. The Ventana cave in Arizona has recently yielded similar evidence of man's association with extinct animals. Bryan (1941) assumes that man entered the Mackenzie River corridor during the interstadial between the second and third Wisconsin advance. These Paleo-Indians were probably all dolichocephalic with rather heavy brow ridges.

The bison appears to have reached the Great Plains shortly before the Middle Pleistocene (Sellards, 1945). The earliest migrants were a giant species like *B. latifrons* in the Kansan formation. The smaller *B. ferox* is found in Yarmouth sediments, and in Sangamon deposits of Nebraska is a still smaller species, but larger than *B. antiquus* Leidy. In a Late Pleistocene (Wisconsin) terrace a species near *B. antiquus* is associated with Folsom and Yuma artifacts. More recent

a. (upper left): Korana girl living in Kimberley Location
b. (upper right): A South African Korana of pseudo-Australoid type, Abraham de Bruyn, age 80. Note the almost Neanderthaloid supra-orbital ridges, the large ear with the large lobe, and the very long jaw
c. (lower): Barotse Negroes at Victoria Falls, Northern Rhodesia

a. (upper left): Two young Basuto women in Pietersburg. Brass rings indicate social status
b. (upper right): Two native workers in the mines near Pretoria
c. (lower left): Young Zulu woman at Durban, East Africa
d. (lower right): A Basuto family near Letaba, Transvaal

XVIII

a. (upper): Typical kraal in Southern Rhodesia
b. (lower): Bavenda women at Pietersburg, northern Transvaal

XIX

a. (upper): Young Sesutos in the northern Transvaal. Their closely cropped hair has a peppercorn appearance

b. (lower): Kaffir women and children in the Game Reserve, northern Transvaal. Note the hair of two women in ringlets

a. Two Matabele women at Bulawayo, Southern Rhodesia

b. Two women near Mafeking, Bechuanaland, showing evidence of Hottentot ancestry

c. Negro children at Letaba in the Kruger National Reserve. Note the "skull-cap" distribution of the hair, and two with peppercorn hair

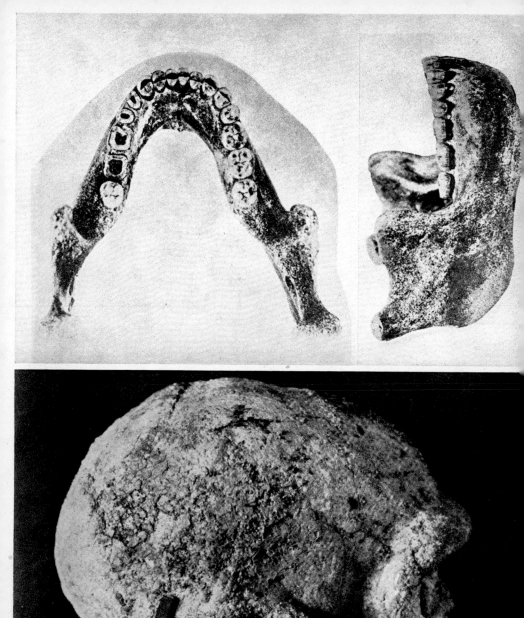

a. and b. (upper): The Heidelberg or Mauer jaw
c. (lower): The Steinheim skull, showing a heavy supra-orbital torus but the occipital region rounded as in modern man

a. (upper): A Tibetan woman photographed at Darjeeling, northern India, showing a remarkable resemblance to Amerinds
b. (lower): Three Slave Indians at Simpson on the Mackenzie River

XXIV
a. (upper left): An Ona woman
b. (upper right): An Ona child
c. (lower left): An Alakaluf man. The nose is broad, the lips somewhat thick, and the eyes intensely dark
d. (lower right): A Yamana woman

deposits all contain *B. bison* Linn. There has evidently been a definite retrogression in size of the bison throughout Pleistocene time, possibly a phenomenon of over-population as the numbers increased in the plains area.[1] Sellards describes a fossil bison quarry on the High Plains of northwest Texas, excavated in 1945. The bone-bed was one and one-half feet thick and contained many whole skeletons together, probably from a stampede. Excavation of five hundred square feet yielded skeletons of fifty or more bison together with twenty-seven artifacts. The latter included nineteen arrow heads, one end scraper, five side scrapers (knives), and two flake scrapers. The points were of Folsom-Yuma type but distinctive. The bison was notably larger than the modern species. There was also a large wolf, and near localities yielded *Elephas columbi* and a species of horse.

Another significant discovery, in Minnesota in 1931, is the skeleton of a girl about fifteen years old, believed to be of Pleistocene age. In road-making on the floor of a glacial lake now known as Pelican Lake, about 150 miles northwest of Minneapolis, in a cutting where the road was being deepened two feet by the grader, a human skull and mussel-shells were exhumed (Jenks, 1932, 1933, 1936). Unfortunately, although the importance of the find was recognized by the men in charge, nearly the whole skeleton had been removed before Professor Jenks was able to see and study the conditions. In further excavations a few remaining bones were found in place, but opinion is divided as to whether the possibility of this being a burial was wholly excluded. Jenks believes this to be the case, and that the girl was drowned in Pelican Lake. The skeleton was in horizontal varves (rock flour) from the glacier, consisting of silt and white sand, over two feet below the old road level and nearly ten feet below the original ground level. The fragments of mussel-shells may have been part of a headdress, and other fragments recovered, such as a wolf's tooth, a bone fragment of a loon, and portions of a turtle carapace, may have been the contents of a medicine bag. A crude dagger of antler and a shell pendant were also recovered. The shell belongs to a species (*Buscyon perversa*) frequently occurring in the

---

[1] This decrease in size of animals during the Pleistocene may be more general than has been realized. The comparisons of Koby (1945) show that there has been a similar decrease in size of teeth and length of limb-bones in the European bears. The Pleistocene cave bear (*Ursus spelaeus*) is largest, a Pleistocene brown bear from Tuscany smaller, and the modern brown bear (*Ursus arctos*) smallest.

Cahokia mounds, but living only in Florida waters now and in Pleistocene times. This implies a long line of exchange by human hands.

The skeleton has several interesting features. The skull is more Eskimoid than Indian, mesaticephalic (C.I. 77, close to that of Alaskan Eskimo). Mongoloid affinities are recognized in the supra-orbital and frontal regions, the shallow supra-orbital fossae, the low nasal bridge, and the malars. The extremely narrow nasal aperture, as in the Eskimo, may be attributed to the glacial conditions. This girl was therefore a generalized Mongoloid rather than a specialized American type. The teeth and palate were large, the upper incisors shovel-shaped, the mandible robust, and the maxilla showed alveolar prognathism as in some Sioux Indians. There is no platycnemia of the femur, a condition found in modern Indians. One unique feature is that the second molars were larger than the first or third, another is the presence of two upright U-shaped grooves in the occiput, which was primitive, large, and projecting. This proto-Eskimo seems to represent a type which might be expected in America in the last phase of the Pleistocene, 20,000 years ago. Whether the makers of Folsom points were of this type is at present unknown. A careful study of this skeleton by Hrdlička (1937) led him to the conclusion that it represents "the characteristic type of Sioux, which differs substantially from that of other North American Indians," but Hooton (1946, p. 405) accepts it as late Pleistocene.

Recent studies in Oregon show that the Paleo-Indian lived there some 15,000 years ago. The post-glacial climate is in three phases, a dry period preceded and followed by times which were cooler and moister. During the dry periods the glaciers on the Sierras disappeared and with increasing moisture were reformed about 4,000 years ago, as indicated by small fresh moraines. In a study of the cave deposits, pollen profiles in the bed of extinct post-glacial lakes, and the geological conditions, particularly the pumice from volcanic eruptions of Mount Mazama and Newberry extinct volcano, Cressman, Hansen, and Allison (1946) show the presence of man in Oregon in early post-glacial time. Two of the caves in Summer Lake valley show occupation interrupted by deposition of pumice from the last eruption of Mount Mazama. Cave No. 3 contained deposits seven feet deep. At the base were evidences of a campfire on a lake beach, with obsidian implements. Bones, broken for their marrow, belonged to extinct species of Equus, camel, and bison. Over this was a bed of fine dust with bat guano and rocks, above this a pumice bed from the eruption of Mount Mazama, and on top evidence of a later occupation. Cave No. 1 showed a long occupation before the eruption, and another afterwards.

On Lower Klamath Lake were three human occupations. The earliest was associated with extinct Equus, camel, and elephant. A portion of a human mandible was found here and long bone arrow points. The second occupation was associated with the refilling of the lake, and the third was by Modoc Indians in historic times. This fauna probably became extinct some 7500 years ago.

The more recent developments of Pueblo pottery and agriculture in the Southwest are being definitely dated by the tree-ring method (Bryan, 1942). There were probably three cycles of erosion and alluviation, due to climatic changes, the last erosion period being about A.D. 1100–1400. The growth of Mayan civilization in Yucatan, mainly within the Christian period, is ably described by Morley (1946).

In concluding this sketch of the migrations of man into North America, it appears probable that the first migrants crossed Bering Strait on a land bridge during one phase of the Wisconsin glaciation, which appears to be contemporary with the Würm in Europe. Kirk Bryan believes that phase to be the second interstadial of the Fourth Glacial period. These migrants probably came southwards through the Mackenzie and Peace River valley corridor during the last interglacial stage. This would afterwards be closed for a time while the Paleo-Indians—Folsom and Minnesota man—spread over the southern half of the continent. At the end of the Ice Age there may have been immigration on a larger scale. The archaic type of Indian with heavy brow ridges would seem most likely to represent the Folsom man. In the modern Indian population there is a large brachycephalic element in some tribes. Whether this was brought in with the transmigrants, as a mixed condition or in certain tribes, or whether it has developed autochthonously in America, is not at present clear. On the coast of British Columbia a long-headed population preceded the modern round-heads, but the latter evidently left Asia too early to have the B blood-group (Gates and Darby, 1934). As they are near the Mongoloid source in eastern Asia, it is reasonable to suppose that their Asiatic ancestors were already broad-headed but emigrated before the B blood group spread to them.

Jenness (1933) points out that excavations on St. Lawrence Island in Bering Strait show that the early inhabitants were long-headed, while the later Eskimos in this region were more broad-headed. Similarly, in Arizona and New Mexico the earliest Basketmakers were prevailingly long-headed; and kitchen middens on the coast of British Columbia show a long-headed people, whereas the modern Indian population is round-headed. How this spread of brachycephaly took place requires further analysis, both in America and in many other parts of the world. Jenness emphasizes that many culture elements are common to the Amerinds and the peoples of northern Asia, having originated in Asia and diffused across Bering Strait. These include bark canoes, cradle-boards, certain types of clothing and methods of dressing skins, scapulimancy, and the bear ceremonies which are well known to the Ainu.

Imbelloni (1943) has written a very stimulating account of the peopling of America, based partly on the work of Dixon and of Griffith Taylor and with some genetical background, but without recognizing the importance of parallel variations. Much discussion of the age of the Indians in America, pro and con, is contained in *Early Man* (1937) edited by MacCurdy.

SOUTH AMERICA

We come now to South America. It seems evident that her population was received entirely from the North American continent. Panama was a strait from Eocene to the end of Pliocene time. During this period the unique South American mammal fauna evolved in isolation, most of them to become extinct in Pleistocene time. But man had already established himself early enough to be associated with extinct ground-sloths in South as well as in North America.

One of Hrdlička's many services to physical anthropology was in a first-hand critical comparative appraisal of all the available human material in South America (1912), with full references to the studies of early workers. The earlier human remains came entirely from two sources: the Lagoa Santa caves in the province of Minas Geraes, Brazil, and various localities in the province of Buenos Aires, Argentina. The Lagoa Santa caves were explored by P. W. Lund of Copenhagen in the years 1835–1844, and the human skeletons are mostly in the Zoölogical Museum of the University of Copenhagen. When this work was nearing completion, a short account was published in the translation of a letter to the Historical and Geographical Society of Brazil (Strain, 1844). The Lagoa Santa and the Punin skulls are the earliest human remains from South America. Lund explored over 200 caves in Brazil, finding in them 115 species of mammals, 27 of which were extinct. H. Winge described the mammals in a series of Danish papers between 1887 and 1915.

After six years of search Lund found human bones in a limestone cavern on the border of a lake (Lagoa do Sumiduoro) which also contained the bones of living and extinct mammals. This cave was frequently flooded in the wet season, but the lake drained out through crevices in the limestone rock in the dry season. The bottom of the cave contained soft black earth, carried in during inundations, in which were the bones of man as well as mammals, birds, reptiles, and fishes scattered in disorder. Some were like fresh bones but more fragile, light and reddish-brown in color. Others were heavy, hard and petrified, and there were all gradations between. This applied to the human as well as the animal bones. Records show that the rate of petrifaction can be rapid or very slow according to the conditions. A single artifact was found—a muller for crushing seeds. Human remains were also found in two other caves in Minas Geraes.

In a lower chamber of the cave of Sumiduoro was a large quantity of animal bones scattered through earth, mainly living species of deer, peccaries, and pacas, but also the extinct Megatherium, Smilodon, and others. A pool in the cave contained petrified bones. Some of the human remains were in another part of the cave, near an entrance. Most of the skulls were collected in one place, perhaps by water action; another pile consisted of small bones from fingers and toes, carpals and tarsals. The pool contained a few human bones, but some of the human skeletons were apparently still in the position where they had been deposited. There were quantities of bones of bats and rodents, species of opossum, a large extinct rodent, and a great fossil jaguar (*Felis protopanther*) now extinct, an extinct cave wolf, and the femur of an extinct monkey, as well as a living species of wolf and two of peccaries, the latter mingled with human remains. Species of peccary and tapir larger than those now living were also found, as well as armadillos, and the llama which is not now found in Brazil. An intriguing discovery was a modern horse, larger than the introduced horses of South America. It differed from two other fossil species in Brazil and apparently agreed with the domestic horse. Opinions differ as to whether the human skeletons were contemporaneous with the extinct animals. One must conclude that contemporaneity is possible but has not been definitely established. Lütken, in a memoir in 1888, considered it probable but not certain.

The human remains belonged to at least thirty individuals of all ages, from newborn to decrepit. They were among fallen blocks of stone and were mostly broken, generally after becoming fragile through decomposition. Several of the skulls had an oblong hole in the temple, believed to have been produced by a stone axe. While most of these skeletons went to Copenhagen, a child's skull in the British Museum with seventeen portions of jaws, parietals, and other skull fragments and long bones were published by Blake in 1864. Another skull, deposited in Rio de Janeiro, was described by Lacerda and Peixoto in 1876. They found the front low, the glabella salient, the superciliary arches very prominent, the occiput almost vertical, occipital protuberance broad, plain, and very protruding, the malars prominent, orbits quadrangular, parietal eminences prominent, cephalic index 69.7, cranial capacity 1388 cc. They compared the skull with that of the modern Botocudo Indians. Hrdlička (1912) viewed this skull but found no primitive characters, the supra-orbitals being moderate in development.

After years devoted to the materials in Copenhagen, Reinhardt, in a memoir in 1888, concluded that they represented a tall tribe, somewhat delicate, dolichocephalic, the skull summit elevated, nearly pyramidal, the cheek bones prominent, eyes wide apart, skull very thick, showing no deformation. Lund formed the erroneous opinion that two races were represented, one having a very low vault, the other small and well formed. In a study of four of the Copenhagen skulls in 1884, Kollmann concluded they were all of Indian type, hypsidolichocephalic, with a height index of 80.2 and mean cranial index 72.2, orbits low, moderate prognathism. All the numerous measurements agree in finding the nasal index rather high, ranging from 47.8 to 53.3. In 1885, ten Kate, from an examination of the Copenhagen crania, found them closely like the Indians of Lower

California. Sören Hansen (1888) studied all the material, finding seventeen measurable skulls and thirty broken mandibles. He found the skulls remarkably uniform, the average length 184 mm., C.I. 70.5, N.I. 50.7, except one female skull which was brachycephalic (C.I. 80.95) but otherwise similar to the rest. The skull was very high, very long, oval, and prognathous, and he makes the statement that the "type corresponds perfectly to that of the Papuans."

Hrdlička (1912) concludes that the cave of Sumiduoro served as a place for burial or deposition of the dead but not as a living place, in the absence of ornaments or utensils.

A skeleton has recently been found in the Confins cave, eight miles from Lagoa Santa (Walter, Cathoud, and Mattos, 1937). This cave has its entrance halfway up a limestone rock 140 feet high. Many Indian skeletons were buried under the rock shelter but the cave entrance was sealed by immense rocks. Excavation inside finally revealed a skeleton lying extended under alluvial soil more than two meters deep, sealed in by stalagmite. Other layers of stalagmite accompanied remains of various Pleistocene fossil mammals, with which the skeleton was contemporary. The skull has a low forehead, although hypsicephalic. There is submaxillary prognathism, the C.I. is 69.1, N.I. 48.9, cranial capacity 1281 cc. The incisors are apparently shovel-shaped. The foreheads are not low and sloping, but the Lagoa Santa cranial type agrees fundamentally with that of dolichocephalic American Indians. This includes facial features, nasal aperture, malars, maxilla, skull base, teeth, and vault, any relation to Melanesians or Polynesians being only basal in character. This modern dolichocephalic type of Indian is found in Brazil and other parts of South America as well as the Aztec, Tarasco, Pima, early Pueblos, Iroquois, Algonquian, and other tribes. The Lagoa Santa crania are not differentiated from the early dolichocephalic Indian type.

The Punin skull, found in Ecuador in 1923 by Mr. G. H. H. Tate on the American Museum expedition led by H. E. Anthony, evidently belongs to the same race as the Lagoa Santa crania. It was discovered in a quebrada or ravine at Punin, near Riobamba, within sight of Mount Chimborazo, in the great interandean plateau, a region of extinct volcanoes, at an elevation of over 9000 feet. The Punin ashbeds were a source of fossil mammals as early as 1883. The ash was deposited in a layer which followed the contour of the ravine. Probably a volcanic cataclysm deposited several feet of ashes and killed large numbers of animals in the ravines. The skull was found in an eroded bank in ash compacted to tufa. Horses, camels, mastodons, and other mammals were found in the quebrada—a Pleistocene fauna. The skull was fifty to one hundred feet from the mammal bones, but of the same geological origin as the bone-beds, so that it probably represents man contemporary with these Pleistocene animals, the latter including the mastodon (*Dibelodon andium*), Mylodon sp., Protauchenia, Smilodon, and *Equus andium*.

The skull was described by Sullivan and Hellman (1925). Photographs show its great morphological similarity to the Lagoa Santa crania. The vault is described as low, the orbits low and widely rectangular, the face low and somewhat prognathous, cranium ovoid. The supra-orbital ridges extend slightly beyond the middle of the superior orbital border, there is a well-marked occipital torus, the teeth are unusually large and there are no third molars. The skull length is 186 mm., C.I. 71, N.I. 59.6 (?). The nasal measurements are doubtful because of imperfections in the skull. The cranial measurements and indices fit best the norms of Tasmania, Australia, and New Guinea. The Australoid appearance is suggested strongly by the cranial vault and face, the glabellar region, orbits, and nose. They find "absolutely no basis for excluding [the Punin skull] from a series of Australian or Tasmanian crania and every reason for including it," although they suggest that the decision might be different if the mandible and skeleton were available. The basion-bregma height is markedly lower than in all the Lagoa Santa crania except one, but there appears to be no other important difference.

Both the Punin and the Lagoa Santa crania appear to differ from the Australian skull in having a narrower nose, and the Lagoa Santa skulls have a less-marked glabellar notch. The latter at least are then Australoid but not Australian. Sullivan and Hellman conclude that these skulls are either basically related to the Australians and Melanesians or they represent a parallel development. From the point of view developed in this book, parallelism is the natural and inevitable conclusion. The skulls probably represent the earliest wave of interglacial Americans, retaining in their evolution from Sinanthropus certain features which belong to the Australoid level of evolution, but involve no direct relationship to *Homo australicus*. The fact that several features of face and cranium are involved probably indicates genetic linkage. Imbelloni (1940) described two Uru skulls of "Australoid" type from an island in Lake Poopó, Bolivia, which were brought to the museum at Buenos Aires in 1934. They are very dolichocephalic (C.I. 72.8 and 73.7) and in vertical as well as posterior view closely resemble Lagoa Santa, but the forehead is less flat than in modern Australians. Posnansky remarked, in discussing the paper, that 80 per cent of the Aymara in the altiplano of the Andes are Urus, and that on the Rio Desaguadero live 100,000 Urus who speak their own language.

In 1910 Hrdlička and Bailey Willis, the geologist, made an expedition to Argentina to study the remains of early man in that region. Florentino Ameghino had described two "new genera" and three other "new species" of man from the Pampean formation. His brother Carlos was a great collector of fossils in the province of Buenos Aires, whence all these supposed remains of early man came, except a few from Patagonia and elsewhere. These remains were all from the soft loess Pampean formation which is everywhere immediately under the humus. Whether this Argentine loess, which is of great extent, had been laid down merely by wind, or whether a glacial climate was involved, is not yet determined. The human skeletons found were mainly either on or near the coast. Large numbers of worked stones were collected at Mar del Plata and other places along a four hundred mile stretch of coast south of the River La Plata. They were of two types, a black pebble industry and other implements of white quartzite brought from a distance, but evidently both made by the same Indian tribes. They included mortars, pestles, hammers, bolas, knives, scrapers, drills, and blades, made from water-worn beach stones. All were on the surface, with potsherds in certain places, and were evidently of no great age. The soft cliffs or barrancas were undergoing rapid erosion.

Hrdlička (1912) described in detail about fifteen finds of human bones which had been made. They all belonged to the modern Indian or to the Lagoa Santa type. As early as 1864, the Carcaraña bones were collected by F. Seguin, a dealer in fossils, from a railway excavation twenty-five miles from Rosario. Parts of four human skeletons with stone implements and a Glyptodon carapace were in addition to a fossil horse and a big bear, *Ursus bonaerensis*. Some believe they belong to the Upper Pampean yellow loess, but Hrdlička found no satisfactory evidence of age, the skulls being like the modern Tehuelche.

Ancient cemeteries were found in the valley of the Rio Negro in 1800. One of these skeletons was in alluvium in the river channel at a depth of four meters. Another, at two meters depth, was more modern and showed skull deformation of the Aymara type. At the Arroyo de Frias, Mercedes, Ameghino found scattered human bones in 1875, at a depth of four meters, with charcoal, rhea egg fragments, modern implements, and pottery. The Saladero skeleton, collected by S. Roth in 1876 ten kilometers from Pergamino, was partly protruding from the loess in a gully three meters deep, but well preserved. It was in a sitting position but was friable and fell to pieces through inadequate treatment. Although this was deeper than an ordinary burial, wind or water could have added to the deposits over it. Roth collected here an arrow point under the femur of Scelidotherium, but Hrdlicka concluded that the antiquity of this skeleton was not substantiated.

The Fontezuelas skeleton was found by Roth five years later near Rio Arrecifes. The Upper Pampean formation was five to twenty-four inches thick, containing bones of Glyptodon, Hoplophorus, and Mylodon. Beneath this the Lower Pampean, one to three meters thick, contained Mastodon, Megatherium, and Toxodon—a Pleistocene fauna. The skeleton was practically complete and hence probably a burial. It was partly exposed by denudation and over it was the inverted carapace of a Glyptodon. Hansen considered the skull like those of Lagoa Santa, hypsistenocephalic, C.I. 73.5. The stature was estimated at 151.5 cm., but some Peruvian Indians are equally short. It was named *Homo pliocenicus* by Kobelt, quite unjustifiably. The Arrecifes skull, near Fontezuelas, was reconstructed from twenty-four pieces covered with a thin calcareous incrustation. Lehmann-Nitsche found it mesocephalic (C.I. 75.8) with prominent supraorbital arches and a well-marked occipital torus. Hrdlička regarded the brow ridges as well developed over the median two-thirds and the occipital torus as rather pronounced, but found "absolutely nothing more primitive . . . than in crania of the American Indians."

The Samborombón skeleton, to which Ameghino gave another specific name, was nearly complete except the skull, of which only the occipital and the mandible remained. The skeleton was divided near the middle into two parts, due to faulting. A few meters away was a fragment of Scelidotherium and part of the antler of a large deer. The sternum of this skeleton had a perforation, but ten Kate found this condition in 16 out of 120 Indian sterna in the Museo de la Plata. Hrdlička considered the antiquity of the skeleton unproven. The Chocorí skeleton, found on the surface in 1888, one hundred yards from the beach, on the coast south of Miramar, consisted of a skull with some other bones. Hrdlička considered it like the Arrecifes skull, with moderate supra-orbitals and a low forehead, C.I. 75.

The Overejo remains, consisting of various human skulls and bones, were far inland on the Rio Dulce thirty miles from Gramilla. Hrdlička found here, in half an hour, another skeleton in the pink loess of a barranca (the vertical wall of a gully), six feet below the surface. The loess had been blown from the river bed and later cut into gullies which were not more than fifty years old. In the same gully he found a jaw of the common horse in aeolian loess nine feet from the surface. Its depth is accounted for by Willis through the fact that where the water accumulates in a pool the very fine-grained loess becomes fluid and sinks in, forming a pit. A bone in such a pit would become covered with in-blown loess. A century ago the Indians buried their dead in this neighborhood. Shells of three living species were collected in the barrancas down to a depth of fifteen feet. The six skulls and skeletons were all modern, at least two were mesocephalic and the one found by Hrdlička was a male with pronounced supra-orbital ridges.

The Baradero skeleton was discovered in 1887 in a railway cutting in the loess about one meter from the surface. It was in extended posture, but Indian burials were sometimes made in this way. Although covered with calcareous concretions, all the other evidence indicates a modern Indian. The Arroyo Siasgo skeleton was excavated in 1909 by C. Ameghino and named as a new species of Homo by his brother. Hrdlička found it to be a modern Indian child of twelve years, showing some Aymara deformation of the skull.

At Necochea, near the coast, a double burial near the surface was found in 1909, and Ameghino invented for it another new species of Homo, based on the fact that one skull had practically no chin prominence. Hrdlička pointed out that in modern mandibles the chin may be equally vertical, as an infantile character. A visit to the site in 1910 showed the presence of black and white chipped stone implements,[2] protruding bones of Glyptodon and other animals. The crania were mesocephalic or nearly so and resembled those of California Indians, with only traces of superciliary ridges. The cranial capacity of one skull was ± 1150 cc., but twenty-one female Peruvian skulls had a capacity of only 920–1050 cc. Another skull, a skull cap, and various other remains from this area, referred to as *H. pampaeus* by Ameghino, showed no morphological differences from modern Indians.

A skeleton found at Miramar on the Atlantic coast in 1898 was made the type of Ameghino's *Homo pampaeus*. It came from near the site of the Chocorí skeleton and was at first stated to be of Lower Pliocene age. Although partly fossilized and bearing incrustations of loess cemented by lime, it is of modern type with a narrow forehead, the frontal artificially flattened, as well as light Aymara deformation. A narrow forehead (7.7–9.6 cm. wide) is characteristic of Indian skulls from Utah, Mexico and Massachusetts.

In the valley of the Rio Negro over one hundred skulls have been found by Moreno showing this type of deformation, one of them ten feet deep in the sands. On this river four miles south of Viedma, Hrdlička obtained from the dry mud eleven nearly black skeletons (one a foetus). All were superficial or partly protruding, and the ten skulls all showed more or less Aymara deformation. The pelvis of a cow exhibited the same discoloration, probably from manganese and iron salts. The lower Rio Negro was evidently a center for this form of head deformation, which tends to obliterate the supraorbital ridges.

We come now to Ameghino's two new genera. In excavations for a dry dock at Buenos Aires in 1896, a skull was found by the workmen at a depth of nearly thirteen meters below low-water. Another skull, which was lost, was at the entrance to dock 4, and in dock 3 were recovered a Glyptodon carapace and a Mastodon tusk. Ameghino described the skull, in 1909, as *Diprothomo platensis*. It consists only of a skull cap, and Ameghino's interpretation of it as a Lower Pliocene forerunner of man resulted from his practically horizontal orienta-

[2] These were always in association. The white were from fragments of quartzite, the "black" from pebbles of jasper, etc., ranging in color from black to red, brown, and yellow.

tion of the calotte. Hrdlička showed that the slope of the frontal in correct orientation is almost exactly that of a modern Piegan Indian. The supra-orbital ridges are indeed prominent over the median half of the orbits, but nearly absent in the distal half. The nasion is 6 mm. below a line joining the upper borders of the orbits and the glabella is prominent. The skull apparently belongs to the archaic Indian type represented by the Lagoa Santa and Punin crania, but Hrdlička considered the prominence of the supra-orbital ridges to be due to the large frontal sinuses and not even Australoid in character. Bailey Willis showed that the river bank near Buenos Aires developed deep irregular holes, and that the skull probably dropped into one of these which was later filled with mud; which would account for the depth at which it was found.

As early as 1887, Ameghino announced the discovery of vestiges of an early forerunner of man, giving it the name *Tetraprothomo argentinus*. This was based on the finding, in a low cliff (barranca) facing the sea at Monte Hermoso on the south coast, of burnt soil, hearts, split bones of animals, and worked stones. Later he referred to this "genus" a human atlas of submedian size and unknown provenance found enveloped in yellowish earth in the Museo de la Plata. Later, C. Ameghino encountered at Monte Hermoso a femur which Florentino also referred to his Tetraprothomo. After careful comparative study, Hrdlička shows that the supposed Pliocene atlas is within the range of variation of modern Indians, while the femur belongs to a Carnivore, probably a felid. Ameghino's mistakes evidently arose through attributing excessive age to relatively recent fossil or subfossil remains, combined with a lack of comparative material and an excess of enthusiasm. All that can be concluded from the whole of these finds is that a few of them apparently belong to the archaic type of dolichocephalic Indian with marked brow ridges, the rest representing the modern Indian type and some even being recent enough to show the Aymara head deformation.

Simpson (1940) has revised the Tertiary formations of South America and proposed a series of new names for the various subdivisions that have come to light. He recognizes about eighteen stages and attempts a world correlation. The age of man in these formations depends upon the age to be attached to two of them—the Chapadmalensian and the Montehermosensian. Simpson considers both of these

Pliocene, although the younger Chapadmalensian is possibly early Pleistocene. While the Tertiary beds of North and South America cannot be directly correlated, Simpson points out that the South American fauna of Cretaceous and Early Tertiary age was replaced by North American forms in the Late Tertiary and Pleistocene. Although Hrdlička ruled out the possibility of Pleistocene man in Argentina, the matter is not yet fully decided. Bryan (1945) recognizes that Hrdlička was mistaken in assuming that all humankind must have come through a Neanderthaloid stage. On the other hand, what I call the orangoid lineage of man is only known from Britain and eastern Africa, while there is morphological evidence that some of the characters of gorilloid Sinanthropus have been passed on to the Amerinds. It therefore seems probable that men of orangoid ancestry played a limited part, if any, in the ancestry of the Amerinds. If it is true that man lived in the Chapadmalensian of Argentina, then he would be older than Pithecanthropus.

The frequency of Glyptodon in association with human finds in South America shows that they survived into recent times. Colbert (1942) has noted these associations in North and South America. The animals include not only various ground-sloths, mastodons, mammoths, and peccaries, but tapirs, capybara, Smilodon, *Felis atrox*, the musk ox, and an extinct moose. Many of these fossils survived so late that they are remarkably fresh, sometimes having remnants of hair, skin, ligaments, and even muscles. Horses and camels and the short-faced bear have been found in caves in the southwestern states, a giant beaver in terraces at Bridgeport, Nebraska. Ground-sloth remains, some showing hair, are known not only in South American caves but in Cuba. In the cordilleran region the bunomastodonts, Cuvieronius and Cordillerion, are associated with man, while recent work in Patagonia shows human association "without doubt" with horses, Equus and the peculiar short-legged South American Hippidion. Man may have been responsible for the relatively sudden extinction of many of these animals.

Excavations of Bird (1938) in Patagonia included large middens on the island of Chiloe on the west coast and others farther south. Naturally, the migration lines converge southwards with narrowing of the continent. The low plains of the eastern (Argentinian) part and the mountainous coast and archipelagoes of the western (Chilean) part offer an impressive contrast in climate and topography. On Navarino island in Tierra del Fuego the first occupation showed percussion flaking, bolas, and scrapers; the second, pressure flaking and the bow and arrow. On the shores of this island were found 183 sites of previous habitation. The floor of a rock-shelter was nine feet deep in ashes and refuse. Estimates based on changes in land levels and on the depth of the middens led Bird to the tentative figure of 3000 to 5400 years since man's first arrival here, and 1800 years since the arrival of Indians with birch bark canoes. Two skulls from the earliest burials were found by Shapiro to have cephalic indices of 72 and 74 respectively. They were of Indian type with some resemblances to the Lagoa Santa crania.

The blood groups taken by Rahm in 1931 of Tierra del Fuegians gave puzzling results, as shown in Table 21. The recent results of

TABLE 21
BLOOD GROUPS OF TIERRA DEL FUEGIANS

	No.	O	A	B	AB	
Ona	18	17	1	—	—	Rahm
Yahgan	33	3	—	30	—	Rahm
Ona	20	14	—	—	—	Lipschutz *et al.*
Yahgan	40	31	—	—	—	Lipschutz *et al.*
Alakaluf	17	13	—	—	—	Lipschutz *et al.*
Total	77	58	11	7	1	
Mestizos	43	24	11	7	1	

Lipschutz *et al.* (1946), which can now be added, show that in the three tribes taken together, those of pure ancestry (5 Ona, 20 Yahgan or Yámana, 9 Alakaluf) were all O, while those having A or B all had some European ancestry. As Rahm did not distinguish between pure and mixed bloods, it is evident that all three tribes were originally O, like most other Indians that have been tested. These remnants are stated to be reduced now to about 40 Onas, 60 Yámana, and 80–100 Alakalufs, evidently on the verge of extinction. In a more complete account of these Fuegians which has just appeared (Lipschutz, Mostny, and Robin, 1946) the family pedigrees and photographs are given; Plate XXIV shows a member of each tribe and an Ona child. The latter seems scarcely distinguishable from a Chinese child except in skin color, or from an Eskimo except in the broader nose.

It may be added here that the only clear exception to the rule that Indians of pure ancestry have only O, is the Blackfoot Indians and a derivative tribe, the Bloods, in Montana and Alberta (Gates, 1946, p. 718). They have 75–80 per cent A, but no B. The high frequency of A might have been derived through chance isolation of a family group; it might have come through early migrants from Asia, or possibly by repeated mutations from O to A. Recent tests of many South American tribes show that they are very high in O, the A and B probably derived from crossing.

Human association with extinct animals has recently been determined in the Gruta de Cadonga, a cave in the Province of Cordoba some 400 miles northwest of Buenos Aires. It was nearly closed when discovered by Lieutenant Colonel Anibal Montes in 1917. He revisited it periodically until 1939 when excavation was begun. The results were published by Castellanos, whose account has recently been summarized by Bryan (1945). The cave is in limestone (marble) and contains four layers of deposits besides the top debris which was removed. The two upper layers contain pottery, that in the second layer being attributed to the Aymara. The third layer consists of reddish compact earth with fragments of a Mylodont and freshwater shells of living species. The fourth and lowest of these beds is separated from the beds above by a strong erosional unconformity. It is a calcareous clay like the pampas loess, containing bone points, scrapers, bones of extinct animals, and the skull of a child, as well as calcareous concretions. The skull, at a depth of more than 6.3 meters, shows the Aymara type of deformation, which must then be an older custom than was formerly supposed. Similar bone points were found in the Sandia cave in North America. While the question of the geological age of this deposit remains uncertain, the association of man with extinct animals appears to be proved. There is also evidence of climatic change which may ultimately be correlated with Pleistocene conditions in the Andes.

While the association of man with extinct mammals has thus been proved in South as well as North America, the chronological age of these associations in relation to the Pleistocene remains to be determined.

Recent excavations in the eastern ranges of the Andes near Mendoza are described by Rusconi (1947). There are five strata of burials, with ossuaries containing over 70 skeletons. The lowest (1), with no head deformation, are regarded as pre-Neolithic. Superposed in stratum (2) were skeletons showing a distinct cranial morphology of more Araucanian type, with head deformation; (3) was of late Inca type, with evidence of tombs and citadels, and Chimu pottery; (4) represented a southern expansion of certain groups as sedentary peasants during late prehispanic time, and (5) was post-hispanic, rooted in the Uspallata valley.

THE ESKIMO

The possible origin of the Eskimos from Magdalenian Europe is hypothetical, based on implements such as the harpoon, and the Chancelade skull. Although it remains possible that they traveled eastwards, across the Siberian tundra, the Eskimos have some Mongoloid characters and it would be difficult to account for them in Europe, except by an earlier westward movement from Asia which would be purely hypothetical. On the other hand, the people represented by Chancelade might have picked up Mongoloid characters by hybridization after their eastward migration. The scaphoid cranium and other features of the Chancelade skull can be more plausibly accounted for as parallel developments in Europe. Collins (1943) believes that the ancestors of the Eskimos are to be found somewhere north or east of Lake Baikal. By excavations, Petro, Okladnikov, and others have discovered Neolithic stone implements, especially arrow points, on Lake Baikal, some of which belong to types found in early Eskimo cultures—the Ipiutak, Old Bering Sea, Dorset. One type of blade on Lake Baikal is Yuma-like, but nothing has been found there resembling the earlier Folsom or Sandia points.

We are thus beginning to find the roots of Indian and Eskimo archaeology in the archaeology of Siberia. The Lake Baikal Neolithic is believed to be 6000 years earlier than the earliest Eskimo remains. Okladnikov dates it from the sixth millennium to the first millennium B.C. Another circumstance which may possibly trace Eskimo origins back to Europe is that the Mesolithic Maglemose culture of Denmark, which is assigned to the Boreal Period II, 6800–5000 B.C., shows certain resemblances (side-bladed projectile points and knives) to Eskimo implements.

The modern Eskimos are a nearly uniform type, representing one stock with a culture adapted to severe Arctic conditions and having one language with only dialect variations. All the evidence indicates that they were the most recent of migrants to the American continent before the historical period. It has already been pointed out that once they were established on the Alaska coast the immigration of other peoples by that route would be blocked. They now stretch along a coast-line of some 6000 miles from Alaska to Greenland, including about two-fifths of the circumpolar area, and there is evidence that they formerly lived on the Siberian coast as far west as the Kolyma river. An Eskimo-like culture has been found by Cernecov further west at the mouth of the Ob, preceding the Samoyeds. Morant (1937) made a study of the Eskimo skull from measurements of seven series compared with twenty-six Asiatic series. He finds a relation to the prehistoric Chinese.

The accompanying photographs were taken by the author in 1928 on an expedition down the Mackenzie River which was supported by the Hudson's Bay Company, whose trading posts extend all over the Canadian Arctic. Plate XXV (a and b) shows two sisters, Laura and Doris Tegitkok, aged 14 and 7 years, from the Arctic coast, who were at school at Hay River on Great Slave Lake. By comparison, Plate XXV (c) shows a Loucheux (Tukudh) Indian girl attending the same Mission School. The Loucheux live near the Arctic circle and have probably intermixed to some extent with the Eskimo. Plate XXV (d) is of a young Eskimo, Jacob Nipalarok, 20 years old, from Baillie Island on the Arctic coast. He may have had a white ancestor, as his features are not typically Eskimo. Plate XXVI (b) is an Eskimo, John Kobigon, photographed at Aklavik, an Eskimo village at the head of the Mackenzie River delta. He is wearing a fur-lined parka. This chiefly Eskimo village has a powerful radio station for sending and receiving news. Plate XXVI (a) is Lennei Iglangasak, an Eskimo at Aklavik who married a Loucheux Indian woman. Like many of the Eskimos in this district, they have their own schooner, which they use for fishing and sealing. Plate XXVI (c) shows a group of Loucheux

Indians at Fort MacPherson. With the increase of communications in the Arctic, by aeroplane and otherwise, the Eskimo culture will no doubt undergo some rapid changes. A considerable amount of crossing with whites has already taken place, beginning a century ago in the Arctic whaling period.

We may suppose that the Eskimos did not migrate to the warmer south from Alaska because they had already developed their highly specialized ice culture which adapted them to the Arctic conditions. There is also some evidence of a general kind that the Eskimos may have undergone some physiological adaptation to low temperatures as a result of their long sojourn in the Arctic. In any case, while they extended southwards on the Alaskan coast and probably played a part in the ancestry of the Aleuts,[3] they did not migrate into warmer territories. This might be because their progress down the coast of British Columbia was blocked by Indians already there. At any rate, they continued eastward along the Arctic coast to Hudson Bay and ultimately to Greenland by the northern route and to Labrador. They even crossed the Strait of Belle Isle into the western part of Newfoundland, where they quarried soapstone for their lamps. Everywhere they have remained maritime, except the Nunatagmiut in the interior of Northern Alaska and the Caribou Eskimo west of Hudson Bay. That the main movements of the Eskimo were east and west along the Arctic coast, and not south, is confirmed by the dialects. Jenness finds that the dialects of northern Alaska are much closer to those of Greenland and Labrador than they are to those of the Yukon-Kuskokwim region immediately to the south.

Some writers have indeed regarded the Caribou Eskimo as representing Indian ancestors of the true Eskimo who were driven north to the Arctic coast by other Indian tribes, but this seems to be an inverted reading of the course of events. There have been contacts between Chipewyan Indians and the Caribou Eskimo in the stretch of barrens west of Hudson Bay; also in Labrador, where the Montagnais and Naskapi Indians have adopted some Eskimo culture elements; and on the Yukon and Kuskokwim rivers in the interior of Alaska; but these exchanges do not appear to have fundamentally affected either party. In the short summer season, when the Eskimo

---

[3] Hrdlicka found that the older skulls on Kodiak Island were more Eskimoid than the modern Aleuts, the prehistoric Aleuts having longer and higher heads (see Fig. 8, p. 283).

migrate inland, live in skin tents and hunt the caribou, their life is
very much like that of the Indians; but their winter life in hunting
the marine mammals is an entirely different one which the Indians
cannot share, although the Nootka Indians on the west coast of Vancouver Island developed a whale-hunting industry of their own. The
Loucheux or Tukudh Indians near the Arctic circle in the Mackenzie
River basin differ from at least the great bulk of Indians in having
a lively, jolly disposition, much more like that of their neighbors, the
Eskimo. This may have been acquired through crossing with Eskimo.

Archaeological investigations have added much to our knowledge of
Eskimo history and have produced some surprises. Collins (1940,
1943) has contributed very useful recent discussions on this subject.
If we return to Siberia, there is a time interval between the oldest
Eskimo culture known and the Siberian Neolithic. Collins suggests
that during this interval the Eskimo were somewhere north or east of
Lake Baikal, and that they gradually extended (or were pushed)
northwards to the Arctic coast and east to the Bering Sea, with a
surviving Mesolithic-Neolithic culture long after it had been discontinued in other regions. There is no reason why the Tungus or some
Mongoloid tribe in Siberia should not have played a part in forcing
the ancestral Eskimos northward to the Arctic coast. It seems highly
probable that their very specialized ice culture was developed in that
region. Once they were tied to the ice in this way, their eastward
progress along the Canadian Arctic would naturally follow. The
Chukchi, Koryak, and Yukaghir have the Eskimo type of culture and
the Chukchi are physically very similar. They displaced the Eskimo
between Bering Strait and the Kolyma River. Probably at an earlier
period, before the reindeer breeding tribes arrived, the Eskimo marine
culture extended west to the Ob. It is to be hoped that someone will
test the blood groups of the Chukchi.

Southern and southwestern Siberia were peopled in the Paleolithic,
and the sources of Eskimo culture extend back to this period. Zolotarev (1938) states that the Barabiusky steppes, the Upper Irtysh,
the Ob and Yenesei regions and a narrow strip extending eastwards
to Yakutsk remained an unglaciated area surrounded by ice and
frozen swamps. The population of this pocket was isolated from
Europe and southern Asia, and here developed the original Paleolithic
culture of Siberia, man afterwards spreading northwards as the ice

disappeared. They dwelt on river banks and the seashore, hunting elk and reindeer in summer and fishing through holes in the ice in winter. They had underground lodges, pottery, deer skins, and they were dog breeders, matriarchal, with totemism, sun worship, and shamanism, this general culture being considered ancestral to that of America. The eastern coastal areas of Siberia formerly made pottery, as did the Thule culture Eskimos. Pottery making was only recently lost in western Siberia, as among the Eskimo. All the Siberian tribes formerly had pottery, but pottery-making was abandoned when they became nomadic. It persisted in the Kamchadals, Chukchi and other tribes into the eighteenth century. Dog breeding long antedated reindeer domestication, which is of recent origin, at least in Siberia. The wild dog probably became a camp follower.

The Tungus formerly had an ice-fishing culture like that of the Eskimos and Siberian tribes. This still persists in some Tungus on the northeast shores of Lake Baikal who have few reindeer. The transition from winter fishermen to forest hunters along the sea of Okhotsk only took place in the eighteenth century and was made possible by the invention of snowshoes (and skis). The Ostyaks and Samoyeds were still dog-breeders in the fifteenth century, but in the sixteenth century the Ostyaks borrowed reindeer from the Samoyeds, although some continued as dog-breeders until the nineteenth century. Thus it appears that all the elements of Eskimo culture, including that of pottery-making which they lost, were originally developed among the Siberian coastal or inland tribes. There is some evidence that even whale-hunting was practised by the Proto-Samoyeds, and semi-underground dwellings were widespread in Siberia until a few centuries ago.

The earliest Eskimo probably hunted whales little if at all, but when they reached Bering Strait and found an abundant food supply of seals, whales, and walrus, Eskimo culture experienced a sudden efflorescence, as was recognized by Rink. Their prehistoric culture was more elaborate than today, as shown by the buried Eskimo city of Ipiutak on Point Hope in northwestern Alaska, two hundred miles north of Bering Strait. This remarkable colony (Rainey, 1941) was developed on a sand-bar twenty miles from the mainland and probably had a population of several thousand for many years. It then disappeared completely and was unknown to the present Eskimo.

Excavations by Rasmusson showed rows of permanent square houses with a central hearth for driftwood and blubber, but no stone lamps. The walls were of driftwood, poles or even plank, with a long entryway, and the houses were not half underground.

There were five avenues of houses and cross streets, the whole stretching for nearly a mile. Their art was highly developed, including a double spiral motif comparable with some Ainu art-forms. Fantastic ivory carvings are unique and in great variety, including links of a chain. Stone implements were also beautifully chipped and unlike those of the Eskimo. An ivory knife-holder, with slots in which small flint blades were set, resembles Maglemose implements. This technique survived on Southampton Island in Hudson Bay. Over sixty graves were excavated. Three of the skulls had been given large ivory eyeballs inset with jade pupils.

This Neolithic people probably preceded the Old Eskimo and is claimed to be older than the Aztec or Maya civilizations. Harpoons and hunting equipment are rare and these may have been the summer houses of people who lived inland in winter. Apparently no study of their physical anthropology has yet been published. According to Collins (1943) the Ipiutak culture showed features reminiscent of Chinese post-Shang times, hence scarcely more than 2000 years old. Some Eskimo middens have been continuously occupied from the earliest known until the nineteenth century, and one on Little Diomede Island in Bering Strait is still occupied. The Diomede Islands form stepping stones across the strait. There is other evidence that the Eskimo occupation is recent: (1) there has been no change in sea level, as shown by the position of the middens; (2) the faunal content of the middens shows that the same land and sea mammals existed then as now, but there is an abundant Pleistocene fauna in the frozen muck near Eskimo sites. These are the extinct species which were hunted by Sandia and Folsom man. The Indians clearly crossed the strait long before the Eskimo.

The Thule culture, found by Mathiassen in excavations on Hudson Bay in 1922, was a whaling culture. The Bay became too shallow for whales (except the small white whale), and this may be why the Thule culture withdrew. It must have spread eastwards from Alaska, and it appeared in the central Arctic about a thousand years ago. From it was derived the Inugsuk stage, which made direct contact of Eskimos in western Greenland with the Norse colonists in southern

Greenland in the thirteenth and fourteenth centuries, when the climate was apparently less severe than now. Central Eskimo life was harder than in Alaska or Greenland, and here developed the greater specialization of the snow house and seal hunting through the ice. The Cape Dorset culture, recognized by Jenness in 1925, was already established in the eastern Arctic when the Thule arrived from the west. The Dorset was a primitive culture, lacking many Eskimo elements and based more on fishing than on hunting sea mammals. In 1926, Jenness excavated Prince of Wales Island and Little Diomede in Bering Strait, uncovering evidence of the Old Bering Sea culture, which was ancestral to the Thule and the modern, but different from both. This old culture flourished also on St. Lawrence Island, 150 miles further south. Although the oldest, it is highly developed. Excavations by Collins on St. Lawrence Island and nearby Punuk Island in 1928 revealed an intermediate (Punuk) culture contemporary with the Thule. The Birnirk culture near Point Barrow is a detached unit, contemporary with Punuk, from which the Thule arose. At one point on St. Lawrence Island a midden was found with a sequence of all these cultures. The occupation of the Alaskan coast has then been continuous since the Eskimos first arrived. No trace has been found of a pre-Eskimo population, but since the earlier Indians must have passed this way some indications of their passage may yet be discovered.

Blood-group tests of various Eskimo groups indicate that when of pure ancestry they may have about 40 per cent A but no B, and the most isolated central Eskimos may be all O. Thus while most of the Indians, of earlier origin, are O (except the Blackfoot who have A in very high percentage and possibly belong to the last Indian wave of immigration), the Eskimos have A, as would be expected from their later arrival, but not B, which is so characteristic of the present Mongoloid population of northeast Asia. However, in 1939 Fabricius-Hansen found 11 per cent of B in the supposedly pure Eskimo population around Angmagssalik in east Greenland—a result which requires a special explanation. Fresh blood grouping of the isolated central Eskimos from Chesterfield Inlet to Baffinland (Jordan, 1946) gives the following results:

Number	O	A	B	AB
369	43%	53	2.7	1.6

Of the 10 B and 6 AB, white ancestry is known in 7 B and 4 AB. It is thus highly probable that these Eskimos were devoid of B before crossing with whites began. Similarly 1.08 per cent were Rh negative, but of the five who were negative three were known to have white blood. The Eskimos in any case are like the Chinese in being nearly all Rh positive.

Various aspects of the problem of the interrelations between Eskimos and Indians were discussed at the Sixth Pacific Science Congress. Shapiro (1933) points out that the western Eskimos are taller, becoming shorter and more dolichocephalic as we pass eastward. He regards scaphocephaly, wide zygomatic arch, progressive dentition, mandibular and palatine tori, a massive mandible with wide rami, and everted gonia as adaptational, in connection with heavy demands on the masticatory apparatus. From comparative measurements of different series of Eskimo skulls, Shapiro finds the Eskimos heterogeneous, containing four, or possibly five, different types: (1) The early Thule type formerly uniform from Alaska to Greenland. These are succeeded by (2) the eastern type, in Greenland and Labrador, having a shorter cranium, a shorter and narrower face with a shorter and narrower nose. (3) The western (Alaska) type, with a shorter, broader, and lower head, shorter cranial base and longer face. (4) In southwestern Alaska and St. Lawrence Island, much shorter, broader heads and lower (chamaecephalic) cranium.

Shapiro finds these skull types very similar in measurements to different Indian tribes. He suggests that the eastern Eskimo are a blend of the Thule type and Indians, perhaps eastern Algonkins. The western type (from Alaska to Coronation Gulf), he thinks are best represented by Chipewyan, Cree, and Huron Indians, while the southwestern Eskimos he relates to the Buriats of Siberia. Because of these cranial similarities, which are marked, he derives most of the Eskimo from Indian ancestry or Indian mixture. If the resemblances are as significant as he believes, his conclusion does not necessarily follow. These Indian tribes could equally well represent original Eskimo immigration waves into Alaska, which subsequently spread over the continent and in doing so necessarily changed their culture. This would also account for the absence around Bering Strait of archaeological remains of Indians as distinct from Eskimos.

In a very recent study of prehistoric cultures in the Yukon region,

De Laguna (1947), enters into further details regarding the elements of the various known Eskimo cultures. She concludes that they were more different 2000 years ago than they are today, the modern Eskimo culture pattern having emerged from a number of distinct cultural beginnings which have since grown more and more alike. There was then no single original home of Eskimo culture.

REFERENCES

Andrews, E. 1875. *Am. J. Sci. Arts.* 10:32.

Barbour, E. H. 1907. Evidence of man in the loess of Nebraska. *Science.* 25:110–112.

Barbour, E. H., and H. B. Ward. 1906. Discovery of an early type of man in Nebraska. *Science.* 24:628–629.

Barcena, M. de la. 1885. Notice of some remains found near the City of Mexico. *Amer. Nat.* 19:739–744. Pls. 2.

Berthoud, E. L. 1866. Description of the Hot Springs of Soda Creek, . . . together with the remarkable discovery of a human skeleton and a fossil pine tree in the Boulder and Gravel formation of Soda Bar, Oct. 13th, 1860. *Proc. Acad. Nat. Sci. Phila.* 18:342–345.

Bird, Junius. 1938. Antiquity and migrations of the early inhabitants of Patagonia. *Georg. Rev.* 28:250–275. Figs. 29.

Bonin, G. von, and G. M. Morant. 1938. Indian races in the United States. *Biometrika.* 30:94–129.

Bowden, A. O., and I. A. Lopatin. 1936. Pleistocene man in California. *Science.* 84:507–508.

Bryan, Kirk. 1941. Geologic antiquity of man in America. *Science.* 93:505–514.

———. 1942. Pre-Columbian agriculture in the Southwest as conditioned by periods of alluviation. *Proc. 8th Amer. Sci. Congr.* 2:57–74.

———. 1945. Recent work on early man at the Gruta de Cadonga in the Argentine Republic. *Amer. Antiquity.* 11:58–60.

Bryan, Kirk, and L. L. Ray. 1940. Geologic antiquity of the Lindenmeier site in Colorado. *Smithson. Misc. Coll.* 99:no. 2. pp. 76, Pls. 6.

Bryan, W. A. 1929. The recent bone-cavern find at Bishop's Cap, New Mexico. *Science.* 70:39–41.

Clements, T. 1938. Age of the "Los Angeles man" deposits. *Am. J. Sci.* 36:137–141.

Colbert, E. H. 1942. The association of man with extinct mammals in the Western hemisphere. *Proc. 8th Am. Sci. Cong.* 2:17–29.

Collins, H. B., Jr. 1940. Outline of Eskimo prehistory. *Smithson. Misc. Coll.* 100:533–592. Pls. 6.

———. 1943. Eskimo archaeology and its bearing on the problem of man's antiquity in America. *Proc. Am. Phil. Soc.* 86:220–235. Figs. 5.

Cook, H. J. 1925. Definite evidence of human artifacts in the American Pleistocene. *Science.* 62:459–460.

Cressman, L. S., H. P. Hansen, and I. S. Allison. 1946. Early man in Oregon. *Sci. Monthly.* 62:43–65. Figs. 10.

DeLaguna, F. 1947. The prehistory of Northern North America as seen from the Yukon. *Mem. Soc. Am. Archaeol.* No. 3. pp. 370. Pls. 30. Figs. 33.

Densmore, F. 1928. Uses of plants by the Chippewa Indians. 44th *Ann. Rept. Bur. Amer. Ethnol.* 1926–27. pp. 275–397. Pls. 36.

Dickeson, M. W. 1846. *Proc. Acad. Nat. Sci. Phila.* 3:107.

Dixon, R. B. 1923. *The Racial History of Man.* New York.

Eiseley, L. C. 1946. Men, Mastodons and myth. *Sci. Monthly.* pp. 517–524.

Fabricius-Hansen, V. 1939. Blood groups and MN types of Eskimos in East Greenland. *J. Immunol.* 36:523–530.

Figgins, J. D. 1927. The antiquity of man in America. *Nat. History.* 27:229–239. Figs. 8.

———. 1927. New geological and paleontological evidence bearing on the antiquity of man in America. *Nat. History.* 27:240–247. Figs. 9.

———. 1928. Further evidence concerning man's antiquity at Frederick, Oklahoma. *Science.* 67:371–373.

Furlong, E. L. 1907. Reconnaissance of a recently discovered Quaternary cave deposit near Auburn, California. *Science.* 25:392–394.

Gates, R. R., and G. E. Darby. 1934. Blood groups and physiognomy of British Columbia coastal Indians. *J. Roy. Anthrop. Inst.* 64:23–44. Pls. 5.

Gidley, J. W. 1926. Fossil man associated with the mammoth in Florida. *J. Wash. Acad. Sci.* 16:310.

———. 1927. Investigating evidence of early man in Florida. *Smithson. Misc. Coll.* 78: No. 7. pp. 168–174. Figs. 7.

Gidley, J. W., and F. B. Loomis. 1926. Fossil man in Florida. *Am. J. Sci.* 12:254–264.

Gould, C. N. 1929. On the recent finding of another flint arrowhead in the Pleistocene deposit at Frederick, Oklahoma. *J. Wash. Acad. Sci.* 19:66–68.

Harrington, J. P. 1940. Southern peripheral Athapaskawan origins, divisions, and migrations. *Smithson. Misc. Coll.* 100:503–532. Figs. 4.

Hay, O. P. 1917. On the finding of supposed Pleistocene human remains at Vero, Florida. *J. Wash. Acad. Sci.* 7:358–359.

———. 1923. The Pleistocene of North America . . . east of the Mississippi. *Carnegie Inst. Publ.* No. 322. pp. 499.

———. 1928. On the antiquity of relics of man at Frederick, Oklahoma. *Science.* 67:442–444.

———. 1928. On the recent discovery of a flint arrowhead in early Pleistocene deposits at Frederick, Oklahoma. *J. Wash. Acad. Sci.* 19:93–98.

Hay, P. R. 1871. *Am. Nat.* 5:147–148.

Heilprin, A. 1887. Explorations on the west coast of Florida and in the Okeechobe wilderness. *Trans. Wagner Free Inst. Sci.* pp. 134. Pls. 19.

Hibben, F. C. 1941. Evidences of early occupation in the Sandia cave, New
 Mexico, and other sites in the Sandia-Manzano region. *Smithson. Misc. Coll.*
 Vol. 99, no. 23. pp. 64. Pls. 15. With appendix by Kirk Bryan.
———. 1942. Pleistocene stratification in the Sandia cave, New Mexico.
 Proc. 8th Am. Sci. Congr. 2:45–48.
Hoijer, Harry, and others. 1946. Linguistic structures of native America.
 Viking Fund Publ. in Anthrop. Vol. 6. pp. 423.
Holmes, F. S. 1859. Collection of fossils from the post-Pliocene of South
 Carolina. *Proc. Acad. Nat. Sci. Phila.* 11:177–186.
Howard, E. B. (Ed.). 1936. Early man in America with particular reference
 to the southwestern United States. *Am. Nat.* 70:313–371.
Howard, H., and A. H. Miller. 1939. The avifauna associated with human
 remains at Rancho La Brea, California. *Carnegie Publ.* No. 514. pp. 39–48.
Hrdlička, A. 1907. Skeletal remains suggesting or attributed to early man
 in North America. *Bur. Am. Ethnol., Bull.* 33. pp. 113. Pls. 21. Figs. 16.
———. 1926. *Smithson. Misc. Coll.* 78:61.
———. 1932. The coming of man from Asia in the light of recent discoveries. *Proc. Am. Phil. Soc.* 71:393–402.
———. 1937. Early man in America. In *Early Man* (ed. G. G. McCurdy),
 p. 103.
———. 1942. The problem of man's antiquity in America. *Proc. 8th Am. Sci.
 Congr.* 2:53–55.
Hrdlička, A., and others. 1912. Early man in South America. *Bur. Am.
 Ethnol. Bull.* 52:1–405. Pls. 68. Figs. 51.
Imbelloni, J. 1940. Sobre craneologia de los Urus. Supervivencias de razas
 australoides en los Andes. *Actas y Trabajos Cient. del XXVII Congr.
 Internac. de Americanistas.* Lima. 1:3–22. Figs. 9.
———. 1943. The peopling of America. *Acta Americana.* 1:309–330.
 Figs. 2.
Jenks, A. E. 1932. Pleistocene man in Minnesota. *Science.* 75:607–608.
———. 1933. Minnesota Pleistocene Homo, an interim communication.
 Proc. Nat. Acad. Sci. 19:1–6. Figs. 3.
———. 1936. *Pleistocene Man in Minnesota.* University of Minnesota Press.
 pp. 197. Figs. 89.
Jenness, D. 1933. Origin and antiquity of the American aborigines. *Proc. 5th
 Pacif. Sci. Congr.* 1:739–747.
Jochelson, W. 1926. The ethnological problems of Bering Sea. *Nat. History.*
 26:90–95. Figs. 3.
Jordan, D. 1946. Blood grouping and Rh factor in Eskimos. *Can. Med.
 Assn. J.* 54:429–434.
Koby, F. (Ed.). 1945. Un squelette d'ours brun du pléistocène italien.
 Verh. Naturf. Ges. Basel. 56:58–85. Figs. 5.
Koch, A. 1841. Description of the Missourium, or Missouri leviathan.
 Pamphlet, St. Louis. Later editions in London and Dublin.

Koch, A. C. 1857. Mastodon remains in the State of Missouri, together with evidence of the existence of Man contemporaneously with the Mastodon. *Trans. Acad. Sci. St. Louis.* 1:61–64.

Leidy, J., and J. F. Clew. 1866. *Proc. Acad. Nat. Sci. Phila.* 18:109.

Lipschutz, A., G. Mostny, L. Robin, and A. Santiana. 1946. Blood groups in tribes of Tierra del Fuego and their bearing on ethnic and genetic relationships. *Nature.* 157:696–697.

Lipschutz, A., G. Mostny, and L. Robin. 1946. The bearing of ethnic and genetic conditions on the blood groups of three Fuegian tribes. *Am. J. Phys. Anthrop.* N. S. 4:301–320. Pls. 5.

Merriam, J. C. 1909. Note on the occurrence of human remains in Californian caves. *Science.* 30:531–532.

———. 1934. Present status of knowledge relating to antiquity of man in America. *Rept. XVI Internat. Geol. Congr., Washington.* 1933. pp. 11.

Montagu, M. F. A., and C. B. Peterson. 1944. The earliest account of the association of human artifacts with fossil mammals in North America. *Proc. Am. Phil. Soc.* 87:407–419.

Morant, G. M. 1937. A contribution to Eskimo craniology based on previously published measurements. *Biometrika.* 29:1–20.

Morley, S. G. 1946. *The Ancient Maya.* Stanford University Press.

Pei, W. C. 1939. On the Upper Cave industry. *Peking Nat. Hist. Bull.* 13:175–179. Pls. 3.

Rainey, F. G. 1941. Mystery people of the Arctic. *Nat. History.* 47:148–155. Figs.

Renaud, E. B. 1942. Cultures of the Black's Fork Basin, Southwest Wyoming. *Proc. 8th Amer. Sci. Congr.* 2:49–52.

Roberts, F. H. H., Jr. 1935. Preliminary report on investigations at the Lindenmeier site in Northern Colorado. *Smithson. Misc. Coll.* Vol. 94, No. 4, pp. 35. Pls. 16. Figs. 3.

———. 1936. Additional information on the Folsom complex. *Ibid.* Vol. 95. No. 10. pp. 38. Pls. 12. Figs. 5.

———. 1940. Developments in the problem of the North American Paleo-Indian. *Ibid.* 100:51–116.

———. 1942. Recent evidence relating to an early Indian occupation in North America. *Proc. 8th Am. Sci. Congr.* 2:31–38.

Rousseau, J., and M. Raymond. 1945. Études ethnobotaniques Québécoises. *Contrib. de l'Inst. Bot. de l'Univ. de Montreal.* No. 55. pp. 154.

Rusconi, C. 1947. Ritos funerarios de los indigenas prehistoricos de Mendoza. *Anales Soc. Cient. Argentina* 143:97–114. Figs. 6.

Sauer, C. O. 1944. A geographical sketch of early man in America. *Geogr. Rev.* 34:529–573. Figs. 3.

Sellards, E. H. 1916a. Human remains from the Pleistocene of Florida. *Science.* 44:615–617. Fig. 1.

———. 1916b. On the discovery of fossil human remains in Florida in association with extinct vertebrates. *Am. J. Sci.* 42:1–18. Figs. 12.

———. 1937. The Vero finds in the light of present knowledge. In *Early Man* (ed. G. G. McCurdy). pp. 193–210. Pls. 2. Figs. 4.

———. 1945. Fossil bison and associated artifacts from Texas. *Bull. Geol. Soc. Am.* 56:1196–1197.

Shapiro, H. L. 1933. Some observations on the origin of the Eskimo. *Proc. 5th Pacif. Sci. Congr.* 4:2723–2732.

Simpson, G. G. 1940. Review of the mammal-bearing Tertiary of South America. *Proc. Am. Phil. Soc.* 83:649–709. Figs. 4.

———. 1943. The beginnings of vertebrate paleontology in North America. *Proc. Am. Phil. Soc.* 86:130–188. Figs. 23.

Spinden, H. J. 1942. Time scale for the New World. *Proc. 8th Am. Sci. Congr.* 2:39–44.

Stewart, T. D. 1935. Skeletal remains from Southwestern Texas. *Am. J. Phys. Anthrop.* 20:213–231.

———. 1940. Some historical implications of physical anthropology in North America. *Smithson. Misc. Coll.* 100:15–50. Figs. 3.

———. 1942. The historical significance of relative head height among North American Indians. *Proc. 8th Am. Sci. Congr.* 3:235–236.

———. 1943. Distribution of cranial height in South America. *Am. J. Phys. Anthrop.* N. S. 1:143–155. Figs. 2.

———. 1946. Re-examination of the human skull found by Gidley and Loomis in association with a Pleistocene fauna at Melbourne, Florida. *Am. J. Phys. Anthrop.* N. S. 4:259.

Stock, C. 1924. A recent discovery of ancient human remains in Los Angeles, California. *Science.* 60:2–5.

———. 1931. Problems of antiquity presented in Gypsum Cave, Nevada. *Sci. Monthly.* 32:22–32. Figs. 11.

Strain, I. G. 1844. Translation of a letter from Dr. Lund, Copenhagen, to the Historical and Geographical Society of Brazil on organic remains recently discovered in the calcareous rocks in Minas Geraes, Brazil. *Proc. Acad. Nat. Sci. Phila.* 2:11–14.

Sullivan, L. R., and M. Hellman. 1925. The Punin calvarium. *Anthrop. Papers, Am. Mus. Nat. Hist.* 23:311–337. Figs. 9.

Walter, H. V., A. Cathoud, and A. Mattos. 1937. The Confins man. In *Early Man* (ed. G. G. McCurdy). pp. 341–348. Pls. 2.

Webster, C. L. 1889. Ancient mounds at Floyd, Iowa. *Am. Nat.* 23:185–188. Pl. 1.

Weidenreich, F. 1939. On the earliest representatives of modern mankind recovered on the soil of east Asia. *Peking Nat. Hist. Bull.* 13:161–174. Pls. 6.

Whitney, J. D. 1880. The auriferous gravels of the Sierra Nevada of California: the fossils of the auriferous gravel series. *Mem. Mus. Comp. Zool. Harvard Coll.* 6:219–288. Pl. 1.

Williams, E. T. 1918. The origins of the Chinese. *Am. J. Phys. Anthrop.* 1:183–211.

Winslow, C. F. 1873. *Proc. Boston Nat. Hist. Soc.* 15:257–259.
Witter, F. M. 1891. Notice of arrow points from the loess in the city of Muscatine. *Proc. Iowa Acad. Sci.* 1:66–68.
Zolotarev, A. 1938. The ancient culture of North Asia. *Am. Anthrop.* 40:13–23.

10

Polynesians, Melanesians, and Negroes

Having now surveyed the distribution and history of man in the six continents, there remain the islands of the Pacific, most of which were peopled very late. The present populations are generally classified roughly as Melanesians, Micronesians, and Polynesians. The term Indonesians, frequently used, has a linguistic basis and applies not only to Sumatra, Java, and Bali, but to Borneo, Celebes, and the Philippines, as well as to Madagascar across the Indian Ocean. The Hindus of India, partly through migration, have exerted a strong influence on Java over a long period. Indian expansion and rule in Indonesia began in the second century A.D. or earlier, and Hindu kingdoms were formed in Java (see Cole, 1945). In some parts of Indonesia the Neolithic culture continued until the Christian era and in other parts much later. A considerable Mediterranean (Brown) element is thus recognized in the population, which has (like that of India) increased enormously under European rule.

Before considering the racial adventures of the Polynesians and Indonesians, some of which began about the time of Christ, it will be well to emphasize the enormous increase in human population which has taken place in most parts of the world in recent centuries. The present population of the earth is over 2,000,000,000 and still increasing rapidly owing to improved conditions and increased food supplies. This condition is quite exceptional and cannot go on indefinitely. In man's early history there were doubtless long periods when the population was sparse and stationary, in equilibrium with the environment, as is true of most animal species. Before Neolithic agriculture began, about 6000 B.C. in Iraq and Egypt (it probably began independently in different parts of the world), man had no means of increasing his food supply except by improved methods of hunting, there was little possibility of accumulating reserves, and practically every family lived at the limits of subsistence.

It has been estimated (Howells, 1944) that the world population in A.D. 1600 was 400 million; in other words, it has increased five-fold in less than three and a half centuries, and the rate of increase in the last century has been wholly abnormal. At the end of the Paleolithic it was probably not over 10 million, already scattered in the five continents. This number is comparable with that of many species of mammals, and was the "natural" condition before man began producing his own food through agriculture and the breeding of animal herds. According to the latest chronology, based on varves and pollen content in Switzerland (Deevey, 1946), the Mesolithic began about 4300 B.C., the Neolithic 3200 B.C., and pollen of cereals, indicating cultivation, is found as early as 3000 B.C. That is to say, five thousand years ago man in Europe and parts of the East was passing well out of the conditions of a natural species and beginning his phenomenal rise in numbers. Nevertheless, even during most of human written history populations have remained relatively stable. In prehistory the isolation was such that evolutionary processes could produce their full effect over long periods, only occasionally disturbed by influx and crossing with another racial group. The truly evolutionary changes have happened in widely separated areas such as Java, northern China, and Africa. In the Lower Pleistocene great areas of the world had no human population at all, and the rest contained only widely scattered and sparse populations of hunters.

In modern populations, Michelot estimated in 1845 that the annual increase in world population was 0.7 per cent. At that rate the population would double in less than a century. Carr-Saunders (article "Population" in *Encyclopaedia Brittanica*, 14th Ed., 1929) estimated that in twenty-six advanced countries in the years 1906–1911 the population increased at the rate of 1.1 per cent per annum. At this rate it would double in sixty years. All increases depend ultimately upon food supply, but many other factors such as climate and customs have a decisive effect. The Eskimos number some 30,000, but were probably more numerous fifty or a hundred years ago. In Greenland the death-rate is 33.6 and in New Zealand 9.1 per thousand. The population of Java was estimated at 5,000,000 in 1835, and in 1929 it was 36,403,833, a seven-fold increase in a century. Even in the same country different racial elements multiply at very different rates, and populations are frequently transformed in this way. Thus in India the Hindu and Moslem elements have increased at a much more rapid rate than the jungle tribes. A different birth rate can almost completely replace one racial element of a country by another without any appreciable change in the total population. Such changes by silent replacement are taking place today in many parts of the world.

The blood groups furnish important evidence regarding general racial movements in Indonesia and across the great stretches of the Pacific. The B has become high not only throughout eastern Asia (except for certain primitive surviving tribes) but also in many parts of Indonesia and Melanesia. The Polynesians, on the contrary, have reached a very high frequency of A while having little or no B. In Table 22 it will be seen that natives of Hawaii and Easter Island have 60 per cent of A, the small amount of B probably resulting from crossing. Hawaii, Easter Island, and New Zealand form a vast Pacific triangle over which the Polynesians navigated in their giant canoes with the aid of the stars. The distance from Hawaii to New Zealand is some 5000 miles, and from Java to Hawaii 6800 miles. The New Zealand Maoris came from Hawaii but they have roughly 40 per cent rather than 60 per cent of A, being correspondingly higher in O. Since a few boat loads constituted the whole population landing in New Zealand, probably the chance composition of those immigrants in A and O determined that their descendants would have a lower frequency of A than the Hawaiians. In line four of the table it will be seen that the tribe of Maoris known as Te Arawa are extraordinarily high in AB while very low in B. The Maoris found in New Zealand a preceding people, the Moriori, whom they largely exterminated; but it is stated that the Te Arawa tribe of Maoris intermarried with them. If the Moriori were Melanesians high in B this condition in the Te Arawa would thus be accounted for.

The origin of the Polynesians remains something of a mystery, although its authentication is almost historic. Originating in India or Java, where the most primitive men we know evolved, the proto-Polynesians are believed to have launched into the Pacific from Java about the time of Christ, so their epic is only a few centuries earlier

Table 22
BLOOD GROUPS OF SOME POLYNESIANS, MELANESIANS, AND NEGROES *

	No.	O	A	B	AB	Author
Polynesia, Hawaii, Oahu	413	36.5	60.8	2.2	0.5	Nigg
Polynesia, Easter Island	63	25.4	69.8	3.1	1.6	Rahm
Polynesia, New Zealand Te Arawa	127	38.6	41.7	0.8	18.9	Phillips
Polynesia, other Maori	73	63.0	35.6	1.4	0	Phillips
Polynesia, Samoa	500	58.6	17.0	19.4	5	Stephenson
Melanesians, New Guinea	753	53.7	26.8	16.3	3.2	Heydon and Murphy
Melanesians, Schouten Islands	1359	63.2	16.9	17.6	2.2	Bos
Java, Modjokerto	1126	37.2	24.4	32.	6.4	Buining
Moluccas	450	51.0	27.0	18.7	3.5	Bijlmer
Celebes, Macassar	195	28.7	29.7	30.8	10.8	Lehmann
Celebes, Buginese	217	34.6	30.4	27.6	7.4	Lehmann
Micronesia, Saipan	293	50.5	33.8	14.0	1.7	Takasaki
Micronesia, E. Carolines	111	34.2	44.2	16.2	5.4	Takasaki
Micronesia, Yap, W. Carolines	213	57.7	20.3	17.8	4.2	Takasaki
Yoruba, Negroes, W. Africa	325	52.3	21.5	23.0	3.2	Muller
Negroes, Belgian Congo	500	45.6	22.2	24.2	8.0	Bruynoghe and Walravens
Negroes, French Congo	400	41.0	27.0	26.0	6.0	Liodt and Pojarski
Pygmies, Belgian Congo	2045	28.8	33.1	28.6	9.4	Jadin, Julien
Bantu Pondo	500	42.6	33.2	22.8	1.4	Elsdon-Dew
Bantu Swazi	500	61.6	19.8	17.4	1.2	Elsdon-Dew
Bantu Zulus	1630	52.0	26.0	20.0	2.0	Elsdon-Dew Pirie, Pijper
Hottentots	506	34.8	30.6	29.2	5.3	Pijper, 1935
Bushmen	548	57.1	29.5	6.8	6.6	Pijper
Madagascar (Imerina, Hova)	266	45.5	27.5	22.5	4.5	Hirszfeld and Hirszfeld
Imerina (blacks)	100	42.0	30.0	23.0	5.0	Herivaux and Rahoerson
Imerina (nobles)	61	35.0	31.7	16.7	18.0	Herivaux and Rahoerson

* In several cases where the results of different authors for the same race were very similar they have been added together, as indicated by the authors cited. The results are taken from the list published by Boyd (1939).

than the conquest of England by Angles, Saxons, Danes, Vikings, and Jutes. Their development thus comes entirely within the historic period. The origin of the Caucasoid element, which is evident in their somatology, is by no means clear. Their rather light brown skin color could be Mediterranean, but they are larger and heavier than the usual Mediterranean type, tending to corpulence.

Percy Smith (1921) traces Polynesian history, by means of traditional genealogies and other means, back to the fourth century B.C., when he finds them on the shores of India. The Aryan invasion of India is frequently dated 1400 B.C. but Toynbee (1935) puts it as early as 1900–1700 B.C. There is thus an interval of a thousand to fifteen hundred years during which groups of Aryans could have moved south and become the ancestral stock of the Polynesians. They may have already cultivated the habit of keeping verbal genealogies through the generations, but it would appear that if there is a connection between the proto-Polynesians and the Aryans all tradition of it has been lost.[1] Nevertheless, this seems the most likely origin of the Polynesian race.

A cave near Sampung in Madiun, Java, was discovered by van Es in 1926. In the floor were two buried skeletons with horn and bone arrowheads, fish hooks, and other implements. These people were tall, 170–180 cm. Many Neolithic stone axes were found, but the megalithic monuments are later, belonging to the metal periods.

The archaeology of Indonesia is amply treated by Heine-Geldern (1945), who throws further light on the origin of the Polynesians and their culture. A full account is given, with an extensive bibliography. As late as 1924, Callenfels discovered the first hand-axes in the late Paleolithic of Sumatra. Caves explored here contained Mesolithic and partly Neolithic cultures, and similar cultures have been found in Java, Celebes, and Borneo, which are related to cultures in French Indo-China and the Malay Peninsula. The full Neolithic is less known than this pre-Neolithic, but there is evidence that the Neolithic population of Java was relatively dense. Heine-Geldern believes that the southeastern Asiatics (Austronesians), already having considerable Mongoloid mixture which had come down into Assam and Burma, migrated westwards into India and introduced the tanged adze between 2500 and 1500 B.C., before the Aryan invasion. Percy Smith (1921) recognizes a Gangetic race in northern India before the Aryan invasion. He believes that a Himalayo-Polynesian race, allied to the Chinese and Tibetan, formerly spread over the Gangetic basin from further India.

These Austronesians or Malayo-Polynesians introduced the quad-

---

[1] Rivet (1930) shows many similarities, not only between the languages of Melanesia, Indonesia, and Polynesia, but also with the Munda and Mom-Khmer as well as the Australian and Tasmanian tongues. He connects the languages of Oceania with Sumerian, 5000 B.C.

rangular adze culture from Malaya into Indonesia. Heine-Geldern derives the Polynesians from the coast of Malaya and their quadrangular adze culture from China via Burma and Malaya. The outrigger canoe, which made possible the spread which took place as far as Madagascar in the west and Easter Island in the east, was developed on the Malayan coast from primitive bamboo outriggers such as are still found in the Shan States of northern Burma and the middle reaches of the Mekong river. Heine-Geldern believes that the megaliths of Polynesia were introduced into Indonesia and southwestern Asia with the quadrangular adze culture between 2500 and 1500 B.C. There is evidence of direct contact of Sumatra with China, in the form of Han pottery, probably in the third century B.C. or even earlier.

If Heine-Geldern is correct regarding the racial source of the Polynesians, it is difficult to see why they are so nearly allied to the Caucasians and why they are not more Mongoloid.[2] They should also, on this hypothesis, have the B blood group from their Mongolian ancestry. The Samoans, unlike other Polynesians, are not only more Mongoloid, but they have 19–26 per cent of B (Gates, 1946, p. 715). The evidence indicates that they are Polynesians who had a later accession of blood from southern Chinese. The Polynesians may have begun with a small amount of B which was subsequently lost as they spread across the Pacific, but this will not explain their lack of Mongoloid characters as well.

The genealogies of the Polynesians, though not written down, were kept in memory with remarkable accuracy after leaving Java. Allowing twenty-five years to a generation gives a time-scale for the family genealogies. The Maoris, Rarotongans, Tahitians, and Hawaiians have many ancestors in common, among them the famous voyagers, Hua and Whiro. Genealogies from these two have been preserved down to the nineteenth century. In Hawaii there are twenty-five generations between the death of Hua and 1900, while in Tahiti there are twenty-five from Whiro, in Rarotonga twenty-six from Whiro, and in New Zealand twenty-six from Whiro and twenty-six from Hua. Six centuries of history can thus be traced with reliability back to A.D. 1300.

[2] This could be explained, however, if the Mongoloids first entered this area at a later period.

According to Buck (1922), while the wanderings of the Polynesians began at about the beginning of the Christian era, the first important emigration was about A.D. 600. These had apparently been preceded in Polynesia by a more primitive non-agricultural people. A second migration about A.D. 1000 initiated the great Polynesian expansion—an eastward drift from the Malay Archipelago (Indonesia) and southern Asia. Buck considers that the proper application of the term Indonesian is to the Caucasoid type in the East Indies, from which the Polynesians arose. Howells (1944) points out that Hawai(ki) and Java are really the same word, showing that the tradition of their Javan origin must have been retained until they reached Hawaii several generations later. According to Percy Smith, this was about A.D. 650. During the eleventh century, and for five or six generations thereafter, parties from the Marquesas, Society, and Samoan Islands arrived in Hawaii (A.D. 1100–1325) and maintained intercourse over these long distances. Samoans and Tongans are believed to have been the first to enter the Pacific, before A.D. 450. The Tonga-Fiji migration is dated about A.D. 600. Then, according to Smith, a new wave reached the Marquesas in 675, Tahiti before 850, New Zealand about 850, Rarotonga 875, Paumoto A.D. 1000. The main migration from Tahiti to New Zealand with a halt at Rarotonga was about 1350, and the Moriori had been driven from New Zealand to the Chatham Islands before this date. Earlier immigrants, about A.D. 1150, had multiplied, and there were other drift-voyagers. The larger immigration in the middle of the fourteenth century was an intentional settlement which introduced kumara (the sweetpotato), taro, yams, the dog, and presumably the pig. The Moriori were probably of Melanesian affinity. The remnant on the Chatham Islands were conquered by a war party of Maoris in 1835 and are now nearly extinct. If any remain, even of mixed descent, their blood groups should be taken. In a cave near Christchurch, New Zealand, the remains in the lowest stratum show that the inhabitants, presumably Moriori, hunted the Moa. The coast has since sunk several feet. These Moa-hunters had a higher standard of craftsmanship than the later Maoris of Otago.

There is a tradition that these predecessors of the Maori were darkskinned, with woolly hair and a very flat nose. The nasal index of Moriori skulls from the Chatham Islands was, however, only 46.1, whereas that of New Zealand whites was 62.6 and of Tongans (with a much larger nose) 77.6. The nose of the Maori infant is said to be massaged to make it narrower. Buck measured Maori soldiers in New Zealand in 1919. He found that a dark face may go with a lighter body skin color. The hair is straight or wavy, and in some pre-Maoris it is said to have been frizzy, indicating some Negrito admixture. Fair hair is believed to have been derived from Pipi, a woman who lived twenty-four generations ago. There are legends of persons with red hair and a fair skin in the pre-Maori race. Buck saw three albinos in the Maori tribes. The hair is generally black, but in one full Maori it was reddish-brown. Among 424 soldiers examined, nine had dark-brown hair and all but one of these had also somewhat lighter skin, indicating possible linkage of genes affecting hair and skin color. The eyes were generally dark-brown or medium-brown, but occasionally light-brown or black. The average height was 170.6 cm., the cephalic index 77.7, ranging from 70–85, generally mesaticephalic. In white New Zealanders the C.I. was 78.4. The face of the Maori is broad, the facial index 85.1 (range 73–97).

Handy (1943) concludes that the breadfruit was probably introduced by the Polynesians to the Marquesas directly from Malaysia where it is native. In the Marquesas there are over 250 varieties, which are distinguished by native names. These may have all developed from cuttings in a period of six centuries. There is a clear tradition that breadfruit was introduced from the Marquesas to Rarotonga. There are over fifty varieties in the Society Islands, fewer in the Cook Islands, and only one in Hawaii, which was introduced from the Society Islands.

Megalithic structures, known as marae, are scattered everywhere in Polynesia east of Tonga and Samoa, as in many other parts of the Old World. They are supposed to have been built by ancestors of the historical Polynesians, but some in Hawaii were of late origin. Elliot Smith derived them from the early type of step-pyramid in Egypt, through diffusion, but this is not accepted. They have been studied in Tahiti and Hawaii. Although Pitcairn Island was uninhabited when the mutineers of the "Bounty" landed there in 1789, there is evidence of a former Polynesian settlement. Marae and stone figures on Pitcairn resemble those which are so numerous on Easter Island. Large stone chisels and axes were also found.

We have seen that the Polynesians of Hawaii and Easter Island and the pure Maoris of New Zealand have little or no B blood group. In a few further tests of Easter Island natives, Sandoval (1945) found that 12 were O and 10 were A_1. As regards the MN, there were 1M and 3N to 8MN, and 4 of 12 tested for Rh were negative.

The modern Javanese (Table 22) have over 30 per cent of B, like the Hindus. It would appear then either that the Javanese have developed their B within the last two thousand years or that the Polynesians began with a small amount of B, which they lost and left behind almost entirely in their migrations between Java and Hawaii. In any case there would appear to have been a very rapid increase of B in Javanese during the last two thousand years, when the Polynesians in Java belonged to a different stock. These questions can only be fully answered with further analysis of the races involved. Some writers recognize a Mongoloid element in the Polynesians, but it must be proto-Mongoloid because of the absence of the B blood group.

In their movements across the Pacific archipelagos there has evidently been a variety of intermixtures of Polynesian, Indonesian, and Melanesian elements, many of which have not yet been disentangled. One clear conclusion which emerges is that the Samoans are a mixture of Hawaiians and Mongoloids, perhaps pre-Chinese. This is shown not only by their physical appearance, which is a "blend" of Polynesian and Mongolian, but also is confirmed by their blood groups (Table 22). They have nearly 20 per cent of B. There has long been a movement of Chinese into parts of the western Pacific. In a short analysis of the Polynesian peoples, Sullivan (1923) has considered the records from Samoa, Tonga, Marquesas, and Hawaii. Like Dixon, he recognizes four elements, but he calls them Indonesian, Melanesian, Polynesian, and Polynesian with deformed (flat) heads, rather than Negrito, Melanesian, Caucasian, and Malay. Indonesians may be regarded as a mixture of the Polynesian (Brown) race with Negritos; and the short-headed element would appear to be of Chinese origin, equivalent to the "Proto-Armenoids" of Elliot Smith. The Polynesian element is found in all parts of the Pacific, the Indonesians in Samoa, Tonga, Marquesas, and Hawaii. The Indonesians are believed to have dispersed in the Pacific for the most part later than the Polynesians, while a Melanesian element is present in the southern and western Pacific and in Tonga. Melanesians in New Guinea show about 16 per cent B blood group (Table 22) while those more isolated in the Schouten Islands north of New Guinea have the same amount of B but are extremely high in O (63 per cent) and low in A (17 per cent). New Guinea contains a great variety of types, most of which still require analysis. In the Moluccas B is about 19 per cent, but in Celebes it goes up to 30 per cent. Further north, in the Carolines, B drops to about 16 per cent and in Yap O rises to 58 per cent.

In an excellent recent account of the Polynesians, with full references to the literature, Buck (1945) points out that in journeying eastwards from Malaysia they introduced the pig, dog, and fowl and a number of plants from this region. These included coconut, breadfruit, banana, taro, yam, and arrowroot. The sweet potato alone came from South America, as already mentioned. He believes it was introduced by some Polynesian navigator who was blown to the coast of South America by a gale and carried westwards again by the trade winds. The paper mulberry (Brousonettia), used in making tapa or bark cloth, was also introduced from Indomalaya.

Buck considers the two possible routes by which the Polynesians could migrate into the Pacific when driven from the mainland by Mongoloid pressure. The more northern route would be through Micronesia, the more southerly via Melanesia. He supports the former route because the Micronesians and Polynesians both had the sling, whereas the Melanesians used the bow and arrow which were unknown to the Polynesians. As we have already seen, the lack of B blood group in most Polynesians would also preclude any great amount of contact with the Melanesians. Buck therefore traces the main early migration through the Gilbert group to central Polynesia, with minor streams to Samoa and Tonga. The Indomalayan food plants and animals, which reached Fiji through the Melanesian chain, could have passed to central Polynesia (the Society Islands) later via Samoa. The early migrants into Micronesia would then have subsisted without their own native plants and animals. In any case, the Society Islands apparently became the ancient source and center of the later spread over the vast eastern Pacific triangle. The island named Havaii (now Raiatea) in this group became the culture center, and from here the Polynesians colonized Hawaii in the north, New Zealand in the south, and Easter Island in the east. These navigators took no animals or plants with them, so they probably left the Society

Islands before the Indomalayan plants and animals reached them. By the fourteenth century all Polynesia had been populated. Local plants such as Pandanus were adopted for making mats and baskets. On Easter Island, rushes were sewn together and used as a substitute, and in New Zealand *Phormium tenax* served the same purpose. At a later period Tahiti became more powerful and populated as a center of Polynesian culture than the Society group.

The Indonesians, apparently stimulated by their Polynesian admixture, circumnavigated the Indian Ocean in outrigger canoes to eastern Africa, from whence they finally reached Madagascar. Had they taken a southerly route directly across the Indian Ocean via the Seychelles with its giant tortoises, or Mauritius, home of the flightless dodo, and Reunion, formerly inhabited by the solitaire, these defenceless animals would have been exterminated. These rich tropical islands, once discovered by man, would have remained inhabited. But they had no human inhabitants, when first discovered by Europeans early in the sixteenth century.

In a valuable study of the culture sequences in the great island of Madagascar, which is over a thousand miles long, Linton (1943) finds that no other part of the world can show a comparable mixture of racial types; yet the languages are singularly uniform, with only dialect differences. In the population of Madagascar there are groups with strong Caucasoid, Mongoloid, and Negroid affinities, and others too mixed to analyze their elements. The earliest element was possibly the Negrito from Africa, crossing by a chain of islands. Next came the brown Caucasian Indonesians in their outrigger canoes. The Imerina have a tradition of landing on the Africa side of Madagascar. Indonesians who came later had more Negro blood, picked up during their sojourn in Africa. Natural selection acted on that racial mixture, for the Indonesian type was later bred out by their susceptibility to malaria, which is mild for the blacks. The Indonesian survivors now occupy the central plateau, which is free from malaria, while the Malagasy blacks, being highly resistant, occupy the lower areas. A "fever line" separates the races. This very striking case of a difference in racial susceptibility to malaria is worthy of further study. The blood-group differences between these races have not been determined.

These Indonesian migrations continued over a long period, but the last of them is believed to have taken place soon after the beginning of the Christian era and before Hindu culture had spread much in Indonesia. This westward movement thus preceded, for the most part, the eastward migrations of the Polynesians into the Pacific. The later Indonesian immigrants into Madagascar were more Mongoloid, through the previous infiltration of Mongoloids into the countries from which they set out. These migrants were looking for a land where there was no death (from fever), so their susceptibility to malaria was the stimulus which initiated their migrations. The Imerina dialect has a few words which are probably derived from Sanskrit. In more recent times, beginning about the twelfth century, there has been a steady infiltration of Arabs into eastern Africa (Zanzibar) and northwestern Madagascar. There are a few Arab ruins on the west coast, not earlier than the fourteenth century. Apparently the Indonesians formerly covered most of southern Madagascar. The Hovas are the lower castes of the Imerina. In the southeast there are tribes of Arab origin, which have degenerated to the native culture level, lost their Mohammedanism and assumed totems, only retaining a knowledge of paper-making and some Arabic words.

There appear to have been at least two groups of Indonesian migrants, differing in physique and culture, with much tribal movement in the island after their arrival. The aborigines (Teroandroka) are now absorbed in the Tanala. They were almost black, very short and slight, with a round face, kinky hair, and round, staring eyes. They were presumably a strain of Negritos. Another aboriginal tribe, the Zanakanony, were also short but powerfully built, with a light skin and long wavy or curly hair. They sound like achondroplastic dwarfs, allied to the Congo Pygmies. They were displaced and absorbed by the Tanala, and are only known by tradition.

The Kimosy are another traditional tribe of aborigines. They were dwarfs less than five feet high, but heavy-muscled and with enormous chests (perhaps also achondroplastic). They agreed with the Zanakanony in having a very light skin and long wavy or curly hair, with long, full beards; but the women had "no breasts" and the children were fed on cow's milk. These people lived in fortified villages, kept cattle, grew rice, and wore bark-cloth kilts. They also had iron and used slings. The Mahafaly tribe have a high percentage of infantile midget dwarfs. Probably a recessive gene for ateleiotic dwarfing is present in this tribe.

On the central plateau, which is free from malaria, the people have marked Caucasian characters, a light-brown skin, wavy-curly hair, a beard, no epicanthic fold. The features are reminiscent of the eastern Polynesians. Many of the Imerina are still lighter, with straight hair, a sparse beard, and the epicanthic fold which is presumably of Mongoloid origin. They resemble the southern Chinese, but the skin is lighter, especially in the Andriana (noble) caste.

Outside the plateau of Madagascar the population is predominantly Negroid, but very different from the Bantu people of the adjacent African coast. These black Malagasy have higher noses and thinner lips than the Negro, the hair is heavy and closely crimped rather than spirally curled, which suggests a hybrid condition between kinky (Negrito) and wavy hair. These very dark Negroids most nearly resemble the Galla and Somali of northeast Africa.

The blood groups of the Imerina show about 23 per cent of B (Table 22, p. 338), but those of the more aboriginal tribes appear to be unknown. The Pygmies of the Belgian Congo are remarkably low in O and approach 30 per cent in B, the Congo Negroes having over 40 per cent of O and about 25 per cent of B. The Yoruba Negroes of western Africa have the same amount of B but over 50 per cent of O. The conclusion is justified that the aboriginal population of Madagascar were Negroid Pygmies from Africa, the Indonesian invaders only coming in late historic time. That they entered Madagascar from the northeast African coast is hypothetical, but the later Arabs would have obliterated any evidence of Indonesians in east Africa (except archaeological). The dominant gentes in many Malagasy tribes claim Arab descent.

After Captain Cook's visit to Hawaii in 1778, many other races and nationalities came there, until now it probably has the most hybrid population in the world. I hope to deal with studies of this and other hybrid populations in a later book. It is only necessary here to refer to Tozzer's (1921) outline of the later racial history of Hawaii. There is a tradition of a Spanish ship from the Mexican coast being wrecked here in 1528, only the captain and his sister being saved. They are said to have become the ancestors of certain well-known families of chiefs. By 1791 there were fourteen Chinese merchants in the islands. An American mission was opened in 1821. In 1851 a group of 180 Chinese coolies arrived and by 1886 there were

21,000 Chinese in Hawaii. Their number has remained more or less constant ever since. The American whalers calling there in the early nineteenth century frequently had Cape Verde islanders in their crews, who introduced a Negro strain. In 1878, 180 Portuguese came from Funchal, Madeira, and by 1885 there were 10,000 Portuguese, mostly from the Azores and Madeira. Porto Ricans have also come, and many other nationalities in small numbers. Hawaii is not only at the crossroads of the Pacific, but all these human strains have intermingled to produce a degree of hybridity unequaled elsewhere. The Japanese began coming late, as indentured labor, and later reached 40 per cent of the population, but they intermarry relatively little with the other races.

The Polynesians in their migrations transported many economic plants from Malaya to Hawaii and other islands. These can be traced by the name-variations in the languages. Kumara, with many variations in Indonesia and Polynesia, is derived from the Sanskrit names for the blue, white, and red lotus-lily which has a floury tuber and was long cultivated in India along with taro and yams. In Ecuador and Peru the Quechua name kumar or kumara has been applied to both the sweet potato and the white potato. On the basis of this near-identity in names it has been argued that the Polynesians carried the sweet potato to the coast of South America, and also that the South American Indians in their primitive boats were carried by the currents into Polynesia and so introduced the sweet potato from South America.[3] It seems likely enough that the far-ranging Polynesian argonauts of the Pacific touched the South American coast at some time, but it seems equally clear that they were never in sufficient numbers to have any perceptible effect on either the race or the culture of any Indian tribe.

Hornell (1946) discusses at length the possible methods of introduction of the sweet potato into Polynesia with its Quichuan name, *kumar*. This evidently happened in pre-Columbian time, probably as a single contact between South America and the Marquesas, through an Indian balsa raft with sail being carried westward by the Humboldt current, taking some of the cultivated roots with them for food. This theory was tested by a Scandinavian party of five sailing from

---

[3] For a discussion of the evidence regarding the sweet potato and other plants in Polynesia, see Merrill (1946, pp. 342, 389).

Peru on such a raft. On August 7, 1947, they had already reached the Polynesians in the Tuamotu Archipelago, south of the Marquesas and east of Tahiti. A few days later they were wrecked on one of these reefs.

In an account of the physical differentiation in Polynesia, Shapiro (1943) finds them a hybrid and diversified population. Twenty-six samples of living populations have been measured. There is marked brachycephaly in the Society Islands, Hawaii, and the Tuamotus, Australs, and some of the Cook Islands, these brachycephalic elements representing a later invasion. Shapiro concludes that the Polynesians represent a fundamental unity of type, in other words, successive immigrations from a common source, the waves not profoundly different. Although ultimately of mixed origin, the fusion had taken place before the invasion of Polynesia began. The fundamental element was regarded as Mediterranean (Brown), but the exact nature of the other ingredients in the mixture is not clear. It must have been from the Javanese population of mixed "Malayan" stock. The successive waves were differentiated in head length, head breadth, and cephalic index, and to a lesser extent in minimum frontal diameter. In Indonesia the earlier dolichocephalic type was supplemented by more round-heads. The last wave was brachycephalic but otherwise not greatly different from the proto-Polynesian.

In contrast to the dense population of Java where the population reaches 1,600 per square mile in rich agricultural lands, Sumatra, which is nearly four times as large, has a population of little over 6,000,000. Its people have frequently been regarded as true Malayans, but contain many elements and groups which have become partly mixed through the centuries. Kitchen middens show the former presence of a coastal folk with stone axes and prehistoric human figures in stone. A more complete account of the anthropological history of Java and Sumatra is given by Cole (1945). The population must have come largely from Malaya, across Malacca Straits. Cole suggests that the proto-Malays came down the great rivers Mekong, Salween, and Irrawaddy from the Tibetan highlands. Their straight hair is Mongoloid, but they had some Caucasoid admixture. The oldest inhabitants belong to the Veddas. They are small, with brown or dark-brown skin, dark wavy hair, dark-brown to black eyes, and long limbs. Zondervan (1931) gives an account of them. He includes in them the Lubu in the deepest jungle behind the east coast, the Lubu on the west coast, who became Mohammedan, the Orang Akit at the mouth of the Siak River, and the Orang Laut living mostly on the water at the mouth of the Kampar, opposite Singapore. The Battak are a widespread proto-Malayan people, probably of mixed Mongoloid and Indonesian ancestry, with a peculiar assemblage of primitive and more advanced cultural conditions combined with a form of writing. The Ulus and Nias are other groups. These tribes were conquered by early Malays from across the straits, and before the Christian era a Malay-Mongol mixture now known as Atchinese, Menangka, Palembang, Lampong, in different areas, gradually spread inland and partly mixed with the inhabitants.

A Paleolithic dwelling was found by Tobler in a cave in the Djambi River basin in central Sumatra. It contained remnants of food and skeletal fragments, obsidian implements, knives, points, and scrapers. These resembled the implements found in caves in Celebes, but the paleolithic tools were better made and there was no neolithic influence. Probably the culture was related to the Stone Age culture of Ceylon, which was formerly widespread and practised by the Vedda, Sakai, Toala, and other peoples. There was also a *coup-de-poing* culture in Sumatra, which began here, at the latest, about 5000 B.C. and was succeeded by an iron culture. The Sumatra *coup-de-poing* culture is related to that in the oldest cave strata of Malaya and Tonking.

In historical time the Hindus, Chinese, Arabs, and Europeans have entered Sumatra in succession. The Hindus had already spread into Sumatra, Java, and the Malay Peninsula at the beginning of the Christian era. Java became a center of Hindu power, religion and culture, and Hindu kingdoms were set up. In Sumatra the Hindus first developed on the north coast and then pushed up the rivers, where there are ruins of numerous Hindu temples. The Province of Palembang in eastern Sumatra became a great Buddhist kingdom between the seventh and twelfth centuries. As early as A.D. 400 a Chinese Buddhist monk wrote on his travels in Sumatra, and in modern times large numbers of Chinese coolies were introduced. But the Arabs were a much more important influence than the Chinese. Mohammedan missionaries arrived A.D. 846 and the spread of Islam in the twelfth century led to the formation of many small Mohammedan states. By the fifteenth century Mohammedanism was dominant through Sumatra, Malay, Borneo, and as far as western Mindanao in the Philippines. Chinese traders exchanged with the Arabs at Palembang in the tenth century, and by the twelfth century had reached as far west as Malabar. The Portuguese came to Sumatra about 1500, soon followed by the Dutch. Since the Dutch occupation there has been much intermixture. It is now uncertain whether some of the primitive culture groups are really at an early culture level or have degenerated. Their "primitive" characters are rapidly being lost.

The Melanesian peoples are included in the region from northeast New Guinea through the Solomons, New Hebrides, Loyalty Islands, New Caledonia, and the intervening archipelagoes to Fiji. Some stocks from Indonesia have moved through Melanesia and out into the eastern Pacific. The Melanesians contain a large Papuan element from New Guinea. They are usually dolichocephalic, the nose is generally broad, and the brow ridges not prominent. The hair is characteristically woolly, but sometimes curly or wavy, the skin ranging from very dark to coppery colored. The Papuans in large parts of New Guinea are still ethnologically unknown, but their cultures are very diverse. They have an extraordinary number of languages belonging to different linguistic stocks. Adjacent villages may speak totally different languages. In British New Guinea (the eastern half of the island) there are more than sixty languages, and in the Dutch part many others. The large nose in some resembles the Armenoid, but is doubtless of independent origin.

The Melanesian element in New Guinea is believed to be comparatively recent, and is probably a mixture of Papuan with Indonesian and proto-Malay, being strongest on the north and northeast coast of New Guinea. Papuans early spread into the Melanesian islands. The Negritos, Papuans, and Melanesians all have frizzy hair, which resembles but is not identical with the woolly hair of the Negro. It is possible that too much has been made of this resemblance, and the propriety of the term "Pacific Negroes" may be questioned. They do not have the thick lips of the African Negro, and the nose, though broad, is not the same shape. The Pulaya on the Malabar coast of India, on the other hand, have rather thick lips and broad nose, like the Negro type much diluted (see Plate XII, a).

In the Snow Mountains of Dutch New Guinea, at the source of the

Mimika River, dwell the Tapiro, who are regarded as typical Negrito Pygmies. Many authors regard all peoples of short stature, dark skin, and woolly or kinky hair as derived from a single (Negrito) stock. Considering the large number of dwarf races in the world, many of which must be of quite independent origin, there seems no necessity to conclude that the Congo Pygmies are related in any direct way to those of New Guinea and other regions in the east. The Andaman Islanders, for instance, are generally classed as typical Negritos, but they appear to have little in common with the Ituri Forest Pygmies, who are achondroplastic dwarfs, which the Jurua of the Andamans are not (Gates, 1940).

Plate XXVII shows a wild Jurua woman, four of her children (one recently born) and another child. All were of the O blood group, and they are typical Negrito dwarfs with woolly hair. Speiser (1943) quotes F. Sarasin as stating that the Negritos of the Andamans and elsewhere are brachycephalic or mesocephalic, and regarding it as more than probable that they gave rise to the Negroes and Melanesians. It is well known that the Negroes and Melanesians are mostly dolichocephalic. There is much evidence that evolution can proceed from dolichocephaly to brachycephaly but none that it has ever proceeded in the reverse direction. This again favors consideration of the Negrito dwarfs as a derived and not an ancestral race.

If the southeastern Asiatic and Papuan peoples of short stature had an origin independent of the Congo Pygmies, then it seems at least possible that the woolly-haired African is independent of the Papuan and Melanesian with kinky hair. In discussion of the Negroes in Chapter 7 we concluded that they must be derived from the same stock as the Bushmen and Hottentots, that is to say, from a Boskop ancestry in Africa. There is no evidence of autochthonous Negroid peoples in the great region between Africa and Melanesia,[4] but there are several islands of Negritos in between, and it seems not unreasonable to suppose that the kinky hair and other features of Melanesians and Papuans, which resemble but do not agree with those of the

[4] A modern exception to this would be the Negro slaves (of African origin) in Arabia, and perhaps the Pulayas of southern India who might have been derived from a few stray Africans in relatively recent times. Moreover, some of the Bhils and Korkus, jungle tribes of central India, combine some Australoid features with thick Negroid lips (see Macfarlane, 1940, 1941b). There are, of course, remnants of Negrito peoples in this area, but their relation to the Negroes is at present uncertain.

Negro, may have arisen independently as mutations from a wavy-haired proto-Australoid stock. An admitted difficulty with this view is that the Tasmanian hair appears to have been more woolly than kinky.

Howells (1943), in an interesting discussion of the Melanesians, shows that they have a greater variety of types than the Polynesians. The first comers to New Guinea were, as one would naturally expect, of Australian aboriginal type. This was on their route to Australia, as we have seen in Chapter 6, and one may expect to find Australoid types persisting in parts of New Guinea. Their ancestors must have occupied in early time parts of the whole region from Malaya, India, and Ceylon, to Sumatra, Java, and on to New Guinea. Howells believes that the Negritos came next, and that these were followed by a long series of invasions by Negro peoples of Congo forest type. It seems incredible that the Negro would occupy two such widely sundered areas, with no evidence of his former existence in between. Howells insists that there are probably true Negroes in Melanesia, whatever their origin, but the possibility that their ancestors never came from Africa does not seem to be excluded. As regards blood groups, the African Negroes have about 25 per cent of B (see Table 22) while the Melanesians have only about 17 per cent. The Negroes seem to have evolved in Africa too late to have spread eastwards in the way suggested. Their passage would have been blocked by the presumably ancestral Boskop and other races in the intervening area.

A number of other relationships can be seen from the blood-group table on p. 338. The Negroes differ from the Bantu in being generally lower in O and higher in B, some Bantu being very high in O. The Hottentots and Bushmen differ markedly in that the latter are high in O and very low in B. The Hottentots tested were mainly Korana from southwestern Africa. The blood groups seem to show that the Bushmen and Pygmies cannot be nearly related. Pijper concluded that the Bush must have crossed with a race high in B to produce the Hottentots. The African Negroes are somewhat higher in B than the Melanesians, but the difference is not great. But this degree of similarity does not necessarily mean a common origin. Dart (1937) is probably right in regarding the Negro as an infantile derivative of the Bush-Boskop race, his prognathism, everted lips, and infantile form of skull being specializations.

The Timor Archipelago consists of Lombok, Sumba, Flores, Timor, and smaller islands stretching in a line east and west between Celebes on the north and Australia on the south. In a study of the anthropology of this region, Bijlmer (1929) found the Melanesian population to be of two types, dolichocephalic and brachycephalic, both purest in eastern Flores. The dolichocephalic element here was said to be of highly Negroid appearance, but in the Kroenese the hair was insufficiently crisped and they were too largely Malay to call them Papuans. The Atoni of west Flores were also Melanesian, the Ngadanese having an equal mixture of proto-Malay and Melanesian ancestry. Bijlmer found the Mongolian eyefold [5] present to a moderate degree in both the proto-Malays and the Polynesians. In Sumba there were some Polynesian traits, but he thinks the Sumbanese are best regarded as proto-Malay, which is the substratum of the whole archipelago and accounts for the mesocephalic element in the population. The cephalic index of the Sumbanese males was 77–79. Sumba also contains tall Polynesian types with faces suggesting those of India and Europe.

The Lamontjong caves in Celebes were excavated by Paul and Fritz Sarasin in 1902–1903. In five of them the floor contained animal bones, stone and bone implements, teeth, and shells. The implements were very crude—scrapers, points, knives, and arrowheads—but the bones were of animal species still living in Celebes. Most of the implements resembled the Magdalenian of Europe, but were more primitive in manufacture. Stone arrows are found nowhere else in Indonesia. These people were hunters with a Paleolithic culture having a touch of the Neolithic. In the upper layer of the cave floors were iron tools and fragments of china. The present Toala of Celebes are apparently descended from these cave dwellers. Hindu and Islamic culture have entered in more recent times.

The facial features of sixty-one Melanesian men and thirty-five women from New Caledonia and the Loyalty Islands were measured from photographs (Keiter, 1936) and compared with Europeans. From measurements of thirty features in each photograph he concluded that the same facial types occur in both populations, but with different frequencies. The sex differences were in the same direction in both races, but more marked in the Melanesians. The noses were

---

[5] These eyefolds need a more detailed examination. See Gates (1946), p. 156.

most different in the two races, and the multiformity of facial types was as great in New Caledonia as in central Europe.

The Polynesians and Melanesians considered in this chapter are of such recent origin that the history of their spread becomes a study in genetics rather than paleontology. More detailed investigations will no doubt yield valuable results from this point of view. We see the last great expansion of mankind into vast areas of the Pacific previously unoccupied by man. This involved collisions and crossing between types differing in varying degrees. The unraveling of these crosses should furnish useful material to human genetics. If the Polynesians, who are so obviously Caucasoid in origin, can be linked with the Aryans, the mystery of their genesis would be solved.

The relation of the Melanesians and Papuans to the African Negroes remains for the future to determine. Their identity of type does not appear to be so close as to necessitate a common origin from a single source. The geographical difficulties also are such as to stimulate the search for some other explanation.

Following Hooton (1931), it appears that the early population of India consisted of two very distinct elements—Negrito and proto-Australoid. These mingled to form the pre-Dravidians. The time of entrance of the Brown (Mediterranean) race into India is not clear, but it must have been fairly early. Much later, within historical times, came the Aryan invasion, introducing another Caucasoid element. The numerous tribes of Negritos farther east were presumably from an Indian source. The blood groups of Negritos and Australoids must have been very different then, as they are now, the former being high in B, the latter having A without B. How this blood-group difference arose we have no means of knowing, but the widely varying blood group frequencies in Indian populations may be regarded as the result of mixture of these two fundamentally different types in varying proportions, plus the Caucasian element. The origin of the proto-Australoids can be traced without difficulty back to Pithecanthropus. The racial source of the Negritos is at present conjectural, but their origin was probably later than that of the proto-Australoids. At any rate, their spread appears to have been at a later date. Many facts referred to in earlier chapters indicate that the B blood group is also much later in its spread, and probably also in its origin, than the A. We have seen good reasons for believing that

A was already present in the anthropoid ancestors of man, whereas B may have arisen independently in the evolution of these two groups. Whether B spread from the Negritos to the Mongoloids or whether it arose independently through mutations in the latter awaits further evidence, but some "primitive" peoples such as the Chwan Miao in western China are even higher in B than the Chinese. They are not, however, physically anywhere near the low level of the Australoids or the Negritos.

It is not necessary to enter into the subject of blood groups in India in detail, but a few of the results may be considered here. From Table 23 some of these results may be compared.

TABLE 23
BLOOD GROUPS IN INDIA

	No.	O	A	B	AB	Author
Mixed Bhotias	112	33.9	32.1	22.3	11.6	Macfarlane, 1941a
All Bhotias	165	37.0	32.1	20.6	10.3	Macfarlane, 1941a
Khasis (Assam)	200	33.0	35.0	18.5	13.5	Macfarlane, 1941a
Chinese (Hunan)	1296	31.9	39.0	19.4	9.8	
Angami Naga (Assam)	165	46.0	38.8	11.5	3.6	Mitra, 1936
Lushai (Assam)	141	32.6	44.7	16.3	6.4	Mitra, 1936
Bhils (west central India)	140	18.6	23.6	41.4	16.4	Macfarlane, 1941b
Bhils (west central India)	44	31.8	13.6	52.3	2.3	Macfarlane, 1940
Korkus (west central India)	140	20.0	28.6	37.9	13.6	Macfarlane, 1941b
Maria Gonds, Bastar State	123	28.5	26.0	34.1	11.4	Macfarlane, 1940
Oraon	155	47.1	12.9	34.8	5.2	Macfarlane and Sarkar, 1941
Munda	120	33.3	30.0	29.2	7.5	Macfarlane and Sarkar, 1941
Santal (Bihar)	339	33.0	20.9	34.8	11.2	Sarkar, 1937
Chenchus	100	37.0	37.0	18.0	8.0	Macfarlane, 1940
Izhuvans (Cochin)	132	58.3	24.2	12.2	5.3	Macfarlane, 1936
Kanikars (Travancore)	211	51.2	18.5	29.9	0.5	Macfarlane and Sarkar, 1941
Paniyan (southern India)	250	20.0	62.4	7.6	10.0	Aijappan, 1936

The Bhotias in northern India and the Khasi in Assam are similar in blood groups to the Chinese in Hunan, having about 20 per cent B. The Angami Naga of Assam are much higher in O and lower in B, indicating their more primitive character, while the Lushai are higher in A but lower in O, having absorbed more B. The Bhils and Korkus are very high in B, but some of them show Australoid features

(Macfarlane, 1940) and some have rather thick lips (Macfarlane, 1941 b). The Oraon and Munda can be grouped together, having about 30 per cent of B, but the former are lower in A and higher in O. The Santals do not differ markedly from these two tribes, all three having more B than the more Mongoloid Bhotias and Chinese.

In the four jungle tribes of southern India A is higher than B, indicating their primitive character, except in the Kanikars of Travancore, whose short stature (153.4 cm.) and somewhat kinky hair (see Plate XI, b) indicate Negrito relationships, and this may perhaps account for the high frequency of B. Photographs of the Chenchu (Macfarlane, 1940) show them to be definitely Australoid or perhaps rather Tasmanoid, as the hair is more or less curly. The Izhuvans (Illuvas) of Cochin are high in O, the Paniyans very high in A, but both these jungle tribes are low in B, which has probably been acquired from their neighbors. There seems every reason to believe that the Australoids in India were originally very low in B or altogether devoid of it. Whether the same was true of the Negrito element, or whether they are an original source of the B which has become so frequent in the main racial elements of modern India is not at present clear. If the B gave greater resistance to malaria, as has been suggested, its great increase in modern India might be explained. In any case, it appears that the B blood group, like brachycephaly, has only spread widely in mankind in relatively recent times.

The Nayadis of Malabar are the lowest of the untouchable Hindu castes. In the census of 1931 they numbered only 709, and this included 144 Ulladans of Cochin. They must shout from a distance and not approach within seventy-four to one hundred feet of any other caste. The traditional purification following pollution by a nearer approach consists in bathing in seven streams and seven tanks and then letting a drop of blood out of the finger! The caste is described by Aiyappan (1937). In 1730 they were a jungle tribe, jet black, with bushy hair, features approaching the kaffirs, no houses or clothing, living in trees. They have cross-cousin marriages. The cephalic index is found to be 73.7, nasal index 85.7 in males, 88.3 in females, stature 159.7 cm. in males, 146 cm. in females. They are microcephalic but hypsicephalic, in many characters intermediate between the hill tribes and the plains castes. The hair is deep black and slightly wavy, the eye almond-shaped, the sclera generally not clear.

The lips are more or less thick, the nose less broad than in the Paniyans and Kadars. Many suffer from night-blindness, which might result from the glare or from deficiency in vitamin A.

REFERENCES

Aiyappan, A. 1936. Blood groups of the Pre-Davidians of the Wynaad Plateau. *Current Sci.* 4:493-494.

———. 1937. Social and physical anthropology of the Nayadis of Malabar. *Bull. Madras Govt. Mus.* 2. pp. 141. Pls. 12.

Bijlmer, H. J. T. 1929. Outlines of the anthropology of the Timor Archipelago. Weltevreden, D. E. I. pp. 234.

Boyd, W. C. 1939. Blood Groups. *Tab. Biol.* 17:113-240. Figs. 13.

Buck, P. H. 1922-23. Maori somatology. Racial averages. *J. Polynesian Soc.* 31:37-44, 144-153, 159-170; 32:21-28, 189-199.

———. 1945. An Introduction to Polynesian Anthropology. *Bishop Mus. Bull.* 187. pp. 133.

Cole, Fay-Cooper. 1945. *The Peoples of Malaysia*. New York: Van Nostrand. pp. 354. Illus.

Dart, R. A. 1927. In Schapera, *Bantu-speaking Tribes of South Africa*.

Deevey, E. S., Jr. 1946. An absolute pollen chronology for Switzerland. *Am. J. Sci.* 244:442-447.

Gates, R. R. 1940. Blood groups from the Andamans. *Man.* 40:55-57. Figs. 3.

Handy, E. S. C. 1943. Two unique petroglyphs in the Marquesas which point to Easter Island and Malaysia. *Peabody Mus. Papers.* 20:22-31. Pl. 1. Figs. 3.

Heine-Geldern, Robert von. 1945. Prehistoric research in the Netherlands Indies. In *Science and Scientists in the Netherlands Indies*. (Ed. P. Honig & F. Verdoorn; New York: G. E. Stechert, pp. 491). pp. 129-167. Figs. 53.

Hooton, E. A. 1931. *Up from the Ape*. New York: Macmillan. New edition, 1946. pp. 788. Pl. 39. Figs. 68.

Hornell, J. 1946. How did the sweet potato reach Oceania? *J. Linn. Soc., Botany.* 53:41-62. Figs. 2.

Howells, W. W. 1943. The racial elements of Melanesia. *Papers of Peabody Mus.* 20:38-49.

———. 1944. *Mankind so far*. New York: Doubleday Doran & Co. pp. 319. Pls.

Keiter, F. 1936. Neukaledonien-Mitteleuropa. Vergleich der Gesichtszüge. *Zeits. f. Morph. u. Anthrop.* 35:377-393.

Linton, R. 1943. Culture sequences in Madagascar, in Studies in the Anthropology of Oceania and Asia. *Papers of Peabody Mus.* 20:72-80.

Macfarlane, E. W. E. 1936. Preliminary note on the blood groups of some Cochin castes. *Current Sci.* 4:653-654.

———. 1940. Blood grouping in the Deccan and Eastern Ghats. *J. Roy. Asiatic Soc. Bengal.* 6:39–49. Pls. 2.

———. 1941 a. Tibetan and Bhotia blood group distributions. *J. Roy. Asiatic Soc. Bengal.* 7:1–5.

———. 1941 b. Blood groups among Balahis (weavers), Bhils, Korkus, and Mundas, with a note on Pardhis, and aboriginal blood types. *J. Roy. Asiatic Soc. Bengal.* 7:15–24. Pl. 1.

———, and S. S. Sarkar. Blood groups in India. *Am. J. Phys. Anthrop.* 28:397–410.

Mitra, P. N. 1936. Blood-groups of the Angami Naga and the Lushai tribes. *Ind. J. Med. Res.* 23:685–686.

Pijper, A. 1930. The blood groups of the Bantu. *Trans. Roy. Soc. S. Afr.* 18: 311–315.

Rivet, P. 1930. Sumérian et Océanien. *Proc. 4th Pacif. Sci. Congr.* 3:519–527.

Sandoval, L., and O. Wilhelm. 1945. Communicación preliminar sobre antropología serológica de los pascuenses. *Bol. Soc. Bio. Concepción (Chile).* 20:11–15.

Sarker, S. S. 1937. Blood grouping investigations in India with special reference to Santal Perganas, Bihar. *Trans. Bose Inst. Calcutta* 12:89–101.

Shapiro, H. L. 1943. Physical differentiation in Polynesia. *Papers of Peabody Mus.* 20:3–8.

Smith, S. Percy. 1921. *Hawaiki, the original home of the Maori.* With a sketch of Polynesian history. 4th Ed. Christchurch, New Zealand. pp. 223.

Speiser, F. 1943. Obituary notice of Dr. Fritz Sarasin. *Verh. Naturforsch. Ges. Basel.* 54:222–264.

Sullivan, L. R. 1923. The racial diversity of the Polynesian peoples. *J. Polynesian Soc.* 32:79–84.

Toynbee, A. J. 1934–35. *A Study of History.* 3 Vols. London.

Tozzer, A. 1921. The anthropology of the Hawaiian race. *Proc. 1st Pan-Pacific Congr.* pp. 70–74.

Zondervan, H. 1931. Das Völkergemisch Sumatras. *Zeits. f. Ethnol.* 62:244–248.

11

Some Principles of Speciation in Primates

Having now surveyed the main lines of evolution in man, there remain certain general principles to be considered. There is reason to believe that with increasing paleontological evidence many obscure points on which differences of opinion now exist will be cleared up. Since man is the best known of all organisms, there is reason to expect that further studies of human phylogeny will throw light on evolutionary principles. Although types of culture and styles in the manufacture of human artifacts can be transmitted from one man to another by culture contact, yet the abundant existence of such artifacts in many parts of the world furnishes an additional durable element in the analysis of human evolution which is not present in connection with any animal group. Since few fossil bones of any species survive, these stone artifacts will ultimately provide the basis for a more detailed knowledge of man's previous existence than is possible in any animal species. Man's world-wide spread also provides a wider basis of comparison than in any genus of animals. If particular human characters have arisen and spread in the past through repeated mutations, as suggested in this book, then further researches in human genetics should provide conclusive evidence of that fact. Such evidence already exists (Gates, *Human Genetics*) regarding the occurrence and inheritance of many abnormal mutations, which creates a presumption that normal racial and specific differences can arise in the same way.

In his *Descent of Man,* Darwin (1871) recognized to a remarkable extent the modern genetic point of view when he wrote (p. 152) that in the earlier editions of the *Origin of Species* he had formerly attributed too much to the action of natural selection as a factor in evolution. In the fifth edition this was altered "so as to confine my remarks to adaptive changes of structure." He says further, "I had not formerly sufficiently considered the existence of many structures which appear to be, as far as we can judge, neither beneficial nor injurious; and this I believe to be one of the greatest oversights as yet detected in my work." On the next page he adds, "That all organic beings, including man, present many modifications of structure which are of no service to them at present, nor have been formerly, is, as I can now see, probable." These statements harmonize extraordinarily well with the present genetic point of view. They show that Darwin in his maturer views placed much less emphasis on natural selection than the neo-Darwinians.

Later, Darwin cites the fact that several species of lemur are tailless although most of the group have a long tail. Also, that in some species of Macacus the tail has twenty-four vertebrae and is longer than the body, but in other species it is a scarcely visible stump, consisting of only three or four vertebrae. In these and other facts regarding tails in mammals it is difficult to see that his views differ from those of the modern mutationist.

A fundamental problem in human genetics is then to determine the mutation frequencies for various characters which have a simple genic basis. If this point of view is correct, it clearly follows that new mutations begin to arise at certain times. Whether such changes in mutation rate are gradual or sudden we have no evidence at the present time, but it is clear that evolution could not have taken place without the origin of new mutations from time to time, meaning by *new* those that had never occurred before in that line of descent. This point of view was first put forward in relation to the blood groups (Gates, 1936). There is genetical evidence that a general increase in mutability is under genic control. Thus Demerec (1937) showed that the higher mutation rate of the Florida strain of *Drosophila melanogaster* was due to one or more recessive genes in the second chromosome. As another example, Rhoades (1938) found a gene a_I (producing red dots in the aleurone in maize) which changes from a stable to a highly mutable state in the presence of a dominant gene, Dt, in another chromosome. These genes affecting general mutability of a specific gene, have presumably themselves originated as mutations. They indicate that there are reasons why the general mutability or the mutation rate of a specific character can undergo a sudden change.

It has sometimes been held, for example, by Lotsy (see Lotsy and Goddijn, 1928) that hybridization between strains and species is in itself sufficient to account for evolution. This view has never been seriously accepted and it clearly has definite limitations, yet there are certain cases of apparently new characters in man which may be explained on this basis. Coon's suggestion that since Neanderthal man is frequently brachycerebral (although never brachycephalic), that brachycephaly might then arise in crosses with Upper Paleolithic (dolichocephalic) man, is worthy of further consideration. If, in such a cross, the presence or absence of brow ridges segregated independently of factors for brain shape, or head shape, then some of the second generation might be expected to have a brachencephalic brain and (whether or not brain-shape is a controlling factor in head-shape) a round head, without brow ridges. However, reasons have already been given for concluding that the mutation hypothesis of the origin of brachycephaly is more probable, and this is further supported by the origin of brachycephaly in regions where there were none of the gorilloid type with which crossing might have taken place.

That secondary human races can arise from hybridization between primary races is well recognized by anthropologists. Coon (1939, p. 629) regarded the Armenoid type as a stable hybrid between the Alpine race and the Irano-Afghan branch of the Mediterranean or Brown race, the mixture being in the ratio of one of the former to two of the latter. This view is based on Bunak's (1927) conclusions from a study of Armenian crania and especially on a thesis of Dr. B. O. Hughes on the physical anthropology of 1500 Armenian males, in which the principle was suggested in 1938. The Armenian head is brachycephalic with a flat occiput and a conspicuously large nose, which is nevertheless leptorrhine. The mean cephalic index is 85.4, nasal index 64, the nasion depression usually slight. This relatively stable condition is supposed to arise when the Mediterranean and Alpine races are present in the mixture in the ratio 2:1. If the analysis is correct, the flat occiput and big nose arise as new characters from a cross in which they are absent from both parental races. It is to be hoped that Hughes' thesis will be published, as an important principle is involved.

Coon applied the same principle to the origin of the Dinaric race in Europe, which has the same features—brachycephaly, flat occiput with the foramen magnum set well back, producing a projecting nose which is frequently convex, and a face frequently elongated. The brachycephaly is often greater than that of the Alpines, the pre-auricular part of the head deriving its dimensions from the dolichocephalic Mediterraneans, so that the ear is posterior in position. The narrow face, also from the Mediterraneans, is often even exaggerated. As a result of the shortening of the head, the nose becomes salient, and the face is frequently elongated. The Dinarics are then not a unity but a secondary type arising in the crosses between two primary races.[1] Dinaricization has thus happened to various types, and the Bronze Age Dinaricized Mediterraneans spread from their source in the eastern Mediterranean to central Europe and down the Rhine into Britain.

In a Dinaric population there will always be some Alpines and Mediterraneans, their relative numbers depending on which is in excess in the population. This hypothesis seems to account for much of the head morphology in eastern Europe and parts of Asia Minor. Coon also applies it to the big-nosed type of Papuans in New Guinea, the aristocratic Arii in Polynesia, and the types of American Indians with high convex leptorrhine noses. The subject is worthy of much more detailed genetic investigation. It is, at any rate, a reasonable hypothesis that the Armenoid nose has originated in this manner and that the peculiarities of the Dinaric head are likewise of hybrid origin. However, recent studies of Ehrich (1947) and Ewing (1947) indicate that the occipital flattening of the Dinarics and Armenoids may be a cradle effect.

There is every reason to believe that crossing from time to time has occurred not only in human evolution but also in many groups of animals. We have already seen how frequent hybridization has been between related strains and races, for instance in the modern history and prehistory of South Africa. We have also seen reason to postulate crossing between widely diverse types of man—even in-

---

[1] The term *ethnic group* has recently been revived by some writers as a partial substitute for race. It was copiously used by Deniker in *The Races of Man* (1913).

tergeneric crosses, as between Palaeoanthropus (Neanderthal) and Homo, in the early Upper Paleolithic of Europe and Asia Minor. There is also the evidence from the upper cave of Choukoutien. Man is certainly not unique in this respect, and so we must recognize occasional crosses, not only between races but also between species and even genera, as a contributing factor in the evolution not only of man but of many animals and plants. Because of the hermaphroditism in higher plants, hybridization has no doubt played a more important role in this group than elsewhere, but it can by no means be excluded from animal evolution. This is one reason why interspecific sterility cannot be regarded as a generally applicable criterion of species, but there are other even more cogent reasons (see Chapter 12).

It is worthwhile pointing out that paleontology can offer little or no direct evidence of crossing between related species, however frequent it may have been in certain groups of animals. It is known, for instance, that lions and tigers will cross both ways in captivity, producing "tigons" and "ligers." No one doubts that the lion and tiger are "good species," yet they are difficult to distinguish osteologically, and paleontologists might well group their bones as belonging to one species. The differences are mainly epidermal—color pattern (the lion being concolorous while the tiger is transversely striped), type of hair, mane in the male lion—in these respects resembling the differences between the so-called human "races," which are at least as marked if not more so.

The lion and the tiger differ slightly (according to Flower and Lydekker, 1891) not only in general size but in the proportionate size of the lower teeth, the general form of the cranium, relative length of nasal bones, and the ascending processes of the maxillaries. By these means the skulls of the two species "can be easily discriminated by the practised observer." The skull differences between so-called races of man are evidently much greater. The lion formerly extended across Arabia, and existed in India within historical times. Flower and Lydekker state that there are still lions in the Tigris and Euphrates river valleys. *Felis spelaea*, which abounded in the caves of western Europe in the Pleistocene, is indistinguishable from *F. leo* by any osteological or dental character. Lions were still found in southeastern Europe within the historical period. *F. tigris* extends from Georgia and south of the Caspian to Lake Baikal, Sumatra, Java, and Bali, while man has nearly exterminated the lion except in Africa. Circumstances are therefore conceivable in which these two species could have crossed in the wild condition, all of which goes to show how flimsy is the argument that interspecific fertility can be used as a criterion of specific identity. If the lion and the tiger are "good species," as no one doubts, then the same is true of some five "races" of mankind.

Anthropology has been rather slow to incorporate the genetical point of view into its thinking, and the reasons for this are fairly obvious. Most anthropological characters are quantitative in nature and few cases of normal racial variations are known to have a single-gene basis, although with further analysis many such cases may yet be disclosed. Another reason for the failure of anthropologists to think in genetical terms was the old antagonism between the genetic and the biometric schools. Probably R. A. Fisher has done more than anyone else to convince biologists that these two approaches to the study of variation are complementary and equally necessary.

Hooton (1925) was one of the first to recognize the importance of discontinuous variations in human evolution. In his paper on the asymmetrical character of human evolution, he showed how frequently characters vary independently in evolution. He says, "All fossil types of man hitherto discovered manifest certain morphological inequalities in the development of various cranial parts." This has been emphasized particularly in Chapter 8, in relation to Neanderthal man, the cranium and mandible of Eoanthropus being probably the most extreme case; but the gorilloid skulls of Sinanthropus and Pithecanthropus combined with a purely modern femur constitute an equally marked example. Hooton showed that similar disharmonic features characterize modern racial types as well as the anthropoid apes. His statement that "one such humanoid development may have attained an almost completely erect posture without any consequent reduction of the jaws or any vast increase in the size of the brain," fits in remarkably well with the characters of Australopithecus, discovered in the same year.

Hooton points out the asymmetrical character of growth and development in various parts of the human body. In one month the stature increases while the weight is stationary. In another month the head broadens and in the next it lengthens. As regards human phylogeny, he postulates (p. 138) "several distinct stocks whose common ancestry must be sought in a proto-human or very inferior human stage of development [which] have developed along lines roughly parallel, but with many unimportant divergences." More recent experimental work with the teeth of embryo rats and the bones of chick embryos, to cite two striking examples, shows that individual teeth and bones placed in a proper nutrient medium undergo self-differentiation, each taking its characteristic shape and structure. The autonomous development of individual bones being shown in this way, it is not surprising to find them varying and mutating independently, although there is evidence that other genes control or affect the development of *groups* of bones, as well as the relative rate of ossification of particular bones such as those of the wrist. Having recognized the relative independence of individual bones in development, we expect them to vary independently. Thus members of one family may have long and those of another short phalanges, and similar characters may be racial, resulting from mutational changes. Other mutations of the same general character may produce inherited abnormalities such as polydactylism, brachydactylism, or apical dystrophy.

Over a period of many years the writer has gradually been obliged to espouse the view, as briefly indicated elsewhere (Gates, 1944), that the main primary "races" of mankind should be recognized as species. The chief argument against this procedure has been that of interfertility, but we have already seen that this argument breaks down, and much more will be said on this subject later. Apart from this, the great objection to the continued acceptance of *Homo sapiens* as a single species including all living races of man is that it gives a false perspective. We recognize different species, and even genera, of man coexisting in Pleistocene times and giving rise to such different types as the Australian, the Bushman, and the Caucasian from independent lines of descent, yet as a convention arising from man's self-conceit we try to crowd them together into one species, implying simple divergence from one ancestry. The evidence is clear, however, that the primary so-called races of living man have arisen independently from different ancestral species in different continents at different times. They have shown some parallel developments, as witness the Australoid type in Australia, in the Koranas of South Africa, and in some Amerind skulls. Yet there is no evidence of convergence, but on the contrary the Australian aborigines and the Bush-Hottentots, for example, have diverged very widely; even in their sex organs, which are regarded by many as specific characters par excellence. Consistency therefore necessitates the recognition of *Homo australicus, H. capensis, H. africanus* (the Negro race), *H. mongoloideus* (including the Amerinds as a geographical subspecies), and *H. caucasicus,* as species, each having its own geographical expression despite the migrations and intercrossing which have taken place especially within historical time. *Homo sapiens* can then be retained as a super-species, including all living species, if desired.

A few quotations from the *Descent of Man* will make Darwin's position clear. He points out that the "races" differ much not only in physical characters but "also in constitution, in acclimatization and in liability to certain diseases.[2] Their mental characteristics are likewise very distinct; chiefly as it would appear in their emotional, but partly in their intellectual faculties." He says everyone is struck with the contrast between the taciturn, even morose, aborigines of South America, and the light-hearted, talkative Negroes. He finds a nearly similar contrast between the Malays and Papuans, who live in the same conditions and are separated only by a narrow space of sea. The North American Indians (except the Loucheux [3]) and the Eskimo show equally contrasting temperamental characteristics. Of course, the old faculty pyschology of Darwin's time has been replaced by other approaches, but the mental differences between races remain and cannot be gainsaid. It has become so much the fashion to decry the existence of mental or even physical differences between the primary races that it is worthwhile quoting Darwin again (p. 109): "The variability or diversity of the mental faculties in men of the same race, not to mention the greater differences between

---

[2] In *Human Genetics* (Gates, 1946) a considerable number of cases are recorded in which a particular inherited disease is found mainly or wholly in a particular race or type of mankind.

[3] My contact with the Loucheux Indians during an expedition down the Mackenzie River in 1928 led me to regard them as temperamentally different from all other Indians, east or west. Living near the Arctic Circle, they had the lively extrovert mentality of the Eskimos. One was left wondering whether this resulted from their Arctic environment or from crosses with the Eskimos.

the men of distinct races, is so notorious that not a word need here be said."

The question of sterility between the human "races" was formerly much discussed (see Broca, 1864) but it is now generally taken for granted that interfertility is universal. However, if a Bushman could be crossed with an Eskimo, it is by no means certain that some degree of intersterility would not be found, if only because one is adapted to a tropical desert and the other to Arctic ice. There is evidence (Gates, 1946, p. 56) of an inversion in a chromosome as between French and Scotch, so there might well be enough inversions between Eskimos and Bushmen to produce a considerable amount of sterility, apart from other causes based on physiological adaptation. Darwin says (p. 222), "Even if it should hereafter be proved that all the races of men were perfectly fertile together, he who was inclined from other reasons to rank them as distinct species, might with justice argue that fertility and sterility are not safe criterions of specific distinctness." The modern work in hybridization not only shows that they are unsafe criteria but that they are completely impossible to apply consistently.

Darwin goes on to say that "with forms which must be ranked as undoubted species, a perfect series exists from those which are absolutely sterile when crossed, to those which are almost or quite fertile." He goes on to remark that "man in many respects may be compared with those animals which have long been domesticated," pointing out the Pallasian doctrine that domestication tends to eliminate the sterility which he regards as general in the crossing of species in nature. He remarks that perfect fertility of intercrossed "races" of man "would not absolutely preclude us from ranking them as distinct species."

Although Darwin would rank as a "lumper" among systematists, yet he says (p. 224), "We have now seen that a naturalist might feel himself fully justified in ranking the races of man as distinct species." Although he finally decides against this procedure, partly because of the difficulty in deciding how many species to recognize, he says further (p. 226), "Every naturalist who has had the misfortune to undertake the description of a group of highly varying organisms, has encountered cases (I speak after experience) precisely like that of man." Darwin recognizes (p. 227) "weighty arguments" for

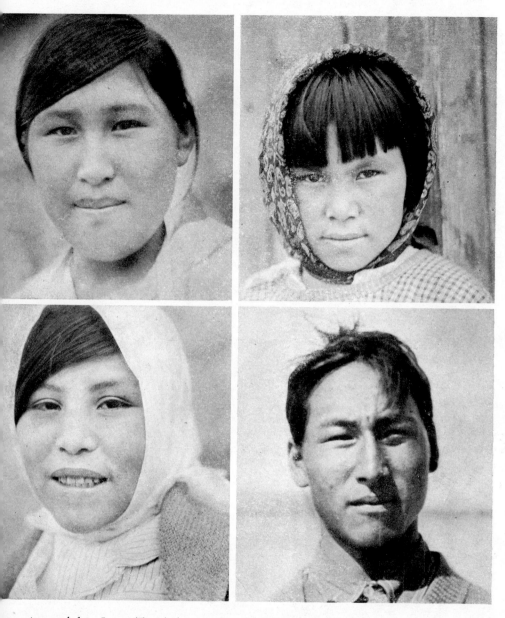

a. (upper left): Laura Tegitkok, 14 years old, Eskimo school girl
b. (upper right): Doris Tegitkok, 7 years old, Laura's sister
c. (lower left): Louisa Alexi, Loucheux Indian girl
d. (lower right): Jacob Nipalarok, Eskimo from Baillie Island

a. (upper left): Lennei Iglangasak, Eskimo at Aklavik. He married a Loucheux Indian woman

b. (upper right): Eskimo at Aklavik, head of the Mackenzie River delta, wearing parka

c. (lower): A group of Loucheux Indians at Fort MacPherson on the Mackenzie River

recognizing human species, but "insuperable difficulties" in defining them. He concludes that the term sub-species might be used "with much propriety," but admits that "from long habit the term 'race' will perhaps always be employed." He points out that complete consistency in the application of the terms species and sub-species is rarely possible, because some species within a genus are frequently more distinct than others. On page 229 he says that those who admit evolution "will feel no doubt that all the races of man are descended from a single primitive stock;[4] whether or not they think fit to designate them as distinct species, for the sake of expressing their amount of difference." Again (p. 235), in discussing mental and language differences, Darwin says, "So again it is almost a matter of indifference whether the so-called races of man are thus designated, or are ranked as species or sub-species; but the latter term appears the most appropriate." Finally he recognizes (p. 240) that "in some cases the crossing of races already distinct has led to the formation of new races."

VARIATION IN ANTHROPOIDS

So much propagandist rubbish has been written about race,[5] sometimes attempting to deny even the existence of human races, that a still broader basis of approach is required. Modern studies of the anthropoids afford much material bearing on their variations and relationships. The orangutans have already been considered from this point of view. We may now briefly examine the variability and distribution of the gorillas and chimpanzees. Regarding the coastal gorilla of the Cameroons and Gaboon, Sanderson (1940) states that the form which he classifies as *Gorilla gorilla gorilla* "varies enormously in colour, countenance, structure and, to a certain extent, in behaviour . . . Marked family likeness in almost every case, just as

---

[4] The later evidence shows that while human ancestry is probably from one stock (with the proviso that the Australopithecinae, if they reached the human level, may be from a separate ancestry) the surviving "races" are by no means all simultaneous divergents from a single ancestry, as was formerly supposed.

[5] This propaganda has reached such a point that a zoologist (Hall, 1946) finds it necessary to point out that the primary races of man can be recognized just as clearly as the geographic races in species of mammals. He takes the conventional view that these are subspecies, whereas we have brought paleontological evidence to show that it is biologically more consistent to regard them as species, although there is no hard-and-fast line between the two categories.

among human beings." The natives, from sight of one individual, could state accurately the number and composition of the rest of the troop to which it belonged. They know all the families by sight. The gorillas pass to and fro in the country, each troop having its distinctive variations. One family has a red top-knot, another has a long face, a third a profuse silvery-white coat. They all become grey in age and the families differ also in habits.[6] Schwarz (1928b) recognizes six subspecies of the gorilla.

Even greater variability is found in the chimpanzees. Yerkes (1940, p. 18) states, "It is said that there is only one species of man, but this does not obscure the striking diversity of so-called human races. Similarly there is diversity of form, mode of life, and habitat among chimpanzees. It seems as logical and appropriate to suppose that they represent a single species as in the case of man. For some reason this point of view has long been unpopular. Nevertheless, I should like to present certain evidences which suggest its reasonableness." It seems probable that the questions of different species in Gorilla, Pan, and Homo will stand or fall together, although the fossil as well as the modern existence of man in Asia, Africa, and Europe (whereas Gorilla and Pan are found only in central Africa) makes the case for species in man stronger than in the anthropoids.

Yerkes goes on to say that chimpanzees range in skin color from Caucasian white to Negro black, from pygmy to giant, from round to long skull; from small to large and erect to lop ear; differing also in many structural and behavioural characters. Some of the structural characters may represent stages of development or aging, sex, health, or physiological status. According to Sanderson (1940) they may also have beetling brows or bland faces. This great variability is expressed by Matschie (1919) in the recognition of fourteen species of Pan—no doubt very excessive. Allen (1939), in his check list of African mammals, recognizes three subspecies of *Pan troglodytes*. These are *troglodytes*, *schweinfurthii* (the long-haired chimpanzee), and *verus* (the western chimpanzee). He recognizes *Pan paniscus*, the dwarf chimpanzee, as a distinct species. To be consistent he also lists the Congo Pygmy as an unnamed species of Homo, with the statement that the "Pygmy of the Congo forests and its desert race, the Bushman, of Southwest Africa form a

---

[6] Variability in the orangutan is considered elsewhere.

basic species." We have already expressed a different view of the relation between the Pygmies and the Bushmen, but it is significant that Allen has at any rate disrupted the hoary dogma of *Homo sapiens* as constituting all living mankind. Sanderson (1940) finds in the chimpanzees not only great individual variation but very different temperaments.

Allen followed Schwarz (1934) in recognizing three local races or subspecies of the chimpanzee, *Pan troglodytes,* but Schwarz (I think rightly) recognized *P. paniscus* as a subspecies only.[7] In earlier work, Rothschild (1904) agreed with Matschie that there were at least four species of chimpanzee living side by side throughout the greater part of the chimpanzee range, in western and central Africa. He maintains five species with twelve geographic subspecies. Yerkes states that all forms of the chimpanzee in captivity interbreed freely, but rivers are a barrier to their migration. Rothschild recognized two groups: *Simia satyrus,* having always a black or brownish face, and *S. pygmaeus* with a pale face in both the adult and the young. On present knowledge it is difficult to say that more than one species exists, especially as they all occur in a limited region and frequently more than one in a given area. Unlike man, there is no paleontological evidence to show independent ancestry of different species in different parts of the world; but on the contrary, all appear to be closely related derivatives from a single common stock. Elliot (1913), in his monograph of the Primates, tentatively recognized some nine forms as having specific rank, but many of these distinctions probably depend on unstable characters. Schwarz (1934) states that newborn chimpanzees are always flesh-colored in their naked parts, their hair long and abundant, with always a white anal tuft. Lönnberg (1917) states that infant gorillas also have a white anal tuft. Schwarz finds that the face color soon changes, becoming dark or even black with age, the same applying to hands and feet and body skin. Specimens occur with reduced pigmentation, and in the young the hair may be reddish instead of black. With age, a bald spot appears on the forehead, more frequently in males than females, and the hair turns grey.

Returning now to the gorillas: Keith (1927), in a study of forty-

---

[7] It is to be hoped that stability will soon be reached, at least in regard to the generic name for the chimpanzee.

two skulls from one locality, found them more variable than in man. They ranged from pronounced dolichocephalic to ultra-brachycephalic (excluding the frontal air sinuses and the occipital protuberance), the cephalic index being 60–79.8 for males and 61.7–80.6 for females. The cranial capacity was 355–620 cc. Coolidge (1929), in a study of some eight hundred skulls in museums, also found great individual variation. Among fifteen specific or subspecific names applied to the gorillas, he recognizes only two subspecies—the coastal form *G. gorilla gorilla* in the forests of the Cameroons and Gaboon, and the mountain form, *G. gorilla beringei,* which is known in the mountainous eastern part of the Belgian Congo and Uganda, from the northern part of Lake Tanganyika to Lake Kivu and Lake Edward, occurring up to altitudes of over seven thousand feet.

G. beringei was described by Matschie in 1903. Its numbers are much smaller than of the coastal species and it is protected by government. These two forms are separated by a forest belt 750 miles wide in which gorillas apparently do not live. Some writers recognize a third species, *G. matschei,* in the coastal region. Coolidge shows that the presence of a chestnut crest and red hair is variable, as is also the shape of the nasals and the superciliary arches, the greatest variation being shown in the development of the sagittal crest. Lönnberg (1917) described a subspecies *mikenensis* of *G. beringei,* found on the slopes of the Mikeno volcano. Its skin is black, hairs black, paler at base. *G. beringei* is distinguished from the western (coastal) gorilla by the great palatal length, generally narrower skull, and the thicker long black fur. It also has a fleshy callosity on the crest of the head.

Schultz (1934) has given good reasons for treating *G. gorilla* and *G. beringei* as distinct species. Having measured and studied some forty skeletons or complete animals, he found various skeletal differences, such as a longer neck, narrower hips, shorter humerus, and longer clavicle in *G. beringei,* the ventral border of the scapula being irregularly concave rather than convex or straight. Among other differences in proportions, *beringei* has a longer trunk, the hand is relatively shorter and broader, the palate longer, face height greater, orbits closer together, the great toe shorter, and a smaller cleft between digits I and II, the other digits being shorter with more extensive webbing, so that *G. beringei* is more or less syndactylous. The webs in some cases extend even to the middle of the toe nails. Some twenty-five differences between the two species are listed, but some of these may prove to be inconstant, after further study. In any case, the morphological differences appear to be ample to give these forms specific rank, but there is no reason to suppose that they would be sterile if crossed. We have already seen that this criterion of species is more honored in the breach than in the observance. These two species differ not only in distribution but in habits, *G. beringei* being almost entirely terrestrial while *G. gorilla* is more arboreal. This difference in habits corresponds with differences in the foot.

The gorilla and chimpanzee have long been recognized as different genera of anthropoids, yet a few animals are on record which are so intermediate or show such a mixture of characters that opinions have differed as to which genus they belonged to. In a detailed anatomical comparison, Keith (1899) showed that there is scarcely a feature in any muscle or bone of the gorilla which is not found in the chimpanzee, though frequently the exception in one is the rule in the other. In both, the muscles of the fifth toe tend to become vestigial, as also in man. Darwin's point (in the ear) is found in 9 per cent of chimpanzees and 26 per cent of gorillas, but the ear of the gorilla is more nearly human. The mean cephalic index of the gorilla is 80 and of the chimpanzee 86 (without the tori). Keith summarizes the main distinctions of the gorilla (apart from its larger size) as, (1) its sullen, untamable, ferocious nature, (2) the long nasal bones, (3) the great alar nasal folds, (4) peculiar molars, premolars, and canines, (5) broad, short, thick, webbed hands and feet, (6) a long heel, very long upper arm, and smaller development of the forearm. There seems no reason why the male gorilla should not cross with the female chimpanzee.

Yerkes (1929) discusses the occurrence of intermediate forms. In 1881, von Koppenfels observed the association of coastal gorillas with chimpanzees in their native habitat. He says, "I believe it is proved that there are crosses between the male *Troglodytes gorilla* and the female *Troglodytes niger,* but for reasons easily understood [the great size difference], there are none in the opposite direction." He believed that he had positive proof of crosses and mentions five supposed cases. Garner denied the possibility, on the ground that they belonged in different genera! Yerkes considers hybridization not definitely established, but agrees that specimens occur which are difficult to classify and for which no other explanation appears to be available. The earliest case is "Mafuka," brought from the Loango coast of western Africa to the Dresden Zoo, in 1874. She was pronounced a chimpanzee by some authorities, a gorilla by others.

Duckworth (1898) described the case of Johanna, an aged female anthropoid from the Gaboon River, which he dissected but was unable to refer to as a true chimpanzee or a genuine gorilla. She had (1) large ears, whereas in gorillas the ear is generally small; (2) there was a lack of supra-orbital prominences, which are present even in the female gorilla. Other points were the interorbital distance, the upper lip, the slender hand and foot, the development of hallux and pollex, and the small size of the teeth. Duckworth places five apes in this intermediate category, including Mafuka and Johanna. It will be noted that Mafuka had heavy brow ridges while Johanna had none. They are not then both likely to be first generation hybrids. Johanna, if a hybrid, as appears likely, would be a later generation segregate. This cross seems no more unlikely than that between Neanderthal and Cro-Magnon man.

Keith (1899) gives a colored plate of Johanna. His analysis of the anatomical and temperamental differences between gorillas and chimpanzees arose from this case. He also considers two other aberrants or intermediates, the Koolookamba of Chaillu and the *aubryi* specimen, regarding them as chimpanzees mistaken for gorillas. Rothschild (1904) treated them as representing two different species of chimpanzee, distinct from *satyrus*, *pygmaeus* and *vellerosus*. It is evident that much more information on the subject is desirable. Apparently the most conspicuous differences between gorillas and chimpanzees are in size and in disposition.

VARIATION IN AFRICAN MONKEYS

A few instances of varietal and specific variation in monkeys as compared with man may be instructive. *Cercopithecus mona* extends across central Africa from Sierra Leone north and south of the Congo rain forest, but not reaching Abyssinia. Schwarz (1928a) classifies ten forms of this species into three groups. He regards Cercopithecus as the most primitive genus of African monkeys, and *C. mona* as probably the most primitive species. *C. mona mona* is found on the coast from upper Guinea to the western Cameroons; *C. m. pogonias* in lower Guinea north of the Congo and west to the Cameroon Mountains. In the latter region the two subspecies overlap. They differ especially in coloring but do not mix. Isolation is believed to account for the origin of the different forms. The denser and wetter the forest, the stronger is the development of both red and black pigmentation. *C. m. wolfi* occurs south of the Congo and on the right bank above the Ubangi tributary. It is regarded as the most primitive, spreading from south of the Congo to the northeast but mainly westwards along the Atlantic coast. *C. m. pogonias* and *wolfi* are much more alike in skull form, face pigmentation, ear tufts, hair on the tail, and color development than either is to the subspecies *mona*. They are also regarded as younger in skull type and more primitive. They produce various mutations, or color forms. The subspecies *pyrogaster* differs from *wolfi* in (1) absence of a black line on the flanks, (2) the red pigmentation being very strong. *Elegans* differs from *wolfi* in (1) the nearly complete absence of red pigmentation, (2) the ear tufts are not red but slightly yellowish white. While the majority of these differences are simply in color pattern, *pogonias* differs from *wolfi* in both color and size, and these two forms both differ from *mona* in skull structure. Schwarz thus finds all stages of differentiation in *C. mona*, from mere modifications to forms which differ in several respects and live in the same area without intercrossing. These local subspecies are very unequal genetically, morphologically, and especially in time value, but all belong to one Formenkreis. Later, Schwarz (1928b) recognized nine species and twenty-five subspecies of Ceropithecus in the Congo Museum.

Sanderson (1940) on the contrary finds that *C. pogonias* and *C. mona* are not subspecies but species. They mingle in the forests but are totally different in appearance, habit, and voice. His first-hand observations of mammals in the tropical forest are of great interest. He found great differences between the wild and preserved skins. Specimens in the field sometimes fell into two clearly divergent groups (Galago species) of which there was no evidence in the preserved material. Conversely, species which appear rather homogeneous in life may show marked subspecific differences in preservation. Sanderson goes on to point out that "the splitting of a Linnaean species into subspecies is only a taxonomic practice of convenience," of purely academic interest. "The mammalian fauna, especially of a large area of tropical forest, may be likened, metaphorically speaking, to a soup with lumps in it. The lumps are Linnaean species, but in nature they are geographically and not taxonomically distributed." In reality there is often a gradation (cline), to a greater or less extent complete. Subspecies similarly "blend hither and thither until the periphery of the whole phylogenetic clot is reached, which in turn often proves to coincide with the natural limits of some vegetational zone."

Skins were found to vary greatly in color according to the method of preservation. *Cercopithecus pogonias* had a bright green back and a yellow belly. The same was true of species of Galago, Anomalurus, and Aethosciurus, also living diurnally in the brilliant sunlight of the upper strata of foliage. These parallel colorations are evidently in adaptation to the conditions. The green color may be due to simple Algae living on the hairs (as in the tree-sloth of South America) or it may be refractive in origin. It fades rapidly after death. In this monkey it goes to bottle-olive green, in Galago it turns rusty. The yellow belly becomes deeper in tone in the monkey, reddish in Anomalurus, slightly paler in Galago. Thus if the original living colors are unknown, dried skins may give quite false impressions.

Sanderson finds that *C. pogonias* is not related to *C. mona*. From Sierra Leone to the Congo he finds a succession of forms— *C. campbelli, lowei, mona mona, pogonias, grayi, nigripes* and *mona wolfi*. *C. nictitans* has white- and black-fronted forms with different distribution. Further west is *C. petaurista* and *C. stampflii*. *C. erythrogaster* has red- and grey-bellied forms. Such dimorphism is found to be not uncommon in mammals, and it also occurs in birds. *C. erythrotis* has three subspecies, *sclateri* in the Niger delta, *cameroonensis* in the Cameroons, and *erythrotis* on the island of Fernando Po and the adjacent mainland. *C. preussi* has its back mainly reddish brown (agouti?), the undersides grey and the iris reddish, whereas in *C. pogonias* the iris is reddish brown. These observations on the color variations of tropical monkeys all deal with characters of less morphological significance than many of the differences between the primary races of man.

In a study of local races in the Colobus monkey, Schwarz (1929) shows that they stretch from Liberia and Sierra Leone in west Africa across central Africa through the Congo basin to Abyssinia, and to the east coast at Mombasa. He finds twenty subspecies or geographic races, only two of which overlap, and these do not interbreed. Lydekker (1905) traced the evolution from a black monkey (*satanus*) with no tail-brush or "mane" in west Africa to various eastern forms with white markings, tail-brush, and long hairs on flanks and shoulders. Schwarz finds two cases in which local forms overlap without crossing. He explains the distribution by the geological history of central Africa. The Congo basin was formerly a vast inland lake, and the lower Congo is of more recent date. Various forms were isolated by this great lake. The twenty subspecies all belong to a single species, *Colobus polykomos*, Zimm. and are divided into four sections. They differ in color markings, shoulder mane, beard, crown-whorl, and tail-tufting. Many of the "racial" differences in man appear to be more fundamental.

Three species of leaf-monkeys of the genus Pithecus are recognized in Borneo. They are large, hairy monkeys weighing twelve to fifteen pounds. Banks (1929) shows that they interbreed. *P. chrysomelas* is black, *P. rubicunda* red, and *P. cruciger* parti-colored. *P. chrysomelas* has close relatives in Malaya and Sumatra. This species normally has a black back with a grey abdomen and a white stripe down the inside of each limb. In one specimen the base of the hairs in some parts of the body were rusty (not black). *P. rubicundus* is chestnut-red all over, and not very variable. At higher altitudes it is more hairy and often darker. In Dutch Borneo there is a tendency to black hands and feet. *P. cruciger* is very variable, scarcely two being alike. The general color is rusty-red (more yellowish than *P. rubicundus*) with a black line down the back, the fore-limbs and tail (forming a cross), the hind feet always black. The young of *chrysomelas* are entirely white at one month except for the black lines forming a cross. The young of *rubicundus* are chestnut-red like the adult, while those of *cruciger* are rusty-reddish (like the adult) with the black cross. Occasionally a single red or black animal will be seen in a troop. The red *rubicundus* is more inland than the black *chrysomelas* and reaches higher altitudes, but the two do not mix. The parti-colored *cruciger* is very local and is found mainly where the other two species overlap. The black and parti-colored species have a blackish face and a light brown iris, the red species a more bluish-grey face with darker brown iris. The voice is also the same in the black and parti-colored species but different in *rubicunda*. One young *cruciger* was white with a black cross like the young *chrysomelas*, but there were reddish hairs in the forehead.

Chrysomelas and *cruciger* are known to interbreed. Banks suggests that *chrysomelas* and *rubicundus* may have been distinct species which interbred to produce *cruciger* where their ranges overlapped. The latter are intermediate in distribution, and the young show the color patterns of all three species. In any case, the convention which reduces them automatically to subspecies when evidence of intercrossing is found adds nothing to our understanding of their relationships, but rather obscures the fact that on the one hand sterility can arise between forms which are very closely related and on the other hand may not arise between forms which show many specific and (as we have seen) even generic differences. It has been pointed out elsewhere (Gates, 1938) that sterility with related forms may arise very early or very late in the differentiation of a species.

The fact that so much local differentiation of type can occur in monkeys shows that they are much more sedentary in their habits than usually supposed. Mayr (1942, p. 240) also remarks, "Very small animals are subject to much involuntary dispersal, but . . . most larger animals are amazingly sedentary throughout their lives and make very little use of their potential dispersal powers." He points out that the same is true in general for birds, and that game fish do not move from one part of a lake to another. The evidence indicates that the primitive races of man were, and are, generally equally sedentary, not moving from their ancestral home unless driven by invading enemies. The Polynesians and other migrating races of the historic period were mentally highly developed and not primitive.

The fact that sterility can arise early or late in the differentiation of species makes it a very poor and indeed often misleading criterion of specific differentiation. For this reason it should, in the writer's opinion, be regarded not as *the* criterion of species, but only as a further consideration in addition to whatever morphological or physiological differences may exist. For this reason the writer prefers Tate Regan's (1925) definition of species. He says (p. 75), "A species is a community, or a number of related communities, whose distinctive morphological characters are, in the opinion of a competent systematist, sufficiently definite to entitle it, or them, to a specific name." It may be pointed out that this definition is in reality only a paraphrase of Darwin's statement that "the opinion of naturalists having sound judgment and wide experience seems the only guide to follow" regarding species. When species can arise in an infinitude of different ways it is folly to try to find one hard and fast criterion which will apply to all. Moreover, it leads to confusion by placing in the same category things which represent very different degrees of differentiation. Surely it is better, as many do, to recognize overtly that species are of many kinds, rather than, by emphasizing a single criterion, to raise the false assumption that all species are equal.

Most systematists appear to be agreed that species are not absolute units but rather conveniences. Species are described in order to deal with nature's multiformity, but every naturalist who is familiar with any group of organisms or has described species knows that the delineation of a particular species from those most nearly related is often a matter of convenience or compromise. When a new species is described the author can hardly ever have full knowledge of just where a break in every series of characters occurs. Often some of the gaps which he believed to exist will be filled in later by fuller knowledge. This is not to say that no gaps in the series exist, but they are bound to be fewer than the gaps in knowledge.

The delimitations of species in polymorphic genera are then partly based on realities (i.e., real gaps), partly on ignorance (supposed gaps), and partly on mere convenience. From this it follows that the number of species recognized in many large genera is not of vast importance, within reasonable limits. There will always be a certain difference of opinion between the lumpers and splitters, except in genera where extinction or larger mutational steps have produced more conspicuous gaps.

The genus, on the other hand, is supposed to record similarities, while species represent a certain aggregate of differences. The present tendency in some groups to create almost as many genera as species by splitting up old and well-recognized genera defeats its own end, because species in the same genus are obviously related; but if they are placed in separate genera that evidence of relationship disappears. As we have already seen in particular cases, differences of opinion frequently exist as to whether particular forms should be ranked as subspecies or species. Mayr (1942, p. 163) says, "Many subspecies are characterized by more striking differences than some 'good' species . . . There is no 'gap' between subspecies and species as far as systematic characters are concerned." This again shows that the decision as to how many forms should rank as species and how many only as subspecies is partly determined by convenience, as the distinction is not absolute. Actual genetical knowledge of intersterility in most wild species is nil, and many surprises develop when an opportunity occurs to check ideas on the subject by actual crosses.

REFERENCES

Allen, G. M. 1939. A checklist of African mammals. *Bull. Mus. Comp. Zool. Harvard.* 83:1–763.

Banks, E. 1929. Interbreeding among some Bornean leaf-monkeys of the genus Pithecus. *Proc. Zool. Soc. London.* 693–695.

Broca, Paul. 1864. *On the Phenomena of Hybridity in the genus Homo.* Ed. C. C. Blake. London: Longmans Green. pp. 76.

Bunak, V. V. 1927. *Crania Armenica.* Moscow. pp. 263. Pls. 25.

Coolidge, H. J., Jr. 1929. A revision of the genus Gorilla. *Mem. Mus. Comp. Zool. Harvard.* 50:292–381. Pls. 21.

Darwin, C. 1871. *The Descent of Man.* 2 Vols. London: Murray.

Demerec, M. 1937. Frequency of spontaneous mutations in certain stocks of Drosophila melanogaster. *Genetics.* 22:469–478.

Deniker, J. 1913. *The Races of Man.* London. (3rd ed.) pp. 611. Figs. 176.

Duckworth, W. L. H. 1898. Note on an anthropoid ape. *Proc. Zool. Soc. London.* 989–994. Figs. 7.

Ehrich, R. W. 1947. Some doubts about the validity of the Dinaric racial classification. *Am. J. Phys. Anthrop.* 5:236.

Elliot, D. G. 1913. *A Review of the Primates.* Monogr. Am. Mus. Nat. Hist. 3 Vols. New York.

Ewing, J. F. 1947. Occipital flattening as a racial diagnostic. *Am. J. Phys. Anthrop.* 5:235.

Flower, W. H., and R. Lydekker. 1891. *An Introduction to the Study of Mammals, living and extinct.* London: Black. pp. 763.

Gates, R. R. 1936. Recent progress in blood group investigations. *Genetica.* 18:47–65.

———. 1938. The species concept in the light of cytology and genetics. *Am. Nat.* 72:340–349.

———. 1944. Phylogeny and classification of hominids and anthropoids. *Am. J. Phys. Anthrop.* N. S. 2:279–292.

Hall, E. R. 1946. Zoological subspecies of man at the peace table. *J. Mammalogy.* 27:358–364. Figs. 2.

Hooton, E. A. 1925. The asymmetrical character of human evolution. *Am. J. Phys. Anthrop.* 8:125–141.

Keith, Arthur. 1899. On the chimpanzees and their relationship to the gorilla. *Proc. Zool. Soc. London.* 296–312. Col. Pl. 1.

Keith, Sir A. 1927. Cranial characteristics of gorillas and chimpanzees. *Nature.* 120:914–915.

Lönnberg, E. 1917. Mammals collected in Central Africa by Captain E. Arrhenius. *K. Svensk. Vetens. Akad. Handl.* 58:No. 2. pp. 110. Pls. 12. Figs. 11.

Lotsy, J. P. and W. A. Goddijn. 1928. Voyages of exploration to judge of the bearing of hybridization upon evolution. *Genetica.* 10:1–315. Figs. 153. Pls. 11.

Lydekker, R. 1905. Colour evolution in Guereza monkeys. *Proc. Zool. Soc. London.* II:325–329. Figs. 5.

Matschie, P. 1919. Neue Ergebnisse der Schimpansenforschung. *Zeits. f. Ethnol.* 51:62–86. Figs. 7.

Mayr, E. 1942. *Systematics and the Origin of Species.* New York: Columbia University Press. pp. 334.

Regan, C. Tate. 1925. Organic evolution. *Brit. Assn. Repts.* pp. 75–86.

Rhoades, M. M. 1938. Effect of the Dt gene on the mutability of the a_1 allele in maize. *Genetics.* 23:377–397. Figs. 5.

Rothschild, Hon. W. 1904. Notes on anthropoid apes. *Proc. Zool. Soc. London.* II:413–440. Figs. 19. Col. pl. 1.

Sanderson, I. T. 1940. The Mammals of the North Cameroons forest area. *Trans. Zool. Soc. London.* 24:623–725. Pls. 22 (4 colored).

Schultz, A. H. 1934. Some distinguishing characters of the mountain gorilla. *J. Mammalogy.* 15:51–61. Figs. 2.

Schwarz, E. 1928a. Studien der Artbildung; die geographischen und biologischen Formen der Mona-Meerkatze (*Cercopithecus mona* Schreber). *Verh. V. Int. Cong. Vereb.* 2:1299–1319.

———. 1928b. Die Sammlung afrikanischer Affen im Congo-Museum. *Rev. de Zool. et Bot. Afr.* 16:105–152.

———. 1929. On the local races and distribution of the black and white Colobus monkeys. *Proc. Zool. Soc. London.* 585–598.

———. 1934. On the local races of the chimpanzee. *Ann. Mag. Nat. Hist.* 13:576–583.

Yerkes, R. M. 1943. *Chimpanzees, a laboratory colony.* New Haven: Yale University Press. pp. 321. Pls. 63.

Yerkes, R. M., and A. W. Yerkes. 1929. *The Great Apes.* New Haven: Yale University Press. pp. 652. Figs. 172.

12

Paleontology, Speciation,[1] and Sterility

It is necessary to realize that time enters as a fourth dimension in evolution as in relativity. Characters are modified or replaced in time, but what we call characters are for the most part only the detritus cast upon the organic beaches of life by the tides of development, both ontogenetic and evolutionary. In previous chapters we have been considering the characters of early types of man and their modern descendants in many parts of the world, thus linking up types into suggested lines of local evolution. The principles involved are of course exactly the same as in ordinary paleontology, so we need to examine further the paleontological method and point of view.

In a valuable study of this subject, Simpson (1945) recognizes the genus as the basic unit of practical and morphological taxonomy, while the species is the basic unit of theoretical and genetic taxonomy. He shows that practical taxonomy in paleontology is, and must be, a balance between a purely vertical or phyletic classification based on time and a horizontal grouping based on space (that is to say, on the geographical distribution of contemporaneous forms). He says (p. 17), "The existence of groups that are ancestral to two or more ultimately quite different phyla and the implication in classification that members of one group are more nearly related to each other than to members of other groups of the same rank give rise to the most difficult problems of classification of fossils." How is the ancestral group

---

[1] The term *speciation* appears to have been first used by Cook (1907, p. 278). His definition is: "The attainment of differential characters by segregated groups of organisms, that is, by subdivisions of older species." It was later used independently by me (Gates, 1917, p. 579) in the general sense of species formation. Osborn (1927, p. 5) afterwards adopted it in the restricted sense of continuous species formation, as contrasted with origin by mutation. It is here used in the general sense. See also Gates (1927, p. 462).

to be classified? In a sequence, is a group more nearly related to its ancestors, its descendants or its contemporaries of like origin? In other words, is a man nearer to his father, his son, or his brother? Where an ancestral unit has two descendant lines, one may (a) include the ancestry in one of the descendant groups (the more conservative course), or (b) give the ancestry a separate name and the same rank as each descendant. Both methods are in use and justifiable in certain circumstances.

Simpson cites the case of the three equid genera Merychippus, Pliohippus, and Hipparion, the first ancestral to the other two. With many examples, he shows that every good classification combines both methods and cannot confine itself to a horizontal or a vertical classification. For instance, the early horses and tapirs were nearly related, but a horizontal classification would unite them while a vertical classification separates them and recognizes the two lines of descent, horses and tapirs. A vertical classification is thus more evolutionary but more subject to modifications with increase in knowledge, whilst a horizontal grouping of species into genera is more Linnaean. Neither can fully express phylogeny, but both can be consistent with phylogeny and both have to be used. If one genus gives rise to another, the last species of the first genus will be most closely related to the first species of the second genus, yet it will be in a different genus. The principle of nearness of affinity is thus violated, but necessarily. The existence of parallel variations will also, of course, increase the difficulties.

In a vertical series of forms, gaps in the record frequently determine where a generic or specific line is to be drawn. If there is no convenient gap, the line will have to be cut somewhere and the position of the "cut" will depend on many factors, including that of convenience. A certain range of variation enters into every character —variation both contemporaneous and in time—and this is an essential part of the character, but no classification can recognize all the affinities and cross-relationships which exist.

Since extended reference has been made to the paleontological views of Simpson, with which the present writer is in agreement, it is only fair to quote his brief reference to man, with some of which the present writer does not agree. Simpson says, "All known hominids, recent and fossil, could well be placed in *Homo*. At most, *Pithecan-*

thropus (with which *Sinanthropus* is clearly synonymous by zoological criteria) and *Eoanthropus* (if the ape-like jaw belongs to it) may be given separate generic rank." The whole of this statement is in harmony with the views expressed in this book, except the first sentence, which appears to the writer to show lack of appreciation of the large differences involved. The second sentence largely makes amends for the first, but the present writer, like many other anthropologists, would add certain other genera, such as *Palaeoanthropus*.

It is well known to taxonomists that a species may be defined by distinguishing it from other related species or by describing its characters and setting limits to the range of variation included in them, these two processes being complementary. Similarly, the analysis practised by "splitters" and the synthesis indulged in by "lumpers" are complementary activities.

Turning now to contemporary species, the term "cline," suggested by Ramsbottom, refers to a series of forms in a species which show variations progressing in particular directions corresponding to their geographic relations at one time, usually in correlation with one or more environmental factors. One of the best examples of this kind of variation was discussed (Gates, 1917) long before the term cline was introduced. *Otus asio* is a screech owl distributed over the greater part of North America and well known for its numerous geographic races or subspecies. In 1917, thirteen subspecies were recognized, with only two cases in which their areas overlapped. Subspecies *aikeni* and *maxwelliae* both occur in central Colorado, this being the northern limit for the former and the southern limit for the latter. But *aikeni* occurs mostly at higher altitudes, so that at Colorado Springs it is found only in summer, while *maxwelliae* occurs there only in winter. Similarly, *gilmani* and *cineraceus* are both found in southwestern Arizona but they can be taken in the same localities only in winter when *cineraceus* comes down from the mountains to the hot valleys occupied by *gilmani*. The most striking cline consists of *kennicottii*, *brewsteri*, *bendirei*, *quercinus*, *cineraceus*, *gilmani*, and *xantusi*, extending in steps from southern Alaska down the Pacific coast to San Francisco Bay (*bendirei*). A break follows, then *cinceraceus* and *gilmani* are found in the desert region of southern California and adjacent Arizona. Finally, *xantusi* occurs at the tip of the peninsula of lower California. The forms in this cline begin in the north with maximum size and darkest vermiculations on the feathers. Proceeding southwards they become progressively smaller and lighter in color, ending with the very small and pale colored *gilmani* in the desert, which builds its nest in the giant cactus.

In Peters' (1940) check-list of birds, the number of recognized subspecies of *Otus asio* is increased from thirteen to twenty-two and the two previous gaps are filled by *clazus* in the mountains of southern California and *cardonensis* on the Pacific coast of lower California. This case is chosen as one of the most extensive clines, extending from Alaska down the whole Pacific coast to the tip of lower California and into the Arizona desert. It exemplifies two well-known laws of geographic variation in animals: 1. That northern forms are large, becoming smaller in lower latitudes; 2. That feathers or fur are darker in moist conditions, becoming palest of all in the desert. Mayr (1942) finds that these ecological rules for mammals and birds have only 25 per cent of exceptions. The other subspecies of *Otus asio* spread over the continent of North America show similar conditions, but the clines are less clearly related to temperature and moisture conditions than in this very striking Pacific coast series.

There are parts of the world where clines may perhaps still be recognized in mankind, but they are largely disrupted by recent man's having more migratory proclivities. Such clines are due to hybridization of neighboring races or to population drift rather than local adaptation. The recognition of clines in the distribution of animals and plants is a significant development in evolutionary study. It is by no means new, although it has only recently received a general name.

Now it is obvious that if there are clines in space, as we have seen, there may also be clines in time, although they will not be quite the same thing. Simpson (1943, p. 174) distinguishes *choroclines* or geographic series of subspecies and *chronoclines* or successive variants in time. This brings us to that "fourth dimensional" aspect of species with which the paleontologist must constantly deal. We may think of choroclines as horizontal in space, on the earth's surface, and chronoclines as vertical, in time. The choroclines at any one time consist of the populations of a species in their geographical arrangement, whereas Simpson regards chronoclines as "purely subjective size groups," not corresponding to any real defined populations that existed in nature and hence not species in any real sense. However, the fossils preserved in the form of chronoclines are samples from successive populations, generally derived by breeding in a particular line of descent; and after all, we can also only see samples of the populations on which choroclines are constructed.

In so far as the populations from later horizons in a series are derived from the earlier, a chronocline is as real as a chorocline, but in case of a shift in type it will be difficult to know in how far it represents a population or chorocline shift resulting from some local climatic change. In this sense chronoclines will generally be subject to a doubt which is not present in choroclines. Probably the study of chronoclines at the subspecific level will only be feasible in the invertebrates, where large numbers of a species are embedded in successive horizons over a considerable period. In man, the remains will always be much too fragmentary to apply this conception below the generic or specific level. Moreover, the serial characters in a chronocline may be often intraspecific and adaptive, whereas in man the "racial" characters appear to be mainly non-adaptive.

The boundary zones between the subspecies in choroclines, such as those of *Otus asio*, require further study to determine the nature of the transitions. Are they generally sharp, with little or no interbreeding where the subspecies meet, or does the whole series form a continuum? If there are sharp gradients between the choroclines, is this because of sterility in the hybrids, aversion to crossing, or some slight ecological difference in habits or habitat? Simpson recognizes that some choroclines have an even slope (in graphic terms) while others have a plateau bounded by steep slopes (narrow transition zones). The character of the boundary between two subspecies will often be determined by the topographic or microclimatic conditions, but the question of interfertility is also involved. In chronoclines a steep gradient may represent rapid evolution or the replacement of one mutant type by another, but generally several characters and not a single one will be affected, producing mutations in the sense of Waagen.

In chronoclines, sterility will not be involved, for it is obvious that they can only be produced by continuous breeding from generation to generation. This is a fundamental point, which has frequently been overlooked by those who rely upon intersterility as the sole criterion of species. Since intersterility cannot exist between the successive elements in a chronocline or the successive species in a paleontological series, why stress intersterility as the sole criterion of distinction between contemporary species, geographic subspecies or members of a chorocline? This is laying undue stress on the presence of one genetical condition and disregarding all the other biological relationships of species. Mayr (1942) recognizes this when he says (p. 119), "A species definition in which sterility is the principal criterion is ... not acceptable, at least not for most groups of animals." This is a point of view with which the present writer heartily agrees. Mayr distinguishes between fertility and crossability, pointing out that two animals may be interfertile but never cross in nature, so that they are reproductively isolated. Disinclination to cross with widely different types is unfortunately a condition which has never developed completely in modern man, although as in all animals, preference for a mate of similar type is quite general in mankind. If it were universal, as in animals, then the frequently unhappy combinations arising from crosses between different human species would never occur.

From what has been said, it is clear that intersterility can arise as an incident or by-product in the differentiation of contemporaneous species, but if it arises in a chronocline the result is extinction and not evolution. It is readily conceivable that some evolution can take place in a group without its separation into contemporary intersterile forms. Just as there is no sterility in the time (vertical) series, there is no necessity for it in the space (horizontal) series. All depends on how widely such contemporaneous groups become differentiated morphologically and physiologically, and how isolating mechanisms develop. In maize it is found that five translocations in the chromosomes will produce 97 per cent of sterile germ cells.

Osborn, whose wealth of experience in mammal paleontology has rarely been equaled, wrote in 1926 (p. 4), "The genus Homo is subdivided into three absolutely distinct stocks, which in zoology would be given the rank of species if not of genera," these three stocks being popularly known as Caucasian, Mongolian, and Negroid. He pointed out that the argument from fertility "is now known to be invalid." He goes further and states (p. 3) that "if an unbiassed zoologist were to descend upon the earth from Mars and study the races of man with the same impartiality as the races of fishes, birds and mammals, he would undoubtedly divide the existing races of man into several genera and into a very large number of species and subspecies." This is an overstatement, but he would no doubt recognize the three primary "races" mentioned above as species, and we may be sure that he would add the Australian aborigines and the Bushmen of South Africa as two more species which are rapidly nearing extinction.

Contrary to the usual view of taxonomists, Dobzhansky (1937) proposes to determine species solely on the basis of their crossability. He recognizes that populations are isolated from each other by different methods in nature, but he defines species as "the stage in the process of evolutionary divergence at which the previously freely interbreeding array of forms splits into two or more separate arrays prevented from interbreeding with each other by some physiological isolating mechanisms." This disregards the fact that intersterility occurs in every possible degree, and in some unexpected ways, as we shall show. In practice it is often impossible to say just at what point exchange of genes between two forms has finally ceased, so that Dobzhansky's criterion is no more absolute than any other. It would seem that the era of search for an absolute criterion of species is about over. They have all broken down, and we shall see that the criterion of intersterility produces more glaring absurdities than any other.

Dobzhansky tacitly excludes from his scheme polymorphic genera of plants; but a genus like Oenothera, with perhaps seventy-five species, is hardly to be classed as polymorphic in the sense of such genera as Rubus, Crataegus, and Salix, yet nearly all the species are interfertile. No botanist of any school would dream, however, of trying to reduce them all to one species; but this procedure, which would make taxonomy a farce, would be necessary under Dobzhansky's scheme.

Notwithstanding Mayr's recognition that intersterility is unsatisfactory as a criterion of species, partly because in taxonomic practice it can only be applied in very few cases, but also because of the extremely contradictory results to which it leads in different genera; yet he says (p. 120) that in many cases in birds, where forms are geographically isolated, the taxonomist must decide (guess) whether two forms would interbreed if their ranges overlapped, and hence whether they are to rank as species or subspecies. This would make a guess about crossability or intersterility of more value than the observable differences between the forms in question! Mayr admits, however (p. 121), "We may have to apply the degree of morphological difference as a yardstick in all those cases in which we cannot determine the presence of reproductive isolation." Surely the former is the normal procedure for a taxonomist; but he merely justifies the use of morphological criteria in "doubtful" cases, thus making the dubious possibility of intersterility more important as a criterion than observation of the phenotypic differences which exist.

Mayr recognizes that "scales of differences" between species and subspecies differ in every family and genus, but modern genetical work shows that the irregularities of intersterility between related forms are just as great and just as unpredictable. On the one hand, forms so closely similar that no taxonomist would give them even varietal rank may be completely intersterile (Drosophila), and on the other hand, a whole subgenus (Onagra) or even a genus (Rubus?) may be interfertile.

It is well recognized that the taxonomy of birds is more advanced than that of any other group of animals, the number of "species" having been reduced from 27,000 to about 8,500, mainly by recognition that many are geographical subspecies. Up to 1870 most of the latter were described as species, but since 1920 nearly all geographic races have been reduced to subspecies. This is no doubt all to the good, as the case of *Otus asio*, already considered, shows; but it does not lessen the importance of morphological criteria in dealing with Linnaean species. The high development of bird taxonomy makes it possible for ornithologists to recognize that most good species of birds are composed of groups of subspecies—showing that the species is not the lowest taxonomic category.

How peculiar can be the conditions which advance or retard crossing between animals is shown in the birds of paradise. This remarkable group, related to the crows, is divided by Mayr (1945) into twenty genera with thirty-nine species and fifty-nine subspecies. Fifteen rare forms turn out to be natural hybrids of well-known species. Hybrids are more frequent in this group because the sexes are generally very unlike, the males with elaborate gaudy plumage while the demure females lack these specializations and are thus more like the ancestral crows. The young females see little or nothing of the male of their species before they are full grown. Not knowing their own father, so to speak, they will accept a male of another species with quite different plumage as readily as one of the species to which they belong. This shows that the species are kept separate, not by sterility but because the female generally meets a male of its own species (which in appearance is quite unlike itself) before that of any other species. Aversion to crossing appears to be lacking, and there is evidently no mutual recognition between males and females of the same species, at any rate at their first meeting.

In man, the only isolating mechanisms (apart from possible intersterility [2] in some of the more widely different types) are space and disinclination. It by no means follows that they should all be collected into one species. Their morphological differences in comparison with those of the anthropoids and monkeys have already been pointed out. We must now justify the statements made in a previous paragraph regarding the incidence of intersterility. A very interesting case has been the two forms now known as *Drosophila pseudoobscura,* races A and B. Both are found in Oregon and Washington on the Pacific coast, and race A has also been found in California and Texas. Lancefield (1929) showed that although phenotypically identical, the reciprocal hybrids, made with some difficulty, produce sterile males and semi-sterile females. The only morphological difference found between these forms is in the Y-chromosome, which is rod-shaped in race A and V-shaped, like the X, in race B. These two races are phenotypically even more alike than *D. melanogaster* and *D. simulans,* which produce sterile hybrids. Most other species of Drosophila will not cross at all, except those most nearly related. Lancefield found evidence of two inversions in the X-chromosome, causing a reduction of crossing-over in the hybrids. Four inverted chromosome sections were subsequently found, the salivary chromosomes pairing in the hybrid except at the breakage points, but the gene-sequences are not constant in either A or B. In chromosome III, twenty-one different inversions have been found —thirteen in race A, seven in race B, and one in both. These in-

[2] Deniker (1913) points out that no crosses have ever taken place between Australian aborigines and Lapps or, for instance, between Bushmen and Patagonians, so it is unknown whether they are interfertile.

versions differ in range of distribution, indicating that some have originated more recently than others. This is confirmed by some of them overlapping in the chromosome. Most populations contain several different gene-sequences and some sequences have discontinuous ranges or occur sporadically. All this goes to show that the conditions in races A and B are still very unstable, in fact that their chromosomes have begun to differentiate and that if these changes continue, they may be expected ultimately to develop into two species.

On the basis of this behavior, Dobzhansky (1941) concludes (p. 307) that "Although these 'races' are practically indistinguishable morphologically, they behave as good species." To call two such forms species because of their intersterility, even though they are phenotypically alike, seems to the writer to be begging the question. Having achieved intersterility, they are in the way to become two species at some future time, provided that they undergo changes which later lead to their differentiation; but to call two forms which are indistinguishable different species is taxonomically absurd. It is, however, the logical result of disregarding the absence of taxonomic distinctions and making sterility an absolute criterion of species. These two races are at the beginning, not the end, of the road to speciation.

Geographically these two races broadly overlap and no hybrids between them have been found in nature. They show certain physiological differences and ecological preferences, and they display sexual isolation, preferring to mate with their own type. All the chromosomes are found to carry genes which affect the interracial sterility, and Dobzhansky concludes that there are at least eight sterility factors in the chromosomes. He finds that the sterility is not due to the differences in the Y— nor to the translocations, but believes it to have a purely genic basis.

In a later study (Dobzhansky and Epling, 1914) the name *D. pseudoobscura* is retained for race A, race B being named *D. persimilis*. Very minor differences have been found in the sex combs and in the wings. Both races are found to have the same gene arrangement in chromosome III, but the other chromosomes show translocations. An attempt is made to justify their taxonomic procedure by the statement (p. 7) that "species exist in nature regardless of whether we can or cannot distinguish them by their structural characters," but it is admitted that the A and B races can only be distinguished by breeding experiments or cytological examination, and even then one form of the Y- is the same in both races. The conclusion is logical enough provided the premises are accepted, but their acceptance would reduce taxonomy to a search for sterility, completely disregarding phenotypes. Every taxonomist knows that such a definition of species is absolutely impractical—an unbalanced attempt to overemphasize a purely genetical in contrast to a broader biological point of view.

In the well-known case of *Drosophila melanogaster* and *D. simulans,* when the latter is the male parent females only are produced. The reciprocal cross produces some males but few or no females. There is a long inversion in chromosome III of simulans, and another in chromosome IV. In certain other species, such as the group *D. virilis, D. americana,* and *D. texana,* crosses produce hybrids of reduced fertility, which varies according to the strains used. There is thus a range in degree of sterility which bears little relation to the degree of specific differentiation. This bears out my view (Gates, 1938) that sterility can arise as the first of the last step in speciation. It seems worthwhile to quote certain earlier statements in this connection. It was pointed out (Gates, 1938, p. 340) regarding the A and B races of *D. pseudoobscura* that the failure to interbreed makes them "the starting point for two new species." Further, "It was supposed until recently that interspecific sterility could only arise gradually over a long period and that it represented the final stage in the production of a true species. Recent experimental work has not only resulted in the immediate production of new species with all the criteria, including those of intersterility and change in chromosome number, but has shown . . . that intersterility is frequently the first stage rather than the last in the production of a new species."

Without going too far into Drosophila genetics we may cite certain other cases which show how variable are the incidence and the expression of intersterility in this genus. Patterson and Crow (1940) describe the hybrid relations of three "species." *D. aldrichi* is found in south central Texas. *D. mulleri* occupies the same area but is also of wider distribution. *D. mojavensis* is not found in Texas but in the California deserts, on Echinocactus. When female *mulleri* is crossed with *aldrichi* the few hybrids produced are sterile, with under-developed gonads. The reciprocal cross gives no offspring. *Mulleri* × *mojavensis* and the reciprocal cross give similar results, and *aldrichi* × *mojavensis* produces a few sterile females. A further study (Crow, 1941) shows that populations of *aldrichi* contain a sex-linked gene (in the X-chromosome which is a dominant semi-lethal in hybrids from *aldrichi* × *mulleri,* but (as in Crepis, see below) produces no effect in *aldrichi*. These three forms and *arizonensis* from Arizona are all now regarded (no doubt rightly) as subspecies of *D. mulleri*. They agree in having five pairs of dot chromosomes, and they have a few major translocations.

Another case, just published by Dobzhansky (1946), goes one stage further than the A and B races of *D. pseudoobscura*. *D. Willistoni* is a species widely known in the American tropics. Dobzhansky describes what he calls a new species, *D. equinoxialis,* found on the upper Amazon with *D. willistoni* at Teffé. It differs from the latter in no respect except that on the average it is a little smaller, but the variations broadly overlap and it is impossible to tell by inspection to which camp any individual belongs. Yet these are both treated as "species" because the male of one will rarely mate with a female of the other even when there is no choice of mates; and no offspring results from insemination, so that the reproductive isolation is complete. This is the perfect species from Dobzhansky's point of view, but from the taxonomic point of view it is nothing more than a basis for the possible development of a future species.[3] To regard forms which are inseparable by any morphological character as "species" is to ignore completely the aims and methods of taxonomy. The possibilities of genetic analysis have no limit, but genetics should not attempt to dictate to taxonomists the adoption of species criteria which would undermine the basis of their science by, on the one hand, naming species where no differences exist, and on the other hand combining hosts of species into one because they can intercross. The resulting taxonomy, if such rules were applied, would be too fantastic to contemplate and would nullify the function of taxonomy, which is to describe the world of organisms in such a way that their forms can be recognized and their relationships to some extent understood.

[3] Even if minor differences are found later between these forms, it will not alter the principle that species should be founded primarily on morphological differences.

The quotation from Mayr (p. 387) indicates that he saw the dangers of sterility as the sole criterion of species, but finally succumbed to the blandishments of the idea or of its advocates. Mayr later (1942, p. 247) defines species as "a reproductively isolated group of populations," thus hinging his definition entirely on the known presence or absence of crossing. Among isolating mechanisms he cites ethological factors (behavior patterns) and concludes that in birds behavior patterns are paramount as an isolating mechanism, thus founding species on the psychology of sex attraction.[4] But he admits that good sympatric species and even genera (occupying the same territory) are interfertile in certain ducks, pheasants, and pigeons. Later still (p. 259) he states again that "it is . . . not admissible to use sterility as a species criterion in animals," yet this appears to be what he has already done. His criterion appears to be that if two subspecies which are brought together by the breakdown of a barrier do not intercross they are species regardless of their degree of differentiation. But (p. 267) the western hedgehog of Europe (*Erinaceus europaeus*) and the eastern hedgehog (*E. roumanicus*) interbreed in the narrow zones where they overlap, yet he accepts them as good species. These are examples of the difficultis which arise when too much stress is laid on interspecific sterility as *the* criterion of species. Stern (1936) has also recognized the dangers of a single criterion of species when he says, "Sterility as a result of crossing has even been made *the*[5] criterion of speciation." He recognizes (p. 124) that "this definition becomes strained on different occasions under the weight of organic manifoldness," but he refrains from attempting to define species and concludes (p. 141) that "the evolutionary origin of interspecific sterility lies not at the beginning of divergent evolution but occurs in the course of it as a by-product." This is in harmony with the views expressed in this book, except that subsequent work has disclosed cases where intersterility arises before any specific differentiation at all has taken place.

That a large amount of gene mutation can take place without disturbing the fertility is shown by the fact that strains of *Drosophila*

---

[4] In *Drosophila subobscura* it has been shown (Rendel, 1945) that a single sex-linked gene produces yellow flies. Normal females will not mate with yellow males, but there is no difficulty in mating yellow × yellow. Sexual aversion has thus arisen through a single gene mutation.

[5] His italics.

melanogaster differing in dozens of genes show no signs of intersterility. It is clear that translocations or inversions of chromosome segments are much more effective than gene mutations in producing intersterility, although occasionally a single gene produces complete intersterility (see below). On the other hand, geographical isolation can lead to a considerable amount of morphological differentiation through different mutations accumulating in the isolated strains, without the development of intersterility. Much therefore depends on the relative frequency of translocations and gene mutations in any isolated race. Sufficient morphological and physiological differentiation must needs lead to the recognition of a "species" whether intersterility has developed or not. This is all the more necessary since in the vast majority of cases intersterility cannot be tested.

Mayr (1940, 1942) has described numerous cases of groups of bird species in various archipelagoes in the Pacific. From the birds of the Solomon Islands and other groups he shows good grounds for concluding that geographical isolation invariably results in speciation in all its stages, but he finds that the time required for species formation is somewhat different in every species and in every district, because of the influence of geographic factors. He concludes with good reason that geographic variation combined with isolation leads to speciation. Dobzhansky (1941) argues that if mutations arise which make their carriers less likely to cross, then these will be selected because they will produce more viable offspring. This would make the development of intersterility dependent upon the selection of such mutations. Dobzhansky himself (p. 287) points out that if physiological isolation developed only by such selective processes, then a group of species isolated on neighbouring islands should not develop intersterility. The fact that they do appears to be fatal to his hypothesis of the origin of intersterility. Certainly more evidence is needed regarding the geographic distribution of isolating genes.

Dobzhansky argues that hybrid sterility by a single gene could only occur if both the homozygotes (AA and aa) were fertile and only the heterozygote (Aa) sterile. He finds such a case in the tail factors in mice. However, in plants the well-known case of *Crepis tectorum* (Hollingshead, 1930) shows that a single dominant gene can be present in homozygous condition and produce no effect on the species, yet it acts as a lethal in the cotyledon stage in crosses with *C. capil-*

laris, C. bursifolia, and *C. leontodontoides* but has no effect in crosses with *C. setosa* and *C. taraxacifolia.* Here is a single very effective isolating gene which is present in some populations of *C. tectorum* in western Europe and central Siberia, but not in others. It is well known that in some species-hybrids in Oenothera, Nicotiana, and other plant genera certain hybrid types die in the seedling stage. There is nothing to show that this gene in Crepis, lethal in the heterozygous condition in one species but of no effect at all in another, has played any part in the differentiation of the species in which it occurs. It appears rather to have arisen incidentally in certain parts of the area of the species and it is in no sense a measure of the degree of differentiation from any other species.

It is necessary to make one more reference to Drosophila and the important work of Patterson (1942), who has made an extensive study of the sterility relationships between various groups of subspecies. He refers to crosses between strains of *D. repleta* which are cross-fertile one way but not in the reciprocal cross. The salivary chromosomes of the hybrids show that sterility is not associated with major chromosome rearrangements but that sexual isolation depends on recessive autosomal factors. He concludes that "despite this isolation the different strains can not be regarded as separate species." In the different species-groups various types of isolating mechanisms are found, all except geographical isolation being due to gene mutations. In the *pseudoobscura* group extreme difference in chromosome structure is accompanied by very little difference in phenotype. In the *mulleri* group phenotypic changes are common but chromosome differences are absent. In the *macrospina* group there is a geographic chain of subspecies, of which only the ends of the series can be regarded as distinct species. The *virilis* group, consisting of five species, *D. virilis, D. americana, D. texana, D. novamexicana,* and *D. montana,* is perhaps of greatest interest. They are geographically and genetically isolated, with little geographic overlap, except *virilis* and *texana,* which have the same distribution but in very different habitats. These five species differ in chromosome morphology. Most interesting of all, *D. americana* is a hybrid species, evolved from crosses between *D. novamexicana* and *texana,* as shown by the chromosome structure.

We may end this discussion of sterility and species with a few ex-

amples from higher animals. In the genus Corvus—the crows—
Corvus corax (the raven) is at home everywhere from Greenland and
the Baltic to the Alps and the Sahara, under conditions as diverse
as those of man himself. The majority of other species have ranges
that partly overlap. *C. corone* L., the carrion crow, and *C. cornix* L.,
the hooded crow, are apposed on a line some 3000 kilometers long in
Europe and Asia (Meise, 1928), the latter being found in the north,
east, and south of Europe. Zones where the two species intercross are
found in Scotland, eastern Europe, and central Siberia, where the
zone is widest in the north. In the Ice Age these species were completely separated, but have since converged, contact taking place
earlier in some areas and later in others. Intergrades between the two
species occur in a zone which varies from a dozen to a hundred
kilometers wide. The zone of contact is narrow where a contact between the species has been longest, and wide where the contact is more
recent, which shows that fresh isolating mechanisms are developing
where the species hybridize. Mayr (1942) regards them as incipient
species, but since they have shown themselves capable of developing
secondary isolating mechanisms where climatic changes have brought
them into contact, they seem to have the essentials of species even
in their sterility relations.

The grackles in eastern North America furnish a similar case.
Chapman (1936, 1939) shows that during the Pleistocene *Quiscalus
quiscala* retreated to Florida and *Q. aeneus* to southern Texas. When
the ice retreated, the former bird took up Alabama and entered
Mississippi further west, while the latter moved eastwards. Where
they met in Louisiana, *aeneus* turned northward, so that they hybridize along an oblique line from Louisiana to New England, producing an intergrading form called *ridgewayi*. This hybridization
zone is only forty miles wide in the south, but in New York and
New England it is much wider. They have, of course, been in contact
longest in the south. *Q. stonei*, found north of Florida, is a mutation
from the Florida bird.

Steiner (1945) has made many species crosses in salamanders and
birds. He concludes that there is no definite line between species and
races. While it is often difficult to say whether a particular cross is
between species or between forms in the same *Formenkreis* (as Mayr
also finds), he reaches the conclusion that the distinction between

good species is easier if they are fully interfertile. Thus, in birds *Poephila acuticauda acuticauda* × *P. a. hecki* is fertile, and in salamanders, *Triturus cristatus cristatus* × *T. c. carnifex* and *T. taeniatus taeniatus* × *T. t. meridionalis* are fully fertile. Also *Pleurodeles waltli* and *P. hagenmuelleri* and certain bird hybrids are fully fertile, but with signs of derangements. These derangements are greater when the geographic ranges overlap, as in *Gallus gallus* × *G. varius*. In this cross, fifteen eggs (three very small and unfertilized) gave in F_2 abnormal embryos which died early. In hybrids of widely separated bird species there is often partial fertility, frequently confined to the males.

The most interesting case is that of *Pleurodeles waltli* × *P. hagenmuelleri*, in which the F_1 is 100 per cent fertile but the F_2 larvae are malformed, all dying within three weeks. In another F_2 experiment, in which 273 eggs were fertilized, all died within six weeks except three which ate and lived twelve weeks. In the backcross, F_1 × *waltli*, of 1247 eggs 92.5 per cent produced normal larvae, but they failed to metamorphose, nearly half dying early and the remainder taking months or years to metamorphose. Steiner concludes that for normal development there must be a harmonic reaction system of genome and cytoplasm. The peculiarity of these crosses is that abnormality first appeared in the F_2, the F_1 being fertile and normal. Newman (1917) found numerous teratological monsters in the F_1 of crosses between the fishes *Fundulus heteroclitus* and *F. majalis;* also in crosses between the minnow and mackerel. These were ascribed to a lowering of the rate of development by the foreign sperm.

Some very interesting studies have been made of the development of intersterility in strains of frogs. Porter (1939) removed the maternal chromatin (second maturation spindle) from the eggs of *Rana pipiens* by microdissection methods after a sperm had entered the egg, and so developed androgenetic haploid individuals. He later (1941) applied these methods to two distinct geographic races of *R. pipiens,* one from Vermont, the other from southern Pennsylvania. The northern form was larger with relatively shorter hind legs, and there were also differences in head shape, vocal sacs, thickness of skin, pigmentation pattern, and other features. Reciprocal ordinary diploid and androgenetic haploid hybrids were studied. It was found in general that when the cytoplasm is from the northern form the hybrids have larger head primordia and smaller post-axial structures than when the egg (cytoplasm) belongs to the southern form. The cytoplasmic differences between the eggs of the northern and southern forms thus have contrasting effects on the developmental processes, and the nuclear differences appear to compensate for the cytoplasmic differences.

In a further study, Porter (1942) finds a gradient of four races of *Rana pipiens* extending from Vermont to Florida. Cytoplasm from the Florida race activated by sperm from Vermont develops much more rapidly than the reciprocal cross. The Pennsylvania form crossed with sperm from Vermont develops normally to the tadpole stage, but in the reciprocal the embryos have abnormally large anterior structures. The cross Pennsylvania × Florida and its reciprocal show less abnormality.

In an extension of this work, Moore (1946) shows that inviability between geographic races of *R. pipiens* develops incidentally to the process of divergence. He used strains of this highly variable species from Vermont and Wisconsin in the north, New Jersey, central and southern Florida, Louisiana, and Texas. Hybrids between adjacent members of the series are normal or nearly so in rate of development and in morphology. Hybrids between progressively more distant members of the series show retardation in development, and morphological defects become more pronounced. Hybrids between the end members show a high degree of inviability, the morphological defects following certain patterns. A female from the south crossed with a male from the north produces hybrids with enlarged heads and defective circulatory system. The reciprocal cross gives a reduced head with fusion of the suckers and olfactory pits, absence of mouth, and abnormalities of the eyes. The extreme geographic races are regarded as nearing the species stage of divergence.

It has already been pointed out that although the white race has crossed with nearly all other races of man, little study has been directed to the possible sterility phenomena involved, and we have no means of knowing what would be the result of crosses between various widely separated primitive races. In *Rana pipiens*, Moore believes that races have been differentiated chiefly by mutational adaptations to temperature differences and other environmental features, the differences being graded and similar to those between northern and southern species. He concludes that in frogs taxonomic species based on morphological differences and "genetic species" based on lack of interbreeding are "usually identical." He points out that if intermediate areas were depopulated of *R. pipiens* and the northern and southern forms afterwards came together again they would cross, with a wastage of gametes owing to the low hybrid viability. Natural selection would then promote the accumulation of additional mechanisms which would prevent gene exchange by increasing the hybrid sterility. This argument in less refined form was much discussed in the Darwin family without any final conclusion being reached regarding its validity.

In a recent further study, Moore (1946a) finds that although intraspecific crosses between the geographically more distant strains of *R. pipiens* (for example, Vermont × Florida strain or Wisconsin × Texas) produce defective offspring, yet these same strains when crossed with *R. palustris* from Massachusetts produce offspring with no defect or inviability. The only exception to this was in the cross with *R. pipiens* from Texas, when very slight defects were found in the offspring. This situation is explained by the assumption that in strains of *R. pipiens* there are complementary lethal genomes. It should give pause to those who try to use intersterility as the sole criterion of species. If this principle were applied, *R. pipiens* and *R. palustris* would have to be united into one species at the same time as *R. pipiens* was divided into several.

A similar study of American toads, by Blair (1941), has several interesting features. There are four species in the eastern States—*B. americana, terrestris, fowleri* and *woodhousi*. In some places where two or more species occur together, intermediates (hybrids) occur. Neither species is exactly the same in any two localities. Some of these characters show clines, some do not. Thus *fowleri* and *americana* show a cline in the body-foot ratio and in the dorsal spots. Intermediates between *B. americana* and *B. woodhousi* are confined to a strip one hundred miles wide, extending from eastern Oklahoma to eastern Nebraska. *B. fowleri* and *B. terrestris* similarly overlap and produce hybrids in the southeastern States. *B. fowleri* and *B. americana* have the same distribution, and intermediates occur throughout their range which are similar to experimentally produced hybrids. The species show different breeding seasons and different breeding

Jurua (Negrito) woman and five children (four of them hers) from the Andaman Islands in the Indian Ocean

sites. Gene exchange between them is restricted by (1) the differences in breeding season, (2) differences in mating calls, (3) ecological preferences as regards breeding sites. Some of the hybrids have intermediate mating calls. The experimentally produced hybrids have (as in man, so far as known) normal viability and fertility. Changes in the countryside produced by man are believed to have increased the gene exchange (crossing of species). In the breeding season males will clasp a female of any species. When the breeding seasons differ, stimulation of the male or female or both by injection of pituitary extract is sometimes necessary to produce copulation. On the whole, the condition as regards species in American toads is very similar to that in the modern species of man.

A very interesting case of natural hybridization in fishes is cited by Hubbs and Miller (1943). *Gila orcuttii* and *Siphateles mohavensis* are two species of minnows—the only native fishes found in the Mohave Desert river system. The former is adapted to streams and the latter to lacustrine conditions. In the pluvial period of the Ice Age they probably had different (complementary) distribution and ecology. As desiccation set in, the river system was reduced to mere creeks in which both species survive; and they now cross extensively, producing 8-9 per cent of hybrids. The hybrids are intermediate in such features as pharyngeal arches, dentition, gill slits, gillrakers, and scales. They have large heads, robust bodies, and large fins, these points being attributed to heterosis. There is a small amount of backcrossing. It is evident that these two species differentiated morphologically and ecologically in the Mohave river system at an earlier period and have since been forced together by the post-glacial drying up of the streams. A similar case is cited, of *Siphateles obesus obesus* and *Leucidius pectinifer*, the former fluviatile, and the latter lacustrine, in connection with the glacial Lake Lahontan system. Intermediate hybrids have been found in the present Lake Tahoe, and as a result *L. pectinifer* is reduced to *S. obesus pectinifer*. This is forcing them under one specific name because they can still intercross although morphologically differentiated at an earlier period into separate species or even genera.

In the flatfishes, six interspecific hybrids have been described in Europe and western North America, all of which Hubbs regards as intergeneric. Hubbs and Kuronuma (1941) add another to the many interspecific fish hybrids now known. The flatfishes, *Kareius bicoloratus* and *Platichthys stellatus* overlap in part of their distribution, on the shores of Japan and the adjacent Asiatic coast. The hybrids are relatively frequent and were at first described as a new genus. Norman combined Kareius with the genus Platichthys, but Hubbs retains the generic rank. In any case there are abundant specific differences between these two forms, which appear to have distinct reaction systems. The hybrids are regarded as intermediate in all respects, but certain characters appear to be dominant, and the hybrids are probably partly fertile.

A species cross in guinea pigs has certain points of interest. *Cobaya aperea* d'Az. and *C. cobaya* Marcg. are found by Pictet (1941) to differ by only two genes, so that in the F_2 from crosses the four body types—F_1 hybrid, *cobaya*, *aperea*, and a new recombination type—appear in the ratio 9:3:3:1. The sex-ratios are also deranged. It hardly needs pointing out that the primary races of man show much more complicated differences.

The extensive studies of mice of the genus *Peromyscus* furnish much material on interfertility. Dice (1933) finds that the species are generally interfertile in the laboratory, but some species and subspecies are only moderately so and a few fail entirely to reproduce in captivity. An adult male and female may fight in a cage, but successful mating may be obtained by placing them together as early as weaning time. The ranges of *P. maniculatus* and *P. polionotus* are now separated by hundreds of miles, but matings between them sometimes succeed and produce fertile offspring. Eleven subspecies of *P. maniculatus* are all more or less interfertile with other subspecies. Sumner produced fertile offspring from all possible crosses of the subspecies, *albifrons, leucocephalus* and *polionotus* of *P. polionotus*. Dice found in *P. leucopus* that five subspecies were interfertile. In the six species, *P. polionotus, maniculatus, leucopus, truei, eremicus,* and *californicus,* all geographic races tested produced fertile offspring when mated with other subspecies of their own species, the interfertile forms thus making a natural biological group, as Mayr finds so frequently in birds.

Dice (1940) finds that in mammals the establishment of reproductive isolation through ethological (behavior) factors precedes, in general, the development of intersterility, and that divergence follows. As in man, no two local populations are exactly the same, distance alone being an important barrier in these small animals, so that there is some inbreeding in every colony. Gilmore and Gregor (1939) suggest the term gamodeme for a more or less isolated interbreeding local population.

Some subspecies of Peromyscus are believed by Dice to be probably polyphyletic in origin, arising from crosses between different subspecies. In *P. maniculatus* the subspecies *artemesiae* is found in forests in Montana, *osgoodi* in grassland, the difference in habitat preventing their interbreeding. *P. maniculatus* and *P. polionotus* differ mainly in size. They cross when brought together, but if *polionotus* is the mother she usually dies in parturition, due to the big foetuses. In the reciprocal cross there is no difficulty and the hybrids are fully fertile. *P. leucopus* and *P. gossypinus* are a similar pair of species, the latter being larger. They are completely interfertile in the laboratory, but seldom cross in nature where they overlap. *P. truei* and *P. nasutus* are species with small morphological differences. In the laboratory they cross with difficulty, producing fertile females and sterile males. In the southwestern States they broadly overlap in distribution.

Dice and Blossom (1937) find that isolation is important in the production of local races of Peromyscus, but that subspecies can occur where there is no physical barrier to dispersal. There is also correlation between pelage color and soil color, a black variety living on lava and pale forms on light, desert soils. The skunks, with a definite black and white pattern, are an exception to this rule.

Many other species of Peromyscus are completely intersterile, including all those which belong in separate species groups. Some species which appear very dissimilar are fully interfertile, while some which appear very much alike are completely intersterile. There is thus little or no correlation between the amount of phenotypic difference and the degree of intersterility. Dice believes that the physiological or psychological (mating) or morphological differences may be the first to diverge in different cases. Man, being "domesticated," appears (like the animals in cages) to have lost in considerable measure his mating aversions. Blair (1943) considers that geographic races usually indicate ecological trends (clines) and hence that they are not *necessarily* incipient species. He cites the case of *Peromyscus maniculatus* and the island form *leucocephalus*. When crossed, many of the

F_1 die soon after birth. There is an excess of females and some of the males are sterile, due to disturbed spermatogenesis.

We may here briefly allude to the fact that self-sterility is a phenomenon of sterility between individuals of a species. It is found in many plants, and group conjugation is a form of self-sterility found in Paramecium and other ciliates. These things show that forms of sterility can develop without any relation to species-formation.

From the results quoted—and many others could be added to them, it is clear that a great variety of conditions exist as regards the relations between species differentiation and intersterility. This of course would be expected if, as we believe, the process of differentiation is a multifarious one, varying from group to group of animals and to some extent from species to species. Views regarding the relative parts played by aversion, sterility, and geographic isolation vary with different writers in the emphasis they would lay on these factors. It seems clear that neither intersterility nor any other condition can be profitably used as an absolute criterion of species. In groups of organisms such as the Cyanophyceae or the Acrasiales, where there is no sexual reproduction, species and genera are as well marked as in sexual organisms, yet intersterility can obviously play no part here in species production or species determination. It is then surely better to recognize that, while intersterility is useful as an additional criterion of species, it can never be satisfactorily applied as a sole criterion, to the exclusion of morphological and other differences.

It must finally be pointed out that Dobzhansky (1944), in applying to anthropology his extreme view that intersterility is the only criterion of species, goes even further than any reasonable application of his theory would demand. He proposes (p. 257) to do away with the genus Pithecanthropus, which all anthropologists with very good reason accept (except perhaps Weidenreich, whose views have been criticized elsewhere (Gates, 1944)). He would reduce Pleistocene Java man to *Homo erectus,* thus doing violence to human phylogeny in a way that it is very doubtful if any biologically minded anthropologist or geneticist would accept. And this in the interest of preserving the phantom of modern *Homo sapiens* as his sole descendant. His distorted perspective in regarding Neanderthal man as only a racial type arises from his dogma that sterility is the only criterion of species, combined with his failure to see that successive species in time must in every case be interfertile because one gives rise to the other. These matters have, however, been pointed out in earlier chapters and need not be repeated here.

When we compare the results of Dobzhansky's ideas as applied to Drosophila and to man we realize that in the name of consistency in adherence to an assumed principle he has achieved the most extreme possible degree of inconsistency. In Drosophila he is such an extreme splitter that he names a "new species" where no morphological difference whatever exists, whereas in man he out-lumps the lumpers, including anatomists and practically all anthropologists, by suppressing genera which every paleontologist and anatomist and practically every anthropologist has recognized since the foundation of the genus Pithecanthropus by Dubois in 1894. Every biologist will recognize that there is something wrong with a principle which leads to such fantastic and contradictory results.

In the light of all the evidence, one can only conclude that the adoption of intersterility as the sole and universal criterion of species is unworkable, leading as it does to complete taxonomic inconsistency and confusion. The attempt to apply it is even self-contradictory in principle since, as we have already seen, sterility cannot arise in the series of forms (chronoclines) which lead from a species to its successor or successors in time.

This being the case, intersterility loses any fundamental character it might otherwise have as a possible criterion of species. The examples already cited amply indicate the multifarious character of species differentiation. Gene mutations and chromosome translocations are but two of these types of change, although possibly the most important. We have seen that in some genera scores of contemporary species can develop without any intersterility arising, whereas in other genera intersterility can arise between forms which show little or no morphological or physiological difference. This confirms the writer's view, expressed a decade ago, that intersterility can arise early or late in the differentiation of species; or it may not arise at all, as is constantly the case when one species gives rise to another in time. All these situations show how impossible it is to subordinate morphological differences to sterility as a criterion of species either in man or any other group of organisms. There is thus no royal road to species determination either in man or any other group of animals or plants. Any approach to uniformity in this matter can only be obtained by recognizing similar morphological criteria in related groups—for example, in man and his relatives the monkeys and apes.

We have shown in this chapter that any attempt to make inter-sterility an absolute criterion of species is highly inconsistent within itself and would lead to utter chaos in taxonomy. There is therefore no danger that it will ever be generally adopted by biologists, except perhaps by a limited number of geneticists who wish to disregard the practical and theoretical aims and functions of classification in plants and animals. Those who have in view the welfare of biology as a whole will refrain from adopting a point of view which is so aberrant and disruptive for biology as a whole, and which is altogether unworkable when applied to widely different genera of plants and animals.

Since sterility fails as the criterion of species, we have to rely on the traditional basis of morphological difference in the discrimination of species, including man. We have already seen that many species and several genera of Hominidae have existed in the past, and it is clear that we must apply to man the same criteria of species that we apply to the apes and monkeys. Consistency in nomenclature and methods of classification thus necessitates the recognition of several species of living man. Those who find this scientific procedure too great a shock can still fall back upon the time-worn *Homo sapiens* as a superspecies embracing all the living species of mankind.

REFERENCES

Blair, W. F. 1941. Variation, isolating mechanisms, and hybridization in certain toads. *Genetics*. 26:398–417. Figs. 6.

―――. 1943. Criteria for species and their subdivisions from the point of view of genetics. *Ann. N. Y. Acad. Sci.* 44:145–178.

Chapman, F. M. 1936. Further remarks on Quiscalus with a report on additional specimens from Louisiana. *Auk*. 53:405–417.

―――. 1939. Quiscalus in Mississippi. *Auk*. 56:28–31.

Cook, O. F. 1907. Aspects of kinetic evolution. *Proc. Wash. Acad. Sci.* 8:197–403.

Crow, J. F. 1941. Studies in Drosophila speciation. II. The *Drosophila mulleri* group. *Genetics*. 26:146.

Deniker, J. 1913. *The Races of Man*. London and New York. pp. 611.

Dice, L. R. 1933. Fertility relationships between some of the species and subspecies in the genus Peromyscus. *J. Mammology.* 14:298–305.

———. 1940. Speciation in Peromyscus. *Am. Nat.* 74:289–298.

Dice, L. R., and P. M. Blossom. 1937. Studies of mammalian ecology in southwestern North America with special attention to the colors of desert animals. *Carneg. Publ.* No. 485. pp. 129. Pls. 8.

Dobzhansky, T. 1937. What is a species? *Scientia.* 61:280–286.

———. 1941. *Genetics and the Origin of Species.* 2nd Ed. New York: Columbia University Press. pp. 446. Figs. 24.

———. 1944. On species and races of living and fossil man. *Am. J. Phys. Anthrop.* N. S. 2:251–265.

———. 1946. Complete reproductive isolation between two morphologically similar species of Drosophila. *Ecology.* 27:205–211.

Dobzhansky, T., and C. Ebling. 1944. Contributions to the Genetics, Taxonomy, and Ecology of *Drosophila pseudoobscura* and its relatives. *Carneg. Publ.* No. 554. Pp. 183.

Gates, R. R. 1917. The mutation theory and the species concept. *Am. Nat.* 51:577–595. Fig. 1.

———. 1927. Mutations: their nature and evolutionary significance. *Am. Nat.* 61:457–465.

———. 1938. The species concept in the light of cytology and genetics. *Am. Nat.* 72:340–349.

Gilmore, J. S. L., and J. W. Gregor. 1939. Demes: a suggested new terminology. *Nature.* 144:333.

Hollinghead, L. 1930. A lethal factor in Crepis effective only in an interspecific hybrid. *Genetics.* 15:114–140. Figs. 6.

Hubbs, C. L., and K. Kuronuma. 1941. Hybridization in nature between two genera of flounders in Japan. *Papers Mich. Acad. Sci.* 27:267–306. Pl. 4. Figs. 5.

Hubbs, C. L., and R. R. Miller. 1943. Mass hybridization between two genera of Cyprinid fishes in the Mohave Desert, California. *Papers of Mich. Acad. Sci.* 28:343–378. Pls. 4.

Lancefield, D. E. 1929. A genetic study of crosses of two races or physiological species in *Drosophila obscura*. *Zeits. f. Abst. u. Vererb.* 52:287–317. Figs. 3.

Mayr, E. 1940. Speciation phenomena in birds. *Am. Nat.* 74:249–278. Figs. 7.

———. 1942. *Systematics and the Origin of Species.* New York: Columbia University Press. pp. 334.

———. 1945. Birds of Paradise. *Nat. History.* 54:264–276. Figs. 22.

Meise, W. 1928. Die Verbreitung der Aaskrähe (Formenkreis *Corvus corone* L.) *J. f. Ornithologie.* 76:1–203.

Moore, J. A. 1946. Incipient intraspecific isolating mechanisms in *Rana pipiens*. *Genetics.* 31:304–326.

———. 1946a. Hybridization between Rana palustris and different geographical forms of Rana pipiens. *Proc. Nat. Acad. Sci.* 32:209–212.

Newman, H. H. 1917. On the production of monsters by hybridization. *Biol. Bull.* 32:306–321. Figs. 14.

Osborn, H. F. 1927. The origin of species. V. Speciation and mutation. *Am. Nat.* 61:1–42.

——. 1926. The evolution of human races. *Nat. Hist.* 26:3–13. Figs. 4.

Patterson, J. T. 1942. Drosophila and speciation. *Science.* 95:153–159.

Patterson, J. T., and J. F. Crow. 1940. Hybridization in the mulleri group of Drosophila. *Univ. Texas Publ.* No. 4032. pp. 251–256.

Peters, J. L. 1940. *Check-List of Birds of the World.* 4 Vols. Cambridge: Harvard University Press.

Pictet, A. 1941. Proportion sexuelle et intersexualité dans la descendance d'un croisement interspécifique de Cobayes. *Actes Soc. Helvét. de Sci. Nat.* 121: 161.

Porter, K. R. 1939. Androgenetic development of the egg of *Rana pipiens*. *Biol. Bull.* 77:233–257. Figs. 40.

——. 1941. Diploid and androgenetic haploid hybridization between two forms of *Rana pipiens*, Schreber. *Biol. Bull.* 80:238–264. Figs. 8.

——. 1942. Developmental variations resulting from various associations of frog cytoplasms and nuclei. *Trans. N. Y. Acad. Sci.* 4:213–217.

Rendel, J. M. 1945. Normal and selective matings in *Drosophila subobscura*. *J. Genetics.* 46:287–302.

Simpson, G. G. 1943. Criteria for genera, species and subspecies in zoology and paleozoology. *Ann. N. Y. Acad. Sci.* 44:145–178.

——. 1945. The principles of classification and a classification of the mammals. *Bull. Am. Mus Nat. Hist.* 85:1–350.

Steiner, H. 1945. Über letale Fehlentwicklung der zweiten Nachkommenschafts-Generation bei tierischen Artbastarden. *Arch. J. Klaus-Stift.* 20: 236–251. Figs. 4.

Stern, Curt. 1936. Interspecific sterility. *Am. Nat.* 70:123–142.

Index

Abel, O., 75
Abel, W., 64, 75
Absolon, 262
Abyssinians, 212, 214
Acrasiales, species in, 404
Adam, W., 148, 162
Africanthropus, 79, 97, 98, 144, 166, 167, 210, 216, 236, 274
Agassiz, Louis, 291
Agriculture, beginnings of, 335
Ainu, 91, 96, 98, 151, 157, 208, 326; bear ceremony, 308; blood groups, 276
Aitape skull, 147, 151
Aiyappan, A., 356, 357, 358
Akkas, 8, 204
Alakaluf, 319
Alaska, 279, 323, 327, 328
Albinism, 151; in Maoris, 342
Aleuts, 278, 279, 282, 323; skull of, 6, 282, 283 (fig.), 284
Alfalou type, 113, 124, 174, 200, 209, 226, 262, 264, 265
Allen, A. L., 184, 230
Allen, G. M., 370, 371, 380
Allison, I. S., 306, 330
Alpine race, 269, 270; origin of, 362
Ambrym island, 148
Ameghino, C., 313, 315, 317
Ameghino, F., 313, 314, 315, 316, 317
American Indians, 8, 9, 92, 274 ff., 363; blood groups, 158, 202, 276; chamaecephaly in, 282; crosses with Eskimos, 328; extermination, 252, 253; hair, 208; head height, 282; head shape, 127, 281, 283, 367; Inca bone in, 93, 113; in South America, 308 ff.; interglacial, 277; languages, 280; Neanderthaloid skulls, 275, 288, 289, 290; nose shape, 284; Sandia culture, 303, 304, 321; skeleton, 288; supraorbital torus, 250; teeth, 55
Americanoids, 279
Amphibians, evolution of, 23
Amphipithecus, 59

Andamans, 352
Andes, excavations in, 321
Andrews, C. W., 16, 17, 41
Andrews, E., 287, 329
Angel, J. L., 249, 271
Angiosperms, evolution in, 23
Anthropoids, variation in, 369 ff.
Arabs: blood groups, 201; in Indonesia, 350; in Madagascar, 346; slaves, 352
Arambourg, C., 209, 230
Armenoid features, 218, 222; nose, 351, 363; skulls, 362; type, 362
Art: Aurignacian, 252, 253; Eskimo, 326; Magdalenian, 207, 261, 264; Neolithic, 209
Aryans, 9, 133, 275, 339, 355
Assam, 356
Asselar man, 257; skeleton, 209; skull, 203
Aurignacian culture, 176, 177, 183, 188, 209, 223, 226, 241, 243, 246, 250, 251, 252, 254, 255, 257, 259, 261, 262
Australian aborigines, 7, 11, 93, 98, 112, 142, 144, 146, 153, 156, 239, 274, 280, 312, 353, 367, 388, 390; blood groups, 126, 154, 157, 201, 204; brain, 159, 160, 250; cranium, 89, 97, 103, 147, 150, 169, 181; tawny hair in, 157; teeth of, 55
Australoids, 148, 151, 152, 154, 179, 180, 181, 182, 183, 192, 216, 263, 274, 284, 312, 352, 355, 356, 367
Australopithecinae, 48, 58, 60 ff., 78, 79, 94, 98, 140, 141, 142, 166, 229, 236, 365, 369
Aversion to crossing, 387, 390
Aye-Aye, 17, 51
Aymara Indians, 313, 320; skull deformation of, 314, 315, 316, 317
Aztecs, 326

Baboons: fossil, 64, 66, 69, 70, 85, 96; modern, 80
Badarians, 125, 134, 137, 139
Bahima, 212
Banks, E., 377, 378, 380

Bañolas mandible, 248
Bantu, 178, 181, 183, 184, 199, 203, 218, 347; blood groups, 201, 338, 353; culture, 192, 195; hybridity in, 132, 179, 189, 194, 195, 196, 197, 219; infant brain, 160, 188; language, 200; origin, 211; skull shapes, 131, 133; somatology, 222; teeth, 193
Barbour, E. H., 290, 329
Barcena, M. de la, 288, 329
Bark cloth, 344, 346
Barma Grande, 259, 263
Barotse, 203 (fig.), 211; origin of, 199
Basketmakers, 302, 308
Basu, 153
Basuto, 194, 196, 203; brain, 160
Bather, F. A., 16, 18, 41, 101
Batrawi, 136
Battak, 350
Bauermeister, W., 203, 230
Baumes-Chaudes, 257, 264, 268
Bavenda, 194, 202
Baxter, H. C., 199, 230
Beaker-folk, 264
Bean, R. B., 151, 162
Bears: in S. America, 313, 318; size of, in Pleistocene, 305
Beattie, J., 50, 75
Bechuana, 178, 181, 189, 193, 194, 199, 203, 211; blood groups, 201
Bell, E. H., 180, 230
Bensley, B. A., 32, 41
Berbers, 266
Berg, L. S., 145, 162
Bernstein, R. E., 193, 230
Berry, E. W., 293
Berry, G. F., 195, 230
Berry, R. J. A., 156, 163
Berthoud, E. L., 288, 329
Bhils, 356
Bhotias, 356, 357
Bijlmer, H. J. T., 338, 354, 358
Bird, J., 318, 319, 329
Birds: hybrids in, 398; species in, 388, 395
Birdsell, J. B., 147, 151, 157, 158, 163, 164
Birds of paradise, 389
Bison, 294, 301, 304; arrow in extinct species of, 299; decrease in size of, 305; marrow from extinct species, 307
Black, Davidson, 79, 86, 87, 88, 89, 115, 253
Blackfoot Indians, 9, 320, 327
Blair, W. F., 400, 403, 406
Blanc, A. C., 110, 115
Bleek, D. F., 197, 230
Bloch, A., 257, 271

Blond element: in Canary Islands, 265; in North Africa, 266; origin of, 257
Blood groups, 9, 154, 157, 347; in Blackfoot Indians, 320; in Bushmen, 180, 204; in Fuegians, 282, 319; in India, 356, 357; in Mongoloids, 276; in the Pacific, 337, 338, 343, 355; mutation in, 122; of Eskimos, 327; of Negroes, 338, 353; of Pygmies, 200, 352; spread of B, 126, 127, 201, 276, 280, 357, 361
Blossom, P. M., 403, 407
Boas, F., 119
Bolas, 81, 102, 229, 319
Bolk, 124
Bonin, G. von, 100, 115, 257, 258, 262, 263, 264, 271, 281, 329
Boomerang, 152, 156
Boskop man, 5, 7, 11, 121, 132, 133, 134, 138, 139, 166, 167, 169 ff., 177, 179, 180, 181, 182, 183, 192, 194, 196, 200, 204, 206, 209, 210, 216, 217, 220, 221, 222, 225, 226, 227, 265, 266, 274, 353; brain of, 160, 188; face, 219; race, 257; skull, 173, 174, 175, 185, 186, 193, 218
Boswell, P. G. H., 224, 233
Botocudo Indians, 310
Boule, M., 107, 209, 230, 243, 246, 254, 255, 257, 269, 271
Bourne, G., 73, 75
Bow, 253, 319, 344
Bowden, A. O., 301, 329
Boyce, Dr., H. H., 298
Boyd, W. C., 158, 163, 358
Brachycephalization, 120, 126
Brachycephaly: dominance of, 121, 267; in Africa, 219, 266; in Amerinds, 281, 283, 284, 290, 307, 311; in Andamans, 352; in Bavaria, 269; in crosses, 265, 363; in Polynesia, 349; origin of, 175, 263, 352, 362; spread of, 4, 120, 122, 123, 124, 125, 130, 135, 179, 195, 197, 210, 227, 257, 260, 262, 264, 268, 270, 308, 357
Brain: and head shape, 123; increase in size of, 145; weights, 162
Breadfruit, in Polynesia, 342, 344
Broca, Paul, 258, 368, 380
Brögger, A. W., 255, 271
Broom, R., 15, 25, 26, 27, 29, 41, 52, 60, 61, 63, 64, 66, 67, 68, 69, 70, 71, 72, 73, 75, 83, 166, 167, 169, 171, 178, 179, 180, 181, 182, 183, 185, 211, 216, 230
Brown, Barnum, 59
Bryan, Kirk, 278, 279, 301, 304, 307, 318, 320, 329
Bryan, W. A., 303, 329

Index

Brythons, 264
Bubalus Bainii, 182, 183
Buck, P. H., 341, 342, 344, 358
Bulldog, type in Andes, 17
Bunak, V. V., 111, 120, 362, 380
Burkitt, A. St. N., 153, 163, 181, 230
Burkitt, M. C., 250, 271
Bury St. Edmunds skull, 241
Bushmen, 10, 11, 100, 112, 126, 132, 133, 134, 139, 166, 167, 169, 171, 172, 177 ff., 184, 187, 188, 207, 274, 368, 370, 390; and Pygmies, 204, 205, 236, 352; blood groups, 201, 202, 280, 338, 353; brain, 159, 160; culture, 192; distribution, 200, 206; foetalization, 121; head shape, 121, 130, 131, 132, 138, 168; origin of, 222; paintings of, 189, 197, 198; pedomorphic degeneration in, 5, 7; race, 196, 197, 209; sex organs, 220, 367; skeletons, 194, 195, 196; skull, 173, 174, 185, 195; somatology, 217, 220; species, 388; taurodont molars, 94
Butler, P. M., 33, 41

Calaveras skull, 295, 298
California: caverns in, 298; early man in, 294
Campbell, T. D., 155, 165
Canary Islands: blondness, 266; skulls, 210, 265, 267
Cape Flats skull, 167, 168 (fig.), 183, 184
Capsian culture, 188, 209, 252
Carabelli cusp, 71, 94
Carpentarians, 151
Carr-Saunders, Sir A. M., 336
Case, E. C., 24, 41
Cassel, J., 174, 230
Castaldi, 163
Catarrhine monkeys, 34, 35, 47, 57, 59, 142
Cathoud, A., 311, 333
Caton-Thompson, Miss G., 194, 230
Cattle, horn distortion in, 214
Caves of Hercules, 109
Celebes, 350, 354
Cephalization, in mammals, 123
Cercopithecidae, 74; variation in Africa, 374
Chancelade man, 64, 254 ff., 258, 321
Chapadmalensian formation, 317, 318
Chapman, F. M., 397, 406
Chatelperronian, 252
Chellean implements, 176, 177, 190, 238, 241, 250
Che Wong, 150
Chimpanzee, 239, 369; cephalic index, 373; dwarf, 370; pigmentation, 371; pygmy, 204; teeth, 240; temperaments, 371, 374; variability, 370, 371
Chimu pottery, 17, 321
Chinese: blood groups, 356; brain, 124, 160; culture, 326; Han pottery, 340; head shape, 126, 343; in East Africa, 7, 197, 198, 199; in Hawaii, 347, 348; in Indonesia, 350, 351; origins, 275, 322; Rh in, 328; skeletons, 87
Choroclines, 385, 387
Choukoutien upper cave, 113, 256, 302, 364
Chromosome numbers, in mammals, 46
Chronoclines, 385, 405
Chubb, E. C., 100, 116
Chukchi, 279, 280, 281, 324, 325
Civilizations, origin of, 199
Clark, W. E., LeGros, 39, 41, 47, 48, 49, 51, 52, 75, 142, 241
Classification, horizontal and vertical, 383
Clements, T., 302, 329
Clew, J. F., 288, 332
Clines, 384, 385, 400, 403
Cohuna skull, 150
Colbert, E. H., 52, 59, 76, 318, 329
Cole, Fay-Cooper, 335, 349, 358
Collins, H. B., Jr., 321, 324, 326, 329
Color-blindness, mutations in, 122
Combe-Capelle, 259, 262, 264
Confins cave, 311
Convergence, 13, 27, 40, 64
Cook, Captain, 347
Cook, H. J., 299, 300, 330
Cook, O. F., 382, 406
Cooke, H. B. S., 183, 196, 230
Coolidge, H. J., Jr., 204, 231, 308, 372
Coon, C. S., 109, 113, 120, 124, 146, 155, 163, 185, 254, 262, 263, 264, 269, 270, 271, 362, 363
Corvus, 397
Cowles, R. B., 26, 41
Crepis: 393; gene lethal in crosses, 395
Cressman, L. S., 306, 330
Cro-Magnon man, 104, 107, 108, 111, 113, 122, 156, 169, 171, 182, 183, 184, 200, 203, 204, 209, 216, 217, 222, 236, 251, 253, 254, 258 ff., 262, 263, 264, 266, 269, 374; origin, 210, 265, 267, 268
Crossopterygian fishes, 22
Crow, J. F., 393, 406, 408
Crowther, W. E. L. H., 156, 163
Culture diffusion rates, 214, 360
Cunningham, D. J., 79, 115
Cyanophyceae, species in, 404

Dachsenbüel, 206
Darby, G. E., 127, 143, 208, 276, 307, 330
Dart, R. A., 60, 61, 62, 66, 70, 71, 76, 79, 121, 125, 126, 133, 134, 135, 136, 138, 139, 143, 166, 169, 192, 196, 198, 199, 204, 206, 216, 217, 218, 219, 220, 221, 222, 231, 266, 267, 268, 353, 358
Darwin, C., 10, 11, 13, 33, 360, 361, 367, 368, 369, 379, 380, 400
Dating, by marine sediments, 106
Dawson, Charles, 237, 238, 271
Dearlove, A. R., 195, 231
Deevey, E. S., Jr., 336, 358
DeLaguna, F., 329, 330
Del Grande, N., 192, 231
Demerec, M., 361, 380
Deniker, J., 11, 363, 380, 390, 406
Densmore, F., 285, 330
Dentition, evolution of: in mammals, 46; in man, 45; of Australopithecines, 65, 66; of Paranthropus, 62, 63, 69, 70; of Sinanthropus, 93
DeSaxe, H., 195, 218, 231
DeTerra, H., 84, 103, 115
Diatryma, 14, 15
Dice, L. R., 402, 403, 407
Dickeson, M. W., 290, 330
Dinaric race, 363
Dingo, 146, 158
Diomede Islands, excavations, 326
Diprothomo platensis, 316
Disease and race, 367
Dixon, R. B., 283, 284, 308, 330, 343
Dobzhansky, T., 388, 391, 393, 395, 404, 405, 407
Dodo, 14, 345
Dog, 285, 325, 341, 344
Dolichocephaly, 120; and pigmentation, 120; and stature, 123; inheritance of, 119
Domestication and sterility, 368
Dravidians, 154, 156, 355; skulls, 134
Drennan, M. R., 80, 115, 167, 183, 184, 185, 186, 192, 194, 195, 200, 203, 220, 221, 222, 231, 232, 234
Dresden Codex, 286
Dreyer, T. F., 132, 143, 167, 169, 173, 186, 188, 200, 232
Drosophila, 12, 389, 405; *aldrichi*, 392; *americana*, 392, 396; *arizonensis*, 393; chromosome inversions in, 390, 392; *equinoxialis*, 393; *macrospina*, 396; *melanogaster*, 361, 390, 395; *montana*, 396; *mojavensis*, 392; *mulleri*, 392, 393, 396; *novamexicana*, 396; *persimilis*, 391; *pseudoobscura*, 390, 391, 392, 393, 396; *repleta*, 396; *simulans*, 390, 391; species hybrids, 392, 393; *subobscura*, 394; *texana*, 392, 396; *virilis*, 392, 396; *Willistoni*, 393
Dru-Drury, E. G., 203, 232
Drury, I., 220, 221, 222, 232
Dryopithecus, 4, 48, 52, 53, 54 (fig.), 55, 58, 59, 61, 65, 72, 80, 82, 85, 94, 225, 229, 240
Dubois, E., 79, 80, 81, 83, 95, 98, 115, 146, 147, 154, 163, 405
Duckworth, W. L. H., 374, 380
Dutch: head shape in, 124; in Sumatra, 351
Dwarfs, 126, 139; achondroplastic, 8, 200, 204, 346; ateleiotic, 8, 204, 346; chimpanzee, 370; in Italy, 169; Neolithic, 208; races, 204 ff., 352

Early Colonial burials, 290
Ears: in gorilla and chimpanzee, 373, 374; small in Negro, 210
East Africa: and civilization, 199; climates in, 176
Easter Island, 340, 342, 344, 345; blood groups, 337, 343
Echinoderms, fossil, 18
Egypt, blondness in, 270; Neolithic in, 335; step-pyramid, 342
Egyptians, 209, 212, 266; blood groups in, 202; brachycephaly in, 125; civilization, 199, 213, 214; pygmy skull, 207; skull shape, 6, 134, 135, 136; types, 210
Ehrich, R. W., 363, 380
Ehringsdorf skull, 104 107, 108, 240, 244
Eickstedt, E. F., 153, 163, 281
Eiseley, L. C., 287, 330
Elephants: evolution in, 35, 176; hairy coat, 98
Elliot, D. G., 371, 380
Elliot Smith, Sir Grafton, 51, 62, 78, 88, 101, 117, 146, 156, 163, 171, 225, 238, 256, 270, 342, 343
Elmenteita: industry, 226; skulls, 177, 183, 223, 227
Elsdon-Dew, R., 201, 202, 232, 338
Endocranial cast: of Boskop, 171; of Bushmen, 160; of Chinese, 89; of Eoanthropus, 238, 240; of La Chapelle-aux-Saints, 246; of Plesianthropus, 70; of Sinanthropus, 88, 91
Eng, R. L., 209, 234
Eoanthropus, 65, 67, 69, 142, 149, 223, 225, 229, 236, 237 ff., 241, 274, 365, 384
Epicanthus, 215
Epling, C., 391, 407
Equidae, evolution of, 16, 17, 34, 383

Equus species, 88, 167, 176
Eskimos, 8, 93, 96, 242, 255, 276, 277, 320 ff., 368; blood groups, 327; Caribou, 323; culture succession, 326, 328; in Alaska, 280, 281, 308; leptorrhiny in, 256, 320; mentality, 367; number, 336; skull, 6, 89, 254, 282, 283 (fig.), 328; taurodont molars, 94; torus mandibularis, 91, 92
Ethnic group, 363
Ethnobotany, 285
Evans, Sir Arthur, 220
Evolution, asymmetrical, 365
Ewing, J. F., 363, 380
Eyefolds, 275, 354

Fabricius-Hansen, V., 327, 330
Fantham, H. B., 211, 232
Femur and erect posture, 79
Fenner, F. J., 147, 163
Field, Dr. Henry, 111
Figgins, J. D., 299, 330
Fire, evidence of, 98, 102
Fischer, E., 212, 232
Fisher, R. A., 365
Fishes, evolution of, 23; hybridization of, 401
Fish Hoek skeleton, 183, 184, 185, 227
FitzSimons, F. W., 171, 172, 173, 197, 232
Flatfishes, interspecific hybrids, 401
Flint, R. F., 106, 115
Florida, early man in, 291 ff.
Florisbad skull, 166, 167, 168 (fig.), 169, 188, 210, 274
Flower, W. H., 364, 380
Foetalization theory, 124, 132
Folsom culture, 286, 303, 304; man, 8, 307; 326; points, 299, 300, 301, 306, 321
Fontezuelas skeleton, 314
Formosa, aborigines, 208; blood groups in, 276
Fouché, L., 194, 232
Fowls, domestic, 214
Franklin, Benjamin, 287
Frassetto, 128, 130
Frets, G. P., 121, 127
Frogs, intraspecific crosses, 398, 400
Frontal torus, reduction in, 141
Fuegians, 93, 114, 282, 390; blood groups of, 319
Furlong, E. L., 298, 330

Galilee skull, 104, 107, 108, 247, 250
Galley Hill skeleton, 149, 241, 242, 250, 264, 268

Galloway, A., 132, 143, 167, 169, 171, 175, 194, 195, 196, 200, 226, 227, 232
Gamble's caves, 226
Gamodeme, 403
Garner, 373
Garrod, Miss Dorothy, 108, 243, 252, 271
Gates, R. R., 4, 13, 15, 18, 41, 46, 63, 76, 87, 93, 94, 107, 113, 119, 122, 124 127, 142, 143, 145, 146, 154, 163, 180, 186, 200, 202, 204, 205, 206, 208, 215, 216, 222, 232, 265, 268, 276, 307, 320, 330, 352, 358, 360, 361, 366, 367, 368, 378, 380, 382, 384, 392, 404, 407
Gear, J. H., 132, 143, 172, 197, 232, 233, 235
Genes: affecting intersterility, 391, 395; affecting mutability, 361; exchange of, 388, 400, 401
Genus, nature of, 379, 382
Geographic variation, laws of, 385
Gibbons, 48, 49, 53, 55, 57, 67, 74, 142
Gibraltar skull, 102, 103, 243, 250
Gidley, J. W., 293, 294, 330
Gifford, E. W., 281
Gigantism, evolution of, 22, 26; in birds, 8, 14; in man, 85; in reptiles, 13
Gigantopithecus, 40, 67, 83, 84, 85, 205
Gilmore, J. S. L., 403, 407
Giuffrida-Ruggeri, 91
Glyptodon, 313, 314, 316, 318
Goddijn, W. A., 362, 381
Golomshtok, E. A., 258, 271
Goodwin, A. J. H., 184, 188, 189, 232, 233
Gorilla: anal tuft in, 371; crosses, 373; described, 369, 373; intermediates with chimpanzee, 373, 374; skulls, 372; species, 372; variation in, 370
Gorilloid line of evolution, 6, 142, 225, 236, 237, 240, 242, 243, 274, 318
Gorjanovic-Kramberger, 244
Gould, C. N., 300, 330
Grackles in eastern states, 397
Granger, W., 14, 42
Graydon, J. J., 158, 163, 233
Greece, population of ancient, 145
Gregor, J. W., 403, 407
Gregory, W. K., 22, 32, 41, 45, 46, 48, 49, 51, 53, 55, 64, 65, 66, 72, 74, 76
Grimaldi man, 209, 254, 256, 257; caves, 259; Grotte des Enfants, 259
Ground-sloth, 291, 294, 298, 303, 308, 318
Gruta de Cadonga cave, 320
Gryphaea, 18
Guanche mummies, 265, 266, 267

Guha, 153
Guinea pigs, species cross, 402

Haas, O., 13, 41
Hadzapi, 197
Hall, E. R., 369, 380
Hambly, W. D., 148, 163
Hamites, 185, 200, 201, 215, 227; origin, 211, 212
Hamy, E. T., 210, 258, 272
Handy, E. S. C., 342, 358
Hansen, H. P., 306, 330
Hansen, Sören, 311, 314
Harappa, 153
Harrington, J. P., 280, 281, 330
Harris, A. C., 174, 230
Harris, D. F., 174, 230
Harris, W. E., 115
Harte, Bret, 296
Haughton, S. H., 169, 233
Hauser, 247, 259
Hawaii, 348, 349
Hawaiians, 340, 343, 347; blood groups in, 337, 338
Hay, O. P., 287, 292, 293, 299, 300, 330
Hay, P. R., 277, 330
Hayata, B., 23, 41
Head shape: genetics of, 5, 121, 127, 128, 362; in Amerinds, 281, 282, 284; in children, 119
Hedgehogs, interbreeding in, 394
Heidelberg jaw, 58, 67, 69, 80, 82, 83, 94, 99, 147, 182, 183, 242, 247
Heilprin, A., 291, 330
Heine-Geldern, Robert von, 339, 340, 358
Hellman, Milo, 53, 55, 61, 64, 65, 66, 76, 150, 163, 312, 333
Hereros, 190, 211
Hibben, F. C., 303, 331
Hill, W. C. Osman, 205, 233
Hilzheimer, M., 17, 41
Hindus, 208, 270, 335, 343, 357; culture, 346, 354; in Africa, 199; in India, 337; in Indonesia, 350
Hirschler, P., 89, 115
Hoffman, A. C., 173, 232
Hoijer, H., 281, 331
Hollingshead, L., 395, 407
Holmes, F. S., 290, 331
Homo: africanus, 367; aurignaciensis, 259; australicus, 11, 225, 312, 367; capensis, 11, 169, 210, 216, 265, 367; caucasicus, 367; erectus, 404; kanamensis, 224, 229, 236, 240, 242, 265; modjokertensis, 81; mongoloideus, 367; mousteriensis, 247;
sapiens, 224, 228, 249, 252, 265, 366, 367, 371, 404, 406
Hooton, E. A., 68, 87, 91, 92, 115, 181, 233, 256, 265, 266, 267, 268, 271, 281, 306, 355, 358, 365, 380
Hopwood, A. T., 57, 76, 102, 233
Hornell, J., 348, 358
Horse, fossil in S. America, 310, 312, 315, 318
Hottentots, 5, 7, 10, 11, 166, 169, 171, 175, 177 ff., 185, 186, 189, 195, 197, 199, 200, 203, 209, 210, 211, 352; apron, 132; blood groups, 180, 201, 202, 216, 338, 353; femur of, 80; measurements, 218; skulls, 133, 173, 174; teeth of, 193
Howard, E. B., 301, 331
Howard, H., 302, 331
Howells, W. W., 156, 163, 336, 341, 353, 358
Hrdlička, A., 8, 61, 63, 76, 100, 116, 157, 163, 180, 230, 239, 240, 242, 243, 249, 250, 251, 252, 253, 271, 276, 277, 279, 281, 288, 289, 290, 294, 295, 306, 308, 310, 311, 313, 314, 315, 316, 317, 318, 323, 331
Hubbs, C. L., 401, 402, 407
Hug, E., 125, 126, 143
Hughes, B. O., 362, 363
Hunter, J. I., 153, 163, 181, 230
Huxley, T. H., 154, 212
Hybridization: and sterility, 368; dysharmony in crossing, 223, 267; hair, 203; in anthropoids, 373; in Okinawa, 208; of human races, 172, 173, 175, 181, 183, 185, 192, 193, 194, 195, 196, 197, 207, 210, 211, 212, 215, 217, 219, 226, 227, 229, 251, 253, 264, 265, 268, 323, 324, 328, 336, 345, 347, 349, 355, 362, 363, 364, 385; origin of secondary races, 369; skulls, 132, 174, 263
Hybrids: in birds, 398; in birds of paradise, 389; in fishes, 398, 401; in frogs, 398, 399; in guinea pigs, 402; in salamanders, 398; in toads, 400
Hybrid vigor, 212, 213, 267
Hyoid bone, 194
Hypsicephaly, 121, 124, 128, 132, 133, 134, 136, 138, 139, 267, 271, 282, 311, 357; origin, 268
Hypsodonty, in horses, 17

Ice Age, 133, 243; cause of, 190; in America, 276, 277; in Europe, 104, 278
Imbelloni, J., 308, 313, 331
Implements, 360; Abbevillian, 86, 241, 251;

Acheulian, 88, 98, 176, 190, 209, 223, 229, 241, 246, 250; Anyathian, 87, 88; Clactonian, 241, 251, 255; in Celebes, 354; Levalloisian, 88, 99, 104, 251, 255; Magdalenian, 207, 243; Monsterian, 102, 182, 191, 192, 213, 241, 243, 246, 250; Soan, 88, 103; Tardenoisian, 257

Inca bone, 89, 92, 93, 113

Incas, 17, 321

India: Aryan invasion, 339; Australoids in, 357; blood groups, 356; Book of Marvels, 198; early inhabitants, 355; Mediterraneans in, 355; Polynesians in, 9, 337, 339; Pulayas, 352

Indonesians, 335, 341, 343, 349, 350, 351; archaeology, 339; blood groups, 337, 338; in Madagascar, 345

Intersterility, 365, 378, 380, 387, 388, 389, 404, 405; origin of, 394, 395; in crows, 397; in Drosophila, 390, 391, 392, 396; in strains of frogs, 398, 399

Ipiutak, 325, 326

Iraq, 335

Iron, introduction into Sudan, 214

Italy, dwarfs in, 206, 208

Iyer, L. A. Krishna, 151, 163

Japanese, 119, 208; in Hawaii, 348

Java: blood groups in, 338, 343; cave in, 339; Polynesians in, 337, 349; population of, 337

Javanthropus, 96, 97, 98, 99, 122, 141, 146, 154

Jenks, A. E., 305, 331

Jenness, D., 308, 323, 327, 331

Jews, 119, 120

Jochelson, W., 279, 331

Jordan, D., 327, 331

Jurua, 8, 352 (fig.)

Kakamas, 173, 174

Kalmucks, 156

Kalomo skeleton, 172, 187, 193

Kamasian pluvial, 176, 224

Kanam man, 142, 223, 224, 226, 250

Kanikars, 151, 152, 356, 357

Kanjera man, 149, 211, 223, 224, 226; endocast, 225

Kappers, Ariëns, 121, 123, 124, 127, 143, 160

Karens, 275

Kedung Brubus, 80

Keen, J. A., 181, 185, 233

Keilor skulls, 147, 148, 149, 150, 151, 157

Keiter, F., 354, 358

Keith, Sir Arthur, 79, 87, 101, 107, 108, 109, 110, 111, 112, 115, 116, 144, 147, 153, 163, 167, 177, 180, 183, 184, 186, 187, 188, 233, 239, 240, 241, 247, 249, 250, 255, 256, 271, 371, 373, 374, 381

Kesslerloch, 206, 207

Kiik-Koba skeleton, 110, 258

King, W. R., 152, 164

Klaatsch, 44, 247, 259

Klopper, A. I. I., 185, 233

Knoche, W., 190, 233

Knysna, 174, 175

Koby, F., 305, 331

Koch, Albert C., 286, 287, 331, 332

Koenigswald, G. H. R., 63, 66, 77, 80, 81, 82, 83, 84, 86, 88, 116

Kohler, W. L., 175, 233

Kollmann, J., 207, 233, 310

Korana, 6, 167, 168, 171, 178, 179, 181, 183, 184, 193, 195, 197, 216, 225, 263, 353, 367

Koryaks, 279, 324

Krapina skeletons, 105, 107, 108, 112, 244, 250

Kumara, 341, 344, 348

Kurgans, 139, 140, 206

Kuronuma, K., 401, 407

Kurumbas, 152

Labyrinthodonts, 24, 25

La Chapelle-aux-Saints, 246

La Ferrasie, 246

Lagoa Santa: caves, 308, 311; skulls, 309 ff., 313, 317, 319

Laing, G. D., 172, 233

Lake Baikal, 321, 324, 325

La Naulette jaw, 248

Lancefield, D. E., 390, 407

Lang, W. D., 21, 41, 171

Languages: evolution of, 282; in New Guinea, 351

Lapouge, 207

Lapps, 91, 92, 121, 265, 280, 390

La Quina, 246, 247

Laugerie-Basse skeletons, 260

Lautsch cave, 262

Leakey, L. S. B., 58, 59, 98, 99, 100, 116, 176, 183, 223, 224, 225, 226, 227, 228, 229, 233, 250

Lebomo mountains, 183

Lehmann, 338

Leidy, Joseph, 288, 291, 332

Lemurs, 17, 28, 34, 35, 40, 47, 49, 51, 52, 59; giant, 86; tailless, 361

Leroi-Gourhan, André, 265, 271

Lethal genes, 12, 395

Lewis, G. E., 52, 53, 55, 76, 77
Lewis, G. N., 234
Libyans, 270
Likasi skeleton, 203
Limnopithecus, 57, 58
Lindenmeier site, 300, 301
Linnaean species, 375, 376, 383, 389
Linnaeus, C., 11, 114
Linton, R., 345, 358
Lion, crosses of, 364
Lipschutz, A., 319, 332
Litopterna, 16
Lolos, 275
London skull, 241
Loomis, F. B., 293, 330
Lönnberg, E., 371, 372, 381
Lopatin, I. A., 301, 329
Lotsy, J. P., 362, 381
Loucheux Indians, 322, 324, 367
Lowe, C. van Riet, 188, 233, 234
Lowe, P. R., 14, 15, 41
Lund, P. W., 308, 309, 310
Lütken, 310
Lydekker, R., 364, 377, 380, 381

Macacus, tail of, 361
McCown, T. D., 107, 108, 109, 111, 112, 116, 249
MacCurdy, G. G., 308, 331
Macfarlane, E. W. E., 352, 356, 357, 358, 359
MacInnes, D. G., 58, 77, 224
Mackay, E. J. H., 153, 164
Mackenzie River valley, 279
MacLennan, G. R., 195, 234
Madagascar, 9, 14, 17, 34, 47, 335, 340, 347; blood groups in, 338; Imerina, 338, 345, 347; Hovas, 346; peoples, 345, 346
Magdalenian: culture, 189, 207, 254, 255, 260, 265, 321; man, 223, 253, 259, 261, 262, 264
Maglemose culture, 322, 326
Mahony, D. J., 148, 149, 150, 164
Maingard, L. F., 189, 193, 234
Maize: culture of, 285; translocations and sterility, 387
Makalian wet phase, 176, 177, 227
Malan, B. D., 183, 196, 230
Malaria and selection, 345, 347, 357
Malarnaud jaw, 248
Mallophaga, 15
Mammae, types of, 220
Mammals: at Alfalou, 210; at Krapina, 244; at Piltdown, 238; evolution of, 26, 27, 29, 30, 204, 277; in Magdalenian cave,

207; in Mauer sands, 242; in Oldoway gorge, 223; origin of, 44, 46, 48, 50; Pleistocene, 245, 246, 248, 260, 286, 290, 291, 292, 294, 298, 299, 303, 304, 306, 309, 311, 314, 318, 320; reproductive isolation in, 402
Man: age in America, 302; epidermal characters of, 98; monkey-like features in, 74; primitive races sedentary, 378; species of living, 368, 370
Maoris, 340; blood groups, 337, 338; hair, 342; nose of, 342
Mapungubwe, 194
Marett, R. R., 245
Marshall, John, 159, 164
Marshall, Sir John, 154, 164
Marsupials, evolution of, 28, 32, 33, 35, 45, 52
Martin, Dr. Henri, 260
Masai, 212, 213, 215
Maška, 261
Mastodon, 286, 287, 288, 290, 291, 294, 297, 304, 312, 314, 318
Mathiassen, T., 326
Matiegka, J., 262, 271
Matjes River people, 132, 173, 185, 186, 187 (fig.), 188, 193, 200, 210
Matoppo Hills, 192
Matschie, P., 370, 371, 372, 381
Matthew, W. D., 14, 16, 27, 28, 34, 42, 45, 50, 77
Mattos, A., 311, 333
Mauer jaw, see Heidelberg jaw
Maya: astronomy, 286; civilization, 199, 307, 326
Mayr, Ernst, 15, 42, 378, 379, 381, 385, 387, 388, 389, 394, 395, 397, 402, 407
Mediterranean race, 201, 206, 212, 218, 219, 220, 222, 257, 264, 267, 269, 270, 284, 335, 338, 349, 362, 363; in India, 355; short, 208
Megalithic culture, 264, 339, 340, 342
Meganthropus, 40, 81, 82, 83, 84, 85, 205, 229, 274
Meiring, A. J. D., 173, 188, 232, 234
Meise, W., 397, 407
Melanesians, 10, 148, 155, 239, 302, 311, 335, 341, 343, 344, 351, 353, 355; blood groups, 337, 338; brachycephalic, 352, 354; languages, 339
Mendes Corrêa, A. A., 257, 271
Meningeal artery in man, 91
Mennell, F. P., 100, 116
Mental differences between races, 367
Meroitic kingdom, 213, 214

Merriam, J. C., 298, 332
Merrill, E. D., 84, 116
Mesembryanthemum, evolution of, 66
Mesolithic, 255, 256, 265, 268, 322, 336, 339
Miao, 275, 356
Mice, interfertility in, 402
Microcephaly, 203, 206, 357
Micronesians, 335, 344; blood groups, 338
Migrations into America, 285, 302, 307, 308
Miller, A. H., 302, 331
Miller, G. S., 16, 42, 239, 272
Miller, R. R., 401, 407
Minnesota, early man in, 305, 307
Miramar skeleton, 316
"Missourium," 286
Mitra, P. N., 356, 359
Mivart, St. George, 80, 116
Mohenjo-Daro, 153, 154
Molars: evolution of, 35; in Carnivora, 33; in Dryopithecus, 61; in lemurs, 34; in man, 53; in Proconsul, 57
Molengraaff, G. A. F., 84, 116
Mongolian, 387; blood groups, 276, 356; eyefold, 275, 354
Mongoloid element: in Amerinds, 307; in Bantu, 218; in Eskimos, 321; in Nilo-Hamites, 215; in South Africa, 197, 203
Mongoloid features, 222; in Malaya, 339; in Polynesians, 340
Monkeys: color changes in, 376; subspecies of, 375, 376; variation in African, 374 ff.
Monotremes, 25, 30, 44
Montagu, M. F. A., 286, 332
Monte Circeo skull, 104, 110, 248
Montehermosensian, 317
Montgomery, Dr. R. B., 123
Moore, J. A., 399, 400, 407
Morant, G. M., 102, 108, 116, 117, 124, 134, 136, 143, 241, 254, 256, 258, 259, 261, 263, 264, 272, 281, 329, 332
Moriori, 337, 341
Morley, S. G., 307, 332
Mostny, G., 319, 332
Mound-builders, 286, 289, 294
Mount Carmel, skeletons, 104, 107, 108, 110 ff., 248, 262; race crossing, 111, 263
Mousterian culture, 81, 103, 108, 110, 156, 167, 176, 177, 183, 192, 209, 223, 229, 230, 244, 245, 248, 251, 252, 253
Movius, H. L., Jr., 87, 108, 117
Moyne, Lord, 205, 234
Muge, 257
Multituberculata, 27, 28, 31, 44, 45
Munda, 356, 357

Murrayians, 151
Mutations, 12, 204, 215, 268; abnormal, 366; adaptational, 399; atavistic, 181; in blood groups, 201, 204; in coloring, 157, 264, 270; in head shape, 4, 120, 124, 126, 180, 263, 269; in stature, 8, 205; of Waagen, 386; rates, 122, 123, 361, 362; spread of, 360; theory, 270
Mydlarski, Jan, 40, 42
Myotragus balearicus, 16

Nabonidus, Chaldean King, 199
Nanyukian, 176
Natural selection, 361, 400
Navaho, 281
Nayadis, 357
Neanderthal man, 6, 8, 63, 64, 72, 78, 91, 95, 96, 98, 102 ff., 144, 147, 149, 171, 183, 184, 204, 217, 243 ff., 251, 404; brain, 145, 238; cranial capacity, 90; femur of, 80, 95; hybridization, 364, 374; skull, 141, 150, 167, 169, 261; teeth of, 94, 186, 249; tools of, 191; variation, 249, 250, 254
Nebarara skull, 195
Nebraska loess man, 290
Nebraska skull, 180
Negrito race, 7, 8, 151, 152, 153, 155, 342, 343, 345, 346, 351, 355, 356, 357; pygmies, 352
Negroes, 5, 9, 93, 98, 114, 132, 134, 167, 172, 177, 184, 199, 203 (figs.), 212, 213, 345, 348, 354, 355, 367, 387; blood groups, 201, 338, 347, 353; hair, 210, 216, 228, 351; head shape, 122, 133, 138, 222; in Egypt, 210, 214; infantile, 353; mental characters, 367; nose and lips, 209, 210; origin of, 217, 352; skeletons of, 194, 209, 256, 257; skulls of, 174, 175, 218, 239; slaves in Arabia, 352; wisdom teeth of, 53, 55, 61
Neolithic, 213, 326, 335; art, 209; beginning of, 336; brachycephaly in, 124, 126; culture, 228, 248, 251; European, 258, 268; implements, 173; in Celebes, 354; in Java, 339; in Portugal, 257; in Siberia, 277, 321; in Switzerland, 206, 269; long barrows, 264; skeletons, 260, 263, 264; in East Africa, 227; times, 269
Nevada, early man in, 303
New Caledonia, 354, 355
Newfoundland, Eskimos in, 323
New Guinea, 148, 343, 344, 351, 353, 363
Newman, H. H., 398, 408
Newman, M. T., 208, 234

Newton, A. and E., 14, 42
Newton, E. T., 250, 272
New Zealand, 337, 340, 341, 342, 344, 345
Nilotes, 184, 212, 213, 214, 223; origin of, 215
Nootka Indians, 324
Nordics, 133, 136, 138, 139, 270; Neolithic, 264, 268, 269; skull, 123, 133, 134, 257
Norman, J. R., 402
Norse, in Greenland, 326
Norway, Paleolithic man in, 255
Notharctus, 49, 50, 51
Notoungulata, 33, 34
Nüesch, J., 206, 207, 234

Oakhurst rock-shelter, 185
Obercassel, 250, 259
Obermaier, 104
Oenothera, 388, 396
Ofnet skulls, 124, 257, 268
Ogilvie, G. S., 150, 164
Okinawa, dwarfs on, 208
Okladnikov, 321
Oldoway man, 176, 183, 222, 223, 241, 250
Olorgesailie Mountain, 229
Olson, E. C., 29, 42
Ona tribe, 263, 282, 319, 320
Ophryonic groove, 103, 150, 226
Oppenoorth, W. F. F., 80, 96, 117
Orangoid line, 6, 142, 211, 236, 237, 250, 252, 264, 274, 318
Oraon, 356, 357
Orangutan, 52, 55, 57, 59, 60, 62, 64, 65, 72, 73, 74, 81, 84, 140, 142; fossil, 67, 85, 94, 239; giant, 205; subspecies, 205; variations, 369, 370
Oregon, Paleo-Indian in, 306
Orford, Marg., 218, 221, 222, 234
Osborn, H. F., 35, 38, 39, 42, 48, 125, 243, 272, 382, 387, 408
Ostriches, 15
Otus asio, 384, 385, 386, 389
Outinequa range, 173, 197
Outrigger canoe, 340, 345
Ovambo, 132, 211
Owen, Sir Richard, 14, 17, 26, 33, 42, 260, 272, 287

Paedopithex rhenanus, 53, 85
Palaeoanthropus, species of, 107, 108, 122, 384
Paleo-Indian, 286, 294, 300, 302, 304, 306; cultures, 303
Paleontology, and classification, 382
Pallasian doctrine, 368

Pampean loess formation, 313
Paniyans, 150, 356, 357, 358
Papuans, 10, 311, 351, 355, 367; nose of, 351, 363; short, 352
Parallel coloration in monkeys, 376
Parallel evolution, principle of, 1, 4, 6, 10, 13, 14, 16, 17, 18, 19, 22, 24, 27, 30, 31, 34, 38, 48, 61, 72, 142, 144, 147, 302, 366; in dentition, 32, 47; in monkeys, 35; in human races, 367; limits of, 44
Parallel mutations, 4, 17, 40, 127, 263, 284; variations, 308, 312
Paramecium, self-sterility in, 404
Paranthropus, 60, 61, 62, 63, 69, 71, 83
Parapithecus, 48, 49
Paricutin volcano, 289
Parsons, F. G., 123, 143, 261
Patagonia, man in, 318, 319
Patterson, J. T., 12, 396, 408
Paterson, T. T., 103, 106, 115, 117
Paumoto archipelago, 341
Paver, F. R., 197, 234
Pearson, Karl, 119, 143
Pecos Pueblo, 281
Pedomorphism, 178, 183, 185, 187, 193, 195, 200, 204, 210, 222
Pei, W. C., 88, 115, 303, 332
Pelvis of Plesianthropus, 68
Pelycosauria, 25, 26, 27
Penguins, giant, 14
Peppercorn hair, 171, 178, 196, 200, 203, 205, 210, 216, 220; on chest, 132
Péringuey, L., 188, 234
Peromyscus, 12; intersterility in, 403, 404; origin of subspecies, 403; species hybrids, 402
Perret, G., 268, 272
Peters, J. L., 385, 408
Peterson, C. B., 286, 332
Petrie, Sir Flinders, 134
Phaup, A. E., 194, 234
Philippines, 151, 355
Phoenicians, in Africa, 198
Phororhachos, 15
Phylogenetic velocity, 39
Physiological isolation, 395
Pickering, S. P., 123, 143
Pictet, A., 408
Pietersburg, Transvaal, 195, 196, 202
Piggott, S., 153, 164
Pijper, A., 180, 338, 353, 359
Piltdown skull, see Eoanthropus
Pitcairn Island, 342
Pithecanthropus, 6, 8, 62, 63, 66, 78, 79 ff., 84, 85, 86, 88, 93, 98, 99, 111, 146, 147,

150, 154, 156, 160, 204, 274, 318, 355, 383, 404, 405; femur, 80, 83, 95; skulls, 89, 90, 92, 97, 109, 114, 122, 140, 181, 225, 365; teeth, 94
Pithecanthropus robustus, 81, 82, 85
Pittard, E., 255, 272
Plattenburg Bay skull, 185
Platyrrhine monkeys, 35, 47, 49, 59
Pleistocene, age of, 73
Pleistocene: and man, 103, 169, 243; correlation, 106; deposits, 176; in America, 277 ff.; in South Africa, 190; length of, 106; strata in Java, 86
Pleurodeles hybrids, 398
Pluvial periods, 176, 177, 190, 212
Pocock, R. I., 49, 77
Podkumok skull, 110, 250, 258
Polymorphic genera, 379
Polynesians, 9, 311, 335 ff., 347, 363; blood groups, 337, 338; Caucasoid element in, 338; genealogies, 340; history, 338; languages, 339; mentality, 378; origin of, 337, 339, 355; Sumbanese, 354
Polyzoa, evolution in, 21
Pondo, 211
Populations: increase of, 335; of world, 336
Port Elizabeth, 172, 173, 175, 182
Porter, K. R., 398, 399, 408
Portuguese, 348
Pottery: Chimu, 321; in America, 285, 286, 307; in Siberia, 325
Predmost, 250, 260, 261, 263
Premaxilla, in anthropoids, 63
Primates, 45, 46, 47, 48, 50, 51, 65, 142; adrenal glands of, 73; giants, 86; origin of, 52, 78; viscera of, 74
Proconsul, 57, 58, 59
Proto-Malays, 349, 350, 351, 354
Pulayas, 152, 351, 352
Pulleine, R. H., 156, 164
Punin skull, 180, 309, 311 ff., 317
Pycraft, W. P., 101, 117
Pygmies, 7, 205, 206, 222; blood groups in, 202, 338, 347; Congo, 5, 8, 204, 346, 370; in New Guinea, 352; Ituri, 200, 201, 219; Neolithic, 207, 208, 236

Quatrefages, A. de, 210, 258, 272

Race: Dinaric, 119, 363; Ethiopian, 212
Races: mental differences between, 367; origin of, 366
Racial characters in man, 386
Rahm, G., 319, 338
Rainey, F. G., 325, 332

Ramsbottom, J., 384
Rancho La Brea, 301, 302
Rasmusson, 326
Ray, 301
Raymond, M., 285, 332
Reck, H., 222, 223, 233, 234
Regan, C. Tate, 378, 381
Reindeer: in Asia, 324; in Europe, 265; domestication, 325
Renaud, E. B., 304, 332
Rendel, J. M., 394, 408
Reptiles, evolution of, 25
Retzius, 124
Rhoades, M. M., 361, 381
Rhodesian man, 79, 86, 96, 107, 108, 147, 166, 167, 181, 211, 216, 225; brain, 160; skull, 6, 7, 67, 69, 80, 92, 97, 99 ff., 102, 140, 150, 160, 169, 180, 183, 236, 263
Rivet, P., 339, 359
Roberts, F. H. H., Jr., 286, 300, 301, 302, 303, 332
Robin, L., 319, 332
Robinson, J. T., 75
Romer, A. S., 17, 25, 27, 42
Rooiberg, 196
Roth, H. L., 156, 164
Roth, S., 314
Rothschild, Hon. W., 371, 374, 381
Rousseau, J., 285, 332
Rusconi, C., 321, 332

Saccopastore, 105, 110, 248
Sagittal crest, 92, 93
Sahara, 7, 113, 133, 177, 209, 211, 212, 216, 257
Sail, introduced, 199
St. Lawrence Island, excavations, 308, 327, 328
Sakai, 150, 350
Saladero skeleton, 314
Salamanders, hybrids in, 12, 398
Saltations, 17, 19
Samborombón skeleton, 314
Samoans, 341, 343, 344; blood groups, 338, 340
Samoyeds, 325
Sanderson, I. T., 369, 370, 371, 375, 376, 381
Sandia culture, 303, 304, 320
Sandoval, L., 343, 359
Sanskrit, 346
Santals, 356, 357
Sapir, 280
Sarkar, S. S., 356, 359
Sauer, C. O., 278, 301, 332

Scelidotherium, 314, 315
Schapera, I., 178, 234
Schebesta, Paul, 204
Scheidt, W., 268, 272
Schepers, G. W. H., 30, 42, 66, 70, 71, 75, 182, 183, 194, 234
Schlosser, 48, 49
Schoetensack, O., 242, 272
Schultz, A. H., 74, 77, 372, 381
Schultze, L., 218, 234
Schwarz, E., 204, 370, 371, 374, 375, 376, 377, 381
Schwarz, E. H. L., 198, 199, 234
Screech owls, geographic variation in, 384
Secondary isolating mechanisms, 397
Secondary races, 12
Selenka, E., 205, 235
Self-differentiation of bones, 366
Self-sterility, 404
Seligman, B. Z., 155, 164, 211, 213, 235
Seligman, C. G., 155, 164, 211, 213, 215, 235, 261
Sellards, E. H., 291, 292, 293, 304, 305, 332, 333
Senyürek, M. S., 77, 109, 117
Sergi, G., 5, 78, 89, 102, 110, 128, 139, 143, 155, 156, 206, 248, 270, 272
Sex attraction, between species, 394
Sexual aversion, 108, 394, 403
Shapiro, H. L., 319, 328, 333, 349, 359
Shaw, J. C. M., 61, 67, 77
Shellshear, J. L., 88, 117, 159, 160, 164
Shovel-shaped incisors, 63, 67, 92, 93, 110, 113, 245, 249, 289, 311
Shrubsall, F. C., 130, 179, 185, 235, 266
Siberia, 279, 282, 285, 304, 324, 325, 328; Neolithic in, 277, 321
Simian shelf, 48, 54 (fig.), 55, 58, 59, 60, 95, 239
Simmons, R. T., 158, 163, 233
Simpson, G. G., 13, 14, 16, 28, 29, 31, 33, 34, 41, 42, 44, 46, 47, 49, 59, 77, 78, 287, 317, 318, 333, 382, 383, 385, 386, 408
Sinanthropus, 6, 8, 55, 63, 64, 67, 69, 79, 80, 81, 83, 84, 85, 86 ff., 113, 114, 144, 204, 225, 274, 275, 318, 384; cannibals, 96; femora, 95; skull, 140, 240, 253, 255, 365
Sipka jaw, 248
Sivapithecus, 63, 65, 70, 72
Skull shapes: genes for, 173, 174; classification of, 283
Skulls, upper Paleolithic, 254
Slome, D., 195, 235
Slome, I., 162, 164

Smithfield industry, 188, 189, 190
Smith, Percy, 338, 339, 341, 359
Smith, S. A., 150, 164
Smith Woodward, Sir Arthur, 100, 117, 237, 238, 239, 240, 271, 272
Smuts, J. C., Jr., 177, 235
Smuts, Rt. Hon. J. C., 190, 235
Soergel, W., 37, 43
Sollas, W. J., 103, 117, 150
Solo man, 80, 86, 96, 107
Solomon Islands, birds of, 395
Solutrean culture, 189, 243, 254, 258, 259, 260, 261
Somali, 212
South Africa: climates of, 191; crossing in, 363; cultures of, 190; map of, 170
Speciation, 12, 360, 382 ff., 391, 395
Species: defined, 378, 387, 388, 392; differentiation of, 379, 404
Species-formation, rate of, 73, 395
Species of mankind, 366 ff., 387, 388; morphological criteria, 406
Speiser, F., 352, 359
Spencer, W. K., 19, 43
Spinden, H. J., 286, 333
Springbok Flats skeleton, 182, 183, 187, 193
Spy skeletons, 243, 245, 246, 247
Stannus, Dr. H. S., 187, 197
Starfishes, evolution of, 19
Stature: of Paleolithic man, 263; sex-ratio, 258, 263
Steatopygia, 132, 196, 197, 206, 207, 208, 222, 257; types of, 220, 221
Steiner, H., 397, 398, 408
Steiner, P. E., 208, 235
Steinheim skull, 104, 107, 108, 109, 149, 169, 242, 245
Sterility, in relation to species, 11, 364, 366, 368, 380 ff., 392, 394, 396
Stern, C., 394, 408
Stewart, T. D., 282, 284, 294, 302, 333
Stillbay culture, 177, 184, 226, 227
Stirton, R. A., 17
Stock, C., 301, 303, 333
Stow, G. W., 178, 235
Strain, I. G., 309, 333
Strandloopers, 130, 172, 173, 174, 175, 178, 179, 185, 186, 192, 195, 200, 227
Straus, W. L., Jr., 57, 65, 74, 77, 142
Sudan, 213, 214, 215, 228
Sulcus lunatus, 60, 71, 89, 160, 162, 225
Sullivan, L. R., 312, 333, 343, 359
Sumatra, inhabitants of, 349, 350
Sumerian, 275, 339
Sumiduoro, cave of, 309, 311

Index

Sundaland emergence, 84, 150, 274
Swanscombe cranium, 80, 104, 149, 240 ff., 250, 264, 268
Sweet potato, 341, 344, 348
Swiss Lake Dwellers, 156, 206

Talgai skull, 146, 150
Tanganyika, terracing in, 198, 200
Tarsioidea, 35
Tarsius spectrum, 50, 51
Takasaki, 338
Tasmanians, 7, 148, 151, 155, 156, 157, 189, 312; blood groups of, 158; hair of, 353
Taubach skull, 104, 245
Taxonomy: and species, 384, 393; in paleontology, 382; of birds, 388, 389
Taylor, Griffith, 308
Teeth: in anteaters, 32; in Carnivora, 33; in Chinese chemists' shops, 83; in Proboscidea, 36; microtuberculate, 27; of Meganthropus, 82; of Parapithecus, 48, 49; of Plesianthropus, 64, 67, 71; taurodont, 87, 94, 108, 109, 147, 186, 195, 245, 247, 249
Teilhard de Chardin, 88, 115
Temagnini, E., 267, 272
Ten Kate, 310, 315
Terraces in East Africa, 198, 200
Teshik-Tash skull, 111
Tetraprothomo argentinus, 317
Texas, early man in, 299, 302, 303, 304
Therapsids, 25, 26, 27, 29, 30
Thule culture, 325, 326, 327, 328
Thylacinus, 32, 146
Tibetan features, 276
Timor, 354
Tindale, N. B., 157, 158, 164
Tippett, L. H. C., 119, 143
Titanotheres, evolution of, 38, 124, 125
Toads, species crossing in, 401
Toala, 150, 350
Tobler, 350
Todas, 152
Torus mandibularis, 89, 91, 255, 256
Torus palatinus, 91, 92
Toynbee, A. J., 199, 235, 339, 359
Tozzer, A., 347, 359
Tree-rings, dating by, 307
Tuamotu Archipelago, 349
Tungus, 265, 324, 325
Turner, Sir William, 155, 164
Turville-Petre, F., 247

Uralis, 152
Uru Indians, 313
Usbekistan, 110

Vallois, H. V., 209, 230, 255, 256, 272
Varves, 265, 336
Vavilov, N. I., 18, 43
Veddas, 150, 153, 155, 156, 157, 349, 350
Verneau, R., 206, 208, 209, 230, 235, 253, 256, 257, 266, 272
Vries, de, mutations of, 38

Waagen, mutations of, 38
Wadjak skull, 72, 146, 147, 148, 150, 151, 154, 169
Wallace's faunal line, 84
Walmsley, T., 79, 117
Walter, H. V., 311, 333
Ward, H. B., 290, 329
Watson, D. M. S., 16, 23, 24, 25, 29, 35, 43
Webster, C. L., 289, 333
Wegner, R. N., 17, 41
Weidenreich, F., 63, 67, 72, 81, 82, 83, 84, 85, 89, 90, 91, 92, 93, 94, 95, 96, 97, 98, 108, 110, 113, 114, 116, 117, 118, 120, 124, 125, 127, 141, 143, 144, 148, 164, 205, 223, 235, 239, 244, 253, 260, 302, 333, 404
Weinert, H., 99, 109, 118, 239, 245, 273
Wells, L. H., 160, 164, 181, 182, 183, 188, 194, 196, 197, 218, 221, 222, 230, 234, 235
Weninger, 215
Werth, E., 48, 77
Whale hunting, 324, 325, 326, 348
White, F., 100, 118
Whitney, J. D., 295, 296, 297, 298, 333
Wilhelm, O., 359
Williams, E. T., 275, 333
Willis, Bailey, 313, 315, 317
Williston, S. W., 25
Wilson, G. E. H., 197, 198, 235
Winge, H., 309
Winslow, C. F., 295, 334
Wirth, A., 208, 235
Witter, F. M., 293, 334
Wood Jones, F., 34, 51, 63, 77, 155, 156, 164, 165
Woollard, H. H., 50, 77, 142, 143, 159, 165
Wright, Sewall, 22
Wunderly, J., 148, 165

Xenopithecus, 57, 58, 59, 60

Yaghan, 282, 319
Yerkes, A. W., 373, 381
Yerkes, R. M., 370, 371, 373, 381
Young, C. C., 88, 115
Young, Matthew, 241, 273

Zanzibar, Arabs in, 346
Zeuner, F. E., 73, 77, 84, 103, 106, 118, 149, 165
Zimbabwe, 194, 198, 199
Zitzikama, 171, 172, 173, 175, 184, 186, 193, 194, 210, 227

Zolotarev, A., 324, 334
Zondervan, H., 349, 359
Zuckerman, S., 154, 165
Zululand, 196
Zulus, 202, 211; blood groups in, 201
Zuurberg, 182, 186